Fertilizers and
Soil Fertility

Fertilizers and Soil Fertility

Ulysses S. Jones

Professor
Department of Agronomy and Soils
Clemson University
Clemson, South Carolina

Reston Publishing Company, Reston, Virginia
A Prentice-Hall Company

Library of Congress Cataloging in Publication Data

Jones, Ulysses S.
   Fertilizers and soil fertility.

   Includes bibliographical references and index.
    1.   Fertilizers and manures.   2.   Soil fertility.
I.   Title.
S633.J76   631.8   79-870
ISBN 0-8359-1960-9

©1979 by
Reston Publishing Company, Inc.
A Prentice-Hall Company
Reston, Virginia 22090

10 9 8 7 6 5 4 3 2 1

Printed in the United States of America.

Contents

chapter 5   MULTINUTRIENT FERTILIZERS   *145*

chapter 6   POTASSIUM—THE CATALYST   *189*

Preface

Fertilizer use accounts for one-third to one-half of the agricultural production in the United States and is rapidly approaching this level of use in the world. Students of soil fertility, plant nutrition and fertilizers are facing more complex problems than ever before. They, as potential agricultural leaders and growers, will rely more and more on highly specialized information.

This text is an introduction to soil fertility, production, marketing and use of fertilizers, agricultural limestone, and other soil amendments. It is written and designed especially for post-high school students of agriculture and the earth sciences, and for fertilizer salesmen and distributors who desire to learn more about their products and those of their competitors.

The ideas and concepts included in this book are the result of thirty years of experience, of which about one-third was spent in the agricultural chemical industry, and two-thirds as a teacher of fertilizers and soils.

In the original concept of this text very little coverage was planned for production, marketing and use of organic wastes and manures as they represent something less than a fraction of one percent of the plant nutrients consumed in the United States. However, because of increasing emphasis on land, water and air quality, a chapter entitled "Protein–Organic Wastes" with emphasis on protein waste management and use has been included.

During the past several years of teaching at Clemson University and abroad, encouragement, suggestions and advice on the subject matter have come from students, faculty, farmers and industry representatives in South Carolina and many other states and nations. Thanks are expressed to the author's co-workers at Clemson and to Drs. C. E. Bardsley and I. M. Wofford who reviewed the nitrogen chapter, to Drs. J. R. Webb and G. G. Williams who supplied data and pictures for the phosphorus and multinutrient chapters, to Dr. S. L. Tisdale who reviewed the sulfur chapter, to Dr. S. E. Younts who reviewed the potassium chapter, to Drs. J. D. Lancaster and R. E. Lucas who supplied data and observations for the micronutrient chapter, to Mr. F. L. McNatt who supplied data on nutrient uptake by plants, and to the late Mr. S. Cottrell of the World Bank who supplied figures and data on manufacturing processes.

Finally, thanks are expressed to Mrs. Linda Breazeale, Mrs. Jacqueline Reardon, Mrs. Eva Suarez, and Mrs. Elaine Teat for the typing and preparation of the manuscript.

ULYSSES S. JONES

chapter 1

The Soil—
Plant Growth
Situation

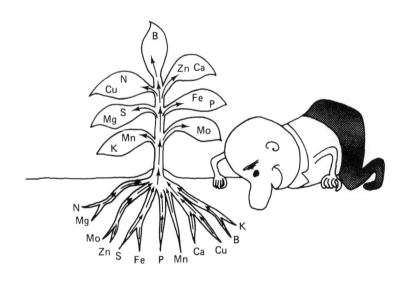

The soil on which plants grow provides a storehouse of minerals, chemicals, water and air.

From a chemical standpoint, the mineral portion of the soil, which makes up about one-half of the volume, is about 93 percent silica (SiO_2) and aluminum and iron oxides (Al_2O_3 and Fe_2O_3) (Table 1.1). Four percent is composed of the oxides of *calcium* (Ca), *magnesium* (Mg) and *potassium* (K). The remaining 3 percent includes titanium (Ti), sodium (Na) and very small amounts of *nitrogen* (N), *sulfur* (S), *phosphorus* (P), *boron* (B), *manganese* (Mn), *zinc* (Zn), *copper* (Cu), *molybdenum* (Mo), *chlorine* (Cl) and many other elements. By far the most prominent chemical constituent in a mineral soil[1] is SiO_2, which accounts for 76 percent or more of the mineral fraction while Al_2O_3 accounts for about 12 percent and Fe_2O_3 for about 5 percent. Of these three elements, silicon, aluminum and iron, only iron is known to be essential for plant growth, and it in only very minute amounts. Thus, the great mass of the soil mineral fraction acts only as an anchor for plant roots and as a skeleton on which to hang the colloidal clay and humus. The essential available mineral nutrients required for plant nutrition are found in the very small colloidal fraction of soil and are those elements italized in the first three sentences of this paragraph; they are also listed in Table 1.2. Refer to Table 1.3 for the chemical symbols and atomic weights of the elements. Useful chemical conversion factors appear in Table 1.4, and quick metric conversions are shown in Table 1.5.

Table 1.1
CHEMICAL COMPOSITION OF A
MINERAL TOPSOIL EXPRESSED AS
OXIDES OF THE ELEMENTS

Oxides	%
SiO_2	76
Al_2O_3	12
Fe_2O_3	5
CaO	1
MgO	1
K_2O	2
All others	3

[1]Soils containing 20 percent or more organic matter are designated organic soils; those with less than 20 percent are referred to as mineral soils.

Table 1.2
ESSENTIAL PLANT NUTRIENT ELEMENTS

Those Used in Large Amounts			Micronutrients	
From Air and Water[1]	From Soil Solids and Other Chemicals		From Soil Solids and Other Chemicals	
Carbon	Nitrogen[2]	Sulfur	Iron	Manganese
Hydrogen	Phosphorus	Calcium	Zinc	Copper
Oxygen	Potassium	Magnesium	Chlorine	Boron
			Molybdenum	Cobalt[3]

[1]From 95 to 99.5 percent of fresh plant tissue is made up of carbon, hydrogen and oxygen, but it is the mineral elements and nitrogen that usually limit plant growth.
[2]Nitrogen may be acquired from soil air indirectly by legumes such as soybeans.
[3]Essential for certain soil microorganisms.

Table 1.3
SYMBOL AND ATOMIC WEIGHTS OF SOME ELEMENTS

Element	Symbol	Atomic Weight	Element	Symbol	Atomic Weight
Aluminum	Al	26.97	Magnesium	Mg	24.32
Boron	B	10.82	Manganese	Mn	54.93
Calcium	Ca	40.08	Molybdenum	Mo	95.95
Carbon	C	12.01	Nitrogen	N	14.008
Chlorine	Cl	35.457	Oxygen	O	16.00
Cobalt	Co	58.94	Phosphorus	P	30.98
Copper	Cu	63.57	Potassium	K	39.096
Fluorine	F	19.00	Silicon	Si	28.06
Hydrogen	H	1.008	Sodium	Na	22.997
Iodine	I	126.92	Sulfur	S	32.06
Iron	Fe	55.85	Zinc	Zn	65.38

Table 1.4
USEFUL CHEMICAL CONVERSION FACTORS

$N \times 1.22 = NH_3$	$HNO_3 \times 0.22 = N$
$P \times 2.29 = P_2O_5$	$H_3PO_4 \times 0.32 = P$
$P_2O_5 \times 0.44 = P$	$Ca_3(PO_4)_2 \times 0.20 = P$
$K \times 1.20 = K_2O$	$KCl \times 0.52 = K$
$K_2O \times 0.83 = K$	$K_2SO_4 \times 0.45 = K$
$Ca \times 1.40 = CaO$	$CaCO_3 \times 0.40 = Ca$
$CaO \times 0.71 = Ca$	$CaSO_4 \times 0.29 = Ca$
$Mg \times 1.67 = MgO$	$MgCO_3 \times 0.28 = Mg$
$MgO \times 0.60 = Mg$	$MgSO_4 \times 0.20 = Mg$
$S \times 3.00 = SO_4$	$H_2SO_4 \times 0.33 = S$
$SO_4 \times 0.33 = S$	$CaSO_4 \times 0.24 = S$

Table 1.5
QUICK METRIC CONVERSIONS

English Unit	Conversion Factor	Metric Unit
VOLUME		
gallon (gal)	× 3.79	liter (l)
foot³ (ft³)	× 0.028	meter³ (m³)
PRESSURE		
pounds per square inch (psi)	÷ 14.7	atmosphere (atm)
LENGTH		
inch (in.)	× 2.54	centimeter (cm)
foot (ft)	÷ 3.3	meter (m)
miles	× 1.61	kilometer (km)
TEMPERATURE		
°F	0.555(°F−32)	°C
AREA		
acre	÷ 2.47	hectare (ha)
foot²/acre	× 23.25	decimeter²/hectare
MASS		
pound (lb)	÷ 2.2	kilogram (kg)
ton	× 0.91	metric ton (m ton)
foot³/ton	× 0.031	meter³/m ton
YIELD OR RATE		
ounce/acre	× 0.07	kg/ha
bushel/acre	× 67.2	kg/ha
ton/acre	× 2240	kg/ha
ton/acre	× 2.24	m ton/ha
lb/acre	× 1.12	kg/ha
lb/ft³	× 16.23	kg/m³
lb/gal	× 0.12	kg/l
lb/ton	× 0.5	kg/m ton
gal/acre	× 9.42	l/ha
gal/ton	× 4.16	l/m ton

Water (H_2O) is a very important food-producing chemical in itself,[2] and it is also the solvent that dissolves carbon dioxide (CO_2) and subsequently the essential plant nutrients into the soil solution. The quantities of the various chemicals dissolved in soil water at any one time are very minute. Yet it is through these chemical changes, with carbonated water serving as the solvent, that the soil mineral and organic fraction is converted into nutrients that are available to plants. Soil organisms supply the CO_2 from which the carbonated water is made. The concentration of CO_2 in soil is approximately one hundred times that in the atmosphere.

Thus, the requirements of a productive soil are a good water-holding capacity, good aeration and a supply of decomposing minerals and humus that are dissolving at a rate rapid enough to meet the requirements of desired plant growth. Unfortunately, most cultivated soils are not this productive, and often minerals and chemicals including water must be added. Obtaining the right combination of water, water-holding capacity, aeration and nutrients economically in the soil is the objective of the food grower. He refers to this as soil management. The science of soil management is included in the profession of agronomy; the word is derived from a Greek noun *agros* meaning field and the verb *nemein*, the English translation of which is to manage.

ESSENTIAL NUTRIENT ELEMENTS

There are sixteen elements considered essential for plant growth and development. They are obtained from air, water and soil solids or from agricultural limestone and other chemicals and minerals that can be processed into fertilizers. About 95 to 99.5 percent of fresh plant tissue is made up of carbon, hydrogen and oxygen. Only about 0.5 to 5.0 percent remains as ash after fresh plant tissue is dried and burned. Although this represents a very small proportion of the total plant, it is usually one or more of the elements found in the ash or one that was lost in the fumes, such as nitrogen, that limits plant growth and development. Plant growth is not seriously retarded by lack of carbon, hydrogen or oxygen as long as H_2O is available and CO_2 in the air is plentiful. Thus, plant nutrition emphasis is placed on the soil's capacity to supply nitrogen and the other essential elements, phosphorus, potassium, calcium, magnesium, sulfur, iron, manganese, zinc, copper, chlorine, boron and molybdenum (see Table 1.2).

The micronutrient elements iron, manganese, zinc, copper, boron and molybdenum are used by higher plants in very small amounts. While

[2]Psalm 65:11. Thou waterest her furrows, thou sendest rain into the little valleys thereof,—and bless the increase of it.

quantities of about 0.5 lb/acre (0.5 kg/ha) are helpful, amounts as small as 4 lb/acre (4.5 kg/ha) can sometimes be damaging to plants. The micronutrients are no less essential than those elements that are needed in large amounts, but they less frequently limit crop production except in certain soils when growing particular crops or where unusually poor soil management has been practiced. Chlorine deficiencies are unknown in field soils, and for this reason chlorine is classified as a micronutrient although root crops may contain as much chlorine as sulfur or phosphorus.

While all of the essential nutrients supplied by the soil must be present and available in order for plants to grow, normally there are six that are of prime importance and are most likely to become limiting factors for plant growth. These six nutrients are nitrogen, phosphorus, potassium, sulfur, calcium and magnesium. The first three are the major plant nutrients, and the latter three are designated as secondary plant nutrients. Collectively these six are designated as macronutrients in contrast to those used in small amounts which are referred to as micronutrients.

When nitrogen, phosphorus, potassium and sulfur are artificially supplied to the soil, they are often added as multinutrient fertilizers. Sulfur is not usually guaranteed as an ingredient of fertilizers but is an incidental ingredient of some multinutrient fertilizers. Nonetheless, if sulfur is deficient in the soil, its addition is very important in soil fertility management; therefore, if sulfur is present in a fertilizer, this should be claimed and guaranteed by the manufacturer.

Calcium and magnesium are usually not guaranteed as ingredients of fertilizer. These two essential elements are supplied when an acid soil is limed with dolomitic limestone.

Thus, if lime[3] and fertilizer are used based on soil chemical test results and on need, the six macronutrients (nitrogen, phosphorus, potassium, sulfur, calcium and magnesium) will be supplied in ample amounts for plant growth.

Soil testing to determine the nutrient needs of the plant to be grown is one of the most practical ways to make decisions about wise use of lime and fertilizer. Soil-testing laboratories are widely distributed. They are located in nearly every state and in many other countries.

In this chapter, the subject of food-producing minerals and chemicals is introduced from the general viewpoint of soils and fertilizers as sources of essential nutrient elements for plant growth. Two major concepts are presented here before moving on to a more complete discussion of plant-soil relationships. First, the plant utilizes essential mineral elements in the form of very simple chemical compounds produced by weathering of the earth's crust. Second, the fertilizers added to assist the soil in producing

---

[3]Chemically pure lime is CaO, but lime, as commonly used, refers also to $CaCO_3$ and $Ca(OH)_2$. *Lime* as used here refers to any of these compounds, with or without magnesium and all other limestone-derived materials added as an amendment to neutralize acid soils.

more and better quality foods are in fact the same simple compounds that nature has been providing from the earth's crust since plants first appeared on the Earth over 300 million years ago.

PLANT-SOIL RELATIONSHIPS

Life depends on the earth's thin layer of soil. This natural body supports plant life in varying degrees of efficiency depending on its productivity potential and nutrient status. Without soil, there would be no plants; without plants there would be no food; without food animals could not survive. Thus, soil was, and is, the beginning of the soil-plant-animal food chain.

For ages and ages, man has searched for fertile, productive soils. Infertile soils produce poor crop yields, despite what else is done. Fertile soils are required for high crop yields. However, many factors, including fertility levels, determine soil productivity. Application of what is known about these factors can improve immensely the use of soils, one of the most valuable, but limited, natural resources. Soil is the basis of man's existence and the measure of his future.

Man's treatment of soils has materially influenced his health, happiness and development. The earliest of the great civilizations were founded upon and nourished by productive, relatively fertile, alluvial flood plain soils along the Euphrates, Tigris, Nile, Ganges and Yellow rivers. These rivers flooded frequently, and the flood plains were thus replenished with soluble nutrients from the watersheds upstream. Only in the last few years of man's evolution from hunter and food gatherer, since about 1850, has he understood the principles of soil fertility that have enabled him to grow two blades of grass where one grew before.

SOIL HORIZONS AND PROFILES

What is a soil? It is a natural body made by nature, not by man. It consists for the most part of minerals and a very little organic matter. It usually has two or more layers called horizons which are more or less obvious from vertical cuts into a soil. The thickness and composition of these horizons account for the differences among soils. Cross sections of the horizons are called soil profiles.

In most cultivated soils, there are at least three, more or less distinct horizons; they are not always immediately obvious to the casual observer. The plow layer is designated the A_p horizon and is normally the one highest in organic matter and easily soluble plant nutrients. Below the plow layer in highly weathered soils, there is usually an A_2 horizon. This

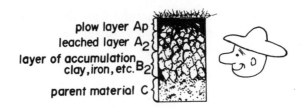

Figure 1.1 Soil profile and horizons (layers).

is a leached layer and is lower in organic matter than the A_p. Below the A_2 there are often one or more B horizons that are designated as the zone of illuviation or accumulation and thus are fairly high in clay content, iron and aluminum (Figure 1.1). The two almost universal properties of soils are:

1. Accumulation of organic matter in the A horizon.
2. Accumulation of clay in the B horizon.

In arid regions, calcium carbonate (lime) and other salts accumulate in the B horizon. Below the B horizon lies the parent material from which the soil developed. This layer is designated the C horizon. The A and B horizons together are designated the solum.

A soil horizon can restrict plant growth (Figure 1.2) because of:

1. Inadequate thickness.
2. Physically impervious layers known as "hard pans."
3. Chemically impervious layers, such as root-damaging aluminum levels in very acid soils.

The careful examination of soil profiles is the first step toward using soils wisely. Profiles reveal many "clues" that by inference identify major problems restricting root growth and development.

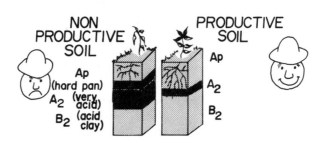

Figure 1.2 Non-productive soil and productive soil.

The soil volume is made up of only four components (Figure 1.3). As an example, a fine sandy loam topsoil is composed of the following: organic matter—0.5 to 5 percent, minerals—45 to 50 percent, water—25 percent and air—25 percent. Water and air occupy the pore space or voids in the soil. Pore space makes up about one-half the volume.

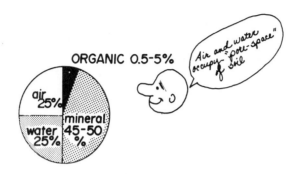

Figure 1.3 The soil composition.

Organic material, which is the residue of dead and dying plant material of macro and microbiological origin, is a reservoir of nitrogen, phosphorus, sulfur and other macro and micronutrients essential for plant growth. A portion of each of these nutrients in topsoils exists in organic forms. For this reason, nitrogen, phosphorus, sulfur, zinc and possibly other nutrient deficiencies occur where soils have been leveled or graded. Organic matter influences favorably the structure and tilth of soils that permit good seedling development and root penetration. One of the most practical methods of maintaining desirable levels of organic matter in mineral soils is to produce adequately fertilized, high-yielding crops and to incorporate the crop residues into the soil.

Soils containing more than 20 percent organic matter are classified as organic and require different management than do mineral soils. For example, organic soils seldom need liming above pH 5.2 to 5.5 because the highly adsorptive organic colloids hold enough calcium, magnesium and other nutrients for plant growth and development at that pH level. Where organic soils have been limed to pH higher than 5.5, instead of their being a reservoir for micronutrients, deficiencies become even more prevalent than in mineral soils.

The mineral portion of soils contains three major substances based on particle size (Figure 1.4). These substances are designated *clay* for particles of colloidal dimension, *silt* for particles of face powder size and *sand* for particles that are gritty. The proportion of sand, silt and clay in a soil determines its *texture* and its capacity to hold water and nutrients. The kind of clay as well as the amount is very important. Several kinds of

9

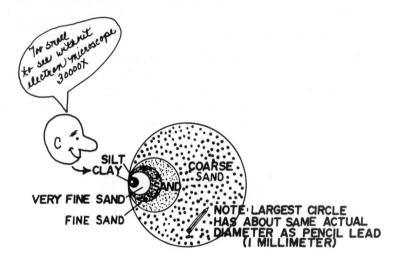

Figure 1.4 Soil minerals.

clay swell or shrink with moisture changes. Some trap potassium and magnesium into a form that is only, with difficulty, available to plants.

The third component of soil is water. Available water is defined as the difference between field capacity and wilting point. *Field capacity* is the soil moisture content after downward water movement has stopped following rainfall or irrigation. *Wilting point* is the soil moisture content at which plants permanently wilt. To a great extent, soil texture and, to a lesser extent, soil structure and organic matter influence the available water content.

Water is the difference between productive soil and a desert. Many of the world's vast desert areas could be made extremely productive with the addition of water and nitrogen fertilizer. Water provides a moist sheath on the surface of roots and has a dissolving effect on soil minerals, producing a soil solution from which plants take up nutrients. Capacity to store available water is one of the key characteristics of productive soils. Soils with low water-holding capacity grow plants subject to drouth. Soil texture (the proportion of sand, silt and clay) is largely responsible for water-holding capacity. Unfortunately, soil texture practically never changes, even in a lifetime, unless rather drastic and usually very expensive procedures are resorted to. Differences in available water contained in three soil textural groups (fine sand, sandy loam and silt loam) may be noted in Figure 1.5.

The fourth component of soil is air. In a dry soil, air occupies nearly all the pore space, which is about one-half the soil volume (Figure 1.3). In flooded soils, water occupies practically all the pores or voids. Under flooded, anaerobic (absence of air) conditions, the lack of oxygen reduces sharply the root growth and development of most crops; hence the importance of good soil drainage. An exception to this is rice, which does

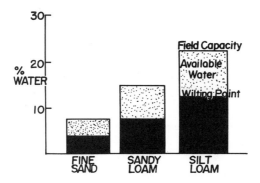

Figure 1.5 The water-holding capacity of soil depends on the soil texture.

well in flooded soils containing very little air. However, corn will wilt and die when soil on which it is growing is flooded for several days.

Anaerobic conditions greatly influence the availability to crops of several nutrients in soil, especially phosphorus. Rice, when growing in water, can flourish on much less soil-available phosphorus than most upland crops, because soil phosphorus becomes more soluble under anaerobic conditions.

CHEMICAL PROPERTIES

Green plants, whether algae in lakes and oceans, peaches in South Carolina, corn in Iowa, oranges in Florida, grapes in France, rice in China, sugar beets in North Dakota or figs in Turkey, convert radiant energy from sunlight into chemical energy of seeds, fruits, foliage and tubers which are basic foods for man and other animals. Producing this chemical energy in foods requires a great deal of carbon dioxide and water plus relatively smaller quantities of simple, inorganic chemical compounds from the soil. These compounds are made up of two or more ions, usually a positive ion balanced by a negative one. In order to understand this basic life process, the student must understand some simple chemistry concepts. The most important of these soil chemistry concepts are referred to as cation exchange capacity, soil acidity and soil alkalinity.

Cation and Anions. Chemical ions in soils have either positive or negative charges. Those with positive charges are designated *cations* and those with negative charges *anions* (Figure 1.6). Soil *colloids*, consisting of clay and humified organic matter, are negatively charged. Consequently, they attract or hold on to positively charged particles called cations. Unlike forces attract, and like forces repel, each other. Therefore, the negatively charged anions are not attracted by the negatively charged soil colloids

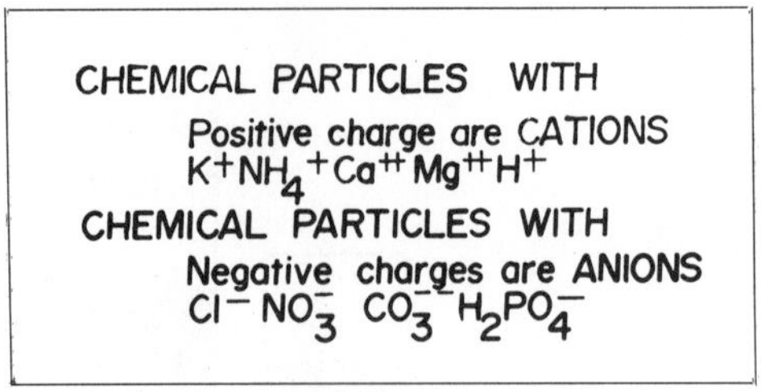

CHEMICAL PARTICLES WITH
Positive charge are CATIONS
$K^+ NH_4^+ Ca^{++} Mg^{++} H^+$
CHEMICAL PARTICLES WITH
Negative charges are ANIONS
$Cl^- NO_3^- CO_3^- H_2PO_4^-$

Figure 1.6 Cations and anions.

and hence move freely with the soil water. An exception to this is phosphate, an anion, which reacts with certain soil clays, notably hydrous oxides of iron and aluminum. Examples of cations are K^+, NH_4^+, Ca^{++}, Mg^{++} and H^+. Examples of anions are Cl^-, NO_3^-, CO_3^{--} and $H_2PO_4^-$.

Cation Exchange Capacity. One of the most practical soil tests for predicting plant nutrient needs involves the determination of the capacity of a soil to hold cations. This soil characteristic is defined as the cation exchange capacity and is frequently abbreviated CEC. The unit of measurement of the cations is the milliequivalent, and the amount of soil involved in the measurement is 100 grams. The CEC is expressed as milliequivalents per 100 grams of oven-dry soil (Figure 1.7). It is the capacity of soils to hold cations expressed as units of one-thousandth of a gram of hydrogen, hence the reason for the prefix milli, which means 1/1000 or 0.001.

One gram of hydrogen is one equivalent weight because the atomic weight of hydrogen is about one. Thus, one milliequivalent of hydrogen is one-thousandth of a gram, or one milligram equivalent weight, which is shortened to milliequivalent.

Each cation element has its own distinct equivalent weight based on how much of it will react with or displace hydrogen from the soil exchange sites. The equivalent weight of an element is its atomic weight divided by

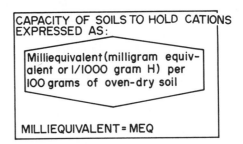

Figure 1.7 Cation exchange capacity (CEC).

its charge. The atomic weight of potassium (K^+) is 39 and its charge is 1; 39 grams of it will displace one gram of H^+. Thus, the equivalent weight of K^+ is 39/1 or 39. Calcium (Ca^{++}) has an atomic weight of about 40 and a charge of 2; the equivalent weight of Ca^{++} is 40/2 or 20. Likewise, magnesium (Mg^{++}) has a charge of 2 and an atomic weight of 24; the equivalent weight of Mg^{++} is 24/2 or 12.

Milliequivalent weight, as explained before, is the milligram equivalent weight. For a given cation, it is the same number as the gram equivalent weight when its unit is expressed as a milligram equivalent weight.

Milliequivalent per 100 grams can be converted to pounds per acre furrow slice by multiplying the equivalent weight of the element involved by 20 (Figure 1.8). For example, the equivalent weight of Ca^{++} is 20 and multiplied by 20 gives 400 pounds of calcium per acre furrow slice (448 kg/ha); the equivalent weight of K^+ is 39 and multiplied by 20 gives 780 pounds of potassium per acre furrow slice (874 kg/ha). Likewise, Mg^{++} has an equivalent weight of 12 and multiplied by 20 gives 240 pounds of magnesium per acre furrow slice (269 kg/ha); hydrogen (H^+) has an equivalent weight of 1 and multiplied by 20 gives 20 pounds of hydrogen per acre furrow slice (22 kg/ha). The factor of 20 is derived from the fact that an acre furrow slice, about 6 in. or 15 cm deep, of an upland mineral soil weighs about 2,000,000 pounds (2,240,000 kg/ha). Similar calculations can be made to convert milliequivalents per 100 grams to kilograms per hectare or other units of weight and area.

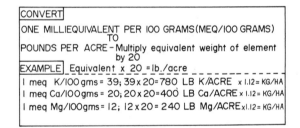

Figure 1.8   Conversion of milliequivalents per 100 grams to lbs. per acre.

Magnitude of CEC.   The amount of CEC of soils is determined by the amount of clay, the kind of clay and the amount of humified organic matter. The magnitude of the capacity varies from 10–20 milliequivalents (meq) per 100 grams for clays such as kaolinite, found in highly weathered soils in the tropics and subtropical areas, to 40 to 80 meq/100 grams for clays such as montmorillonite, found in cooler temperate climates. CEC values for organic matter are much higher, amounting to 100 to 200 meq/100 grams.

The CEC is a very important soil property (Figure 1.9). It affects the:

1. Capacity of the soil to hold nutrients such as Ca^{++}, Mg^{++}, K^+ and NH_4^+.
2. Quantity of a nutrient required to change its relative level in soils. For example, high CEC soils require more potassium (K^+) to raise soil potassium from a low to a high level than do low CEC soils.

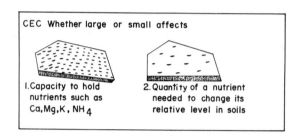

Figure 1.9   Magnitude of CEC.

This brings up the subject of base saturation which is an important concept about the cation exchange capacity. Hydrogen is a cation, but it is acidic and not basic like Ca^{++}, Mg^{++}, K^+, etc. Percent base saturation (Figure 1.10) expresses the percent of the CEC that is composed of basic cations—calcium, magnesium, potassium and sodium. For example, if a soil has a CEC of 18 meq/100 grams and if the total milliequivalents of Ca^{++}, Mg^{++}, K^+ and Na^+ is 15.3, then the soil is 85 percent base saturated, $15.3/18.0 \times 100 = 85$ percent. Availability of these nutrients to plants increases with their percentage of saturation. Also, as the percentage of base saturation increases, the amount of H^+ acid saturation decreases.

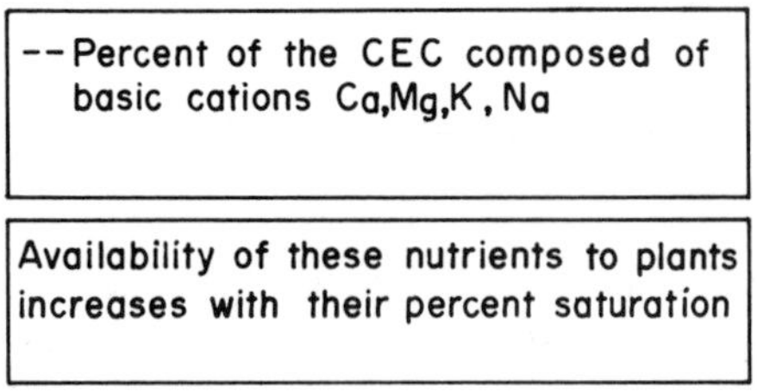

> --Percent of the CEC composed of
> basic cations Ca,Mg,K,Na

> Availability of these nutrients to plants
> increases with their percent saturation

Figure 1.10   Percent base saturation.

Soil Acidity and Alkalinity.   Another important chemical property of soils is their degree of acidity or alkalinity (Figure 1.11). This is expressed by a range of pH units from 0 to 14. A pH of 7 is *neutral*, below 7 is *acidic* and above 7 is *alkaline*. Most soils have a pH value between 4 and 8.5. The pH scale represents concentration or activity, not quantity. Two soils, a sand and a clay, may have the identical pH value and the identical percentage of base saturation, but they would differ greatly in total acidity because of a difference in CEC. Soil pH is measured in soil-water mixtures, usually 1:1 or 2:1 mixtures.

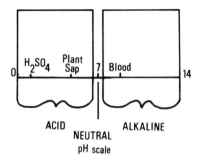

Figure 1.11   Acidity-alkalinity scale—(pH).

There are two causes of soil acidity: hydrogen and aluminum ions (Figure 1.12). Leaching of calcium, magnesium, potassium and other basic elements out of the soil results in a corresponding increase in hydrogen ions occupying adsorption sites on soil colloids or particle surfaces. Thus, acidity is increased. Aluminum ions create acidity as they react with water and form hydrogen ions. Aluminum is a "big bully" in mineral soils below pH 5.5. It nearly saturates the exchange sites of many soils at pH 5.0 and below. At about pH 5.5, it begins to be precipitated in inactive forms. Soluble aluminum is one of the most damaging elements to plant roots; it restricts growth and development of apex cells in the root systems in acid soils below pH 5.5.

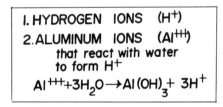

Figure 1.12   Two causes of soil acidity.

Factors contributing to development of soil acidity include:

1. Loss of bases by leaching and erosion.
2. Crop removal of bases such as K^+, Ca^{++} and Mg^{++}.
3. Carbon dioxide (CO_2) released by roots and microorganisms forming carbonic acid (H_2CO_3).
4. Acid-forming plant residues and fertilizers.

Nearly all organic plant residues, nitrogen-phosphorus-potassium fertilizers, and nitrogen materials are acid forming.

Finely ground limestone is the principal soil amendment used to reduce soil acidity and to increase soil alkalinity. The general reaction of dolomitic limestone in soil is:

$$\boxed{\begin{array}{c}\text{soil}\\\text{colloid}\end{array}}\begin{array}{l}\text{H}\\\text{H}\\\text{Al}\end{array} + CaCO_3 \cdot MgCO_3 + H_2O \rightarrow \boxed{\begin{array}{c}\text{soil}\\\text{colloid}\end{array}}\begin{array}{l}\text{H}\\\text{Ca}\\\text{Mg}\end{array} + Al(OH)_3 + CO_2$$

$$\text{(dolomitic lime)}$$

Acid condition   Limed condition

The negatively charged acid soil colloid is surrounded by H^+ and Al^{+3} ions. The finely ground limestone slowly dissolves in soil water forming CO_2 and the cations Ca^{++} and Mg^{++} that displace Al^{+3} and H^+ from the soil particle. This is a neutralization of acidic cations (H^+ and Al^{+3}) by basic cations (Ca^{++} and Mg^{++}). Aluminum hydroxide, $Al(OH)_3$, is an inactive product of the neutralization process.

Plants do best when the pH is right, so soils should be limed with the crop in mind (see Table 7.5 on page 229). Some plants do best in soils with a high pH value, whereas others do well at much lower pH values. Alfalfa has one of the highest pH, calcium and magnesium requirements. Potatoes have one of the lowest pH requirements in soils in which scab is a problem because the activity of actinomycetes, the soil microorganism causing scab, is increased as the pH level increases.

Salts in Soils and Fertilizers. Salts are defined as the products, other than water, of the reaction of an acid with a base. Salts commonly found in soils break up into cations (sodium, calcium, potassium, etc.) and anions (chloride, sulfate, etc.) when dissolved in water.

A *saline soil* is one containing enough soluble salts to impair its productivity for plants but not containing an excess of exchangeable sodium.

A *saline-alkali* soil is one containing sufficient exchangeable sodium to interfere with the growth of most crop plants and containing appreciable quantities of soluble salts. The exchangeable sodium is greater than 15 percent saturation of the CEC, and the electrical conductivity of the saturated soil extract is greater than 4 mmhos per centimeter at 25°C. The pH of the saturated soil is usually less than 8.5.

Reclamation is the process of removing excess soluble salts or excess sodium from soils and restoring lands to productivity.

The *salt index* is an index used to compare solubilities of chemical compounds. Most nitrogen and potassium compounds have high indexes, and phosphate compounds have low indexes (Table 1.6). When applied too close to seed or on foliage, the compounds with high indexes cause plants to wilt or die. Therefore, placement of fertilizer is very important.

Table 1.6
SALT INDEX (RELATIVE EFFECT OF FERTILIZER MATERIALS ON THE SOIL SOLUTION)[1]

Material	Salt Index[2]	Material	Salt Index[2]
Anhydrous ammonia	47.1	Nitrate of soda	100.0
Ammonium nitrate	104.7	Nitrogen solution, 37%	77.8
Ammonium nitrate–lime	61.1	Nitrogen solution, 40%	70.4
Ammonium phosphate (11–48)	26.9	Potassium chloride, 50%	109.4
Ammonium sulfate	69.0	Potassium chloride, 60%	116.3
Calcium carbonate (limestone)	4.7	Potassium chloride, 63%	114.3
Calcium cyanamide	31.0	Potassium nitrate	73.6
Calcium nitrate	52.5	Potassium sulfate	46.1
Calcium sulfate (gypsum)	8.1	Sodium chloride	153.8
Diammonium phosphate	29.9	Sulfate of potash-magnesia	43.2
Dolomite (calcium and magnesium carbonates)	0.8	Superphosphate, 16%	7.8
Kainit, 13.5%	105.9	Superphosphate, 20%	7.8
Kainit, 17.5%	109.4	Superphosphate, 45%	10.1
Manure salts, 20%	112.7	Superphosphate, 48%	10.1
Manure salts, 30%	91.9	Uramon	66.4
Monoammonium phosphate	34.2	Urea	75.4
Monocalcium phosphate	15.4		

[1]After L. F. Rader, Jr., et al., Soil Sci. 55 (1943), 210–218.
[2]Relative values based on the salt index of nitrate of soda as 100.

FIXATION OF PHOSPHORUS

Fixation of phosphorus generally implies the conversion of the applied phosphorus that is water or citrate soluble to a more insoluble form. The phosphate ion, $H_2PO_4^-$, which is in solution, moves from the solution to attach itself to iron and aluminum on clay surfaces or to be precipitated into insoluble forms by iron, aluminum and/or calcium ions, which are also in solution.

Much research has been done to learn how phosphorus behaves in soils. The diagram (Figure 1.13), "Hills and Valleys of Phosphate Fixation," sums up the relationships that exist between ions in the soil solution, the clays of the soil and phosphorus as affected by the different degrees of acidity in the soil (expressed in terms of pH). The diagram is made up of experimental curves that express definite relationships. The higher the hill, the greater the fixation capacity; the lower the hill, or the deeper the valley, the less the relative fixation.

It is obvious that phosphorus has the best chance to be available in the soil at the slightly acidic reaction of pH 6.5. This is one of the reasons why it is so important to correct strong acidity by liming the soil.

Since phosphates are rapidly fixed, they do not leach out readily or

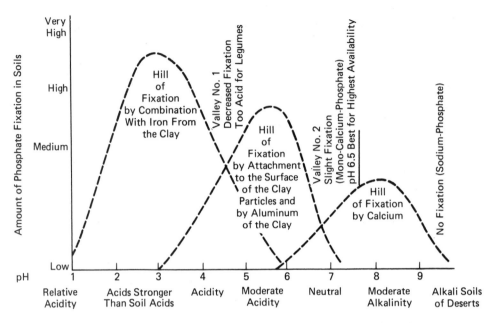

Figure 1.13 Soil reaction range—hills and valleys of phosphate fixation. (*Better Crops with Plant Food*, G.D. Scarseth, vol. 24, no. 3, 1937.)

move much in the soil. To offset this, phosphate fertilizers should be applied frequently (every year) instead of in infrequent, large applications aimed at supplying the phosphorus needs of the plants for three or more years. The available phosphorus-carrying fertilizers should not be mixed too thoroughly with the soil since thorough mixing increases the contact between the phosphorus and the fixation components in the soil. Adding the available phosphates (superphosphates and ammonium phosphates) in narrow bands in the soil at the side of the seed is usually best.

SOIL MICROBES

Microbial life in soils includes animals in addition to plants (Figure 1.14). Protozoa are the smallest forms of animal life and are most plentiful under high water levels. Nematodes are small, eel-like animals that can be serious parasites, especially in cotton on sandy, coarse-textured soils and in sugar beets grown in cooler, temperate climates. Although barely visible to the naked eye, nematodes can be easily identified under a microscope.

Microbial plants in soils include bacteria, actinomycetes, fungi and algae (Figure 1.15). Of these, bacteria are the most important in soil fertility and plant nutrition. Bacteria thrive in soils near a neutral pH. Actinomycetes also are favored in limed soils, and white potatoes under

18

ANIMALS

PROTOZOA -
the smallest forms
of animal life

NEMATODES -
eel-like animals that
can be serious parasites

Figure 1.14   Microbial animal life in soil.

limed conditions develop scab which is a disease caused by actinomycetes. In plant classification, actinomycetes are intermediate between bacteria and fungi. Fungi grow best in acid soils. Algae are a varied group. Some fix nitrogen from the atmosphere, and others carry on photosynthesis like higher plants; blue-green algae do both. Japanese research results indicate that these algae in moist rice soils fix up to 36 pounds of nitrogen per acre (40 kg/ha) from the air.

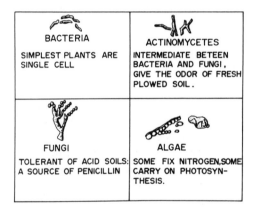

Figure 1.15   Microbial plant life in soil.

The three major roles (Figure 1.16) of bacteria in soil fertility and plant nutrition are:

1. Nitrogen fixation, both symbiotic, which means in partnership with legumes, and nonsymbiotic, which means free living, without legumes. Legume plants provide food energy for the symbiotic group whereas the nonsymbiotic group obtains food energy from decomposing organic matter in soils.
2. Nitrogen transformations, which are:
   a. Mineralization, which is the conversion of organic nitrogen such as protein into ammonium nitrogen.

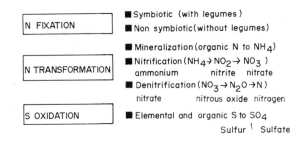

Figure 1.16   Major roles of bacteria in plant nutrition.

    b. Nitrification, which is the conversion of ammonium to nitrite and nitrite to nitrate, the preferable form of nitrogen for higher plants.

    c. Denitrification, which is the conversion of nitrate nitrogen to nitrous oxide (N_2O) and eventually to elemental nitrogen (N_2), which is the form that constitutes about four-fifths of air by volume.

   3. Sulfur oxidation, which is the conversion of elemental sulfur and organic sulfur to sulfate ($SO_4^=$), the preferable form of sulfur for higher plants.

The importance of these transformations for the equilibrium of the environment, as well as for the production of basic food crops, cannot be overemphasized.

    Legume inoculation is essential for symbiotic nitrogen fixation (Figure 1.17). Nature provides the bacteria for legumes, but nature must be assisted when man creates enormous volumes of feed, oil and food grains such as soybeans. The seed should be artificially inoculated every few years or, if the same species of legume is planted every third year on the same land, one inoculation will suffice. This symbiotic relationship accounts for the production of millions of tons of nitrogen annually. But, the nitrogen-fixing bacteria do not work for nothing. Plants must provide them

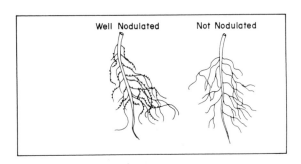

Figure 1.17   Legume inoculation essential for nitrogen fixation.

with enormous quantities of food energy in the form of carbohydrates in order for the bacteria to convert nitrogen in air into forms plants can use.

The quantity of nitrogen fixed by these bacteria, known as *Rhizobia*, depends on the supply of carbohydrates in the legume, the vigor of the microorganism, the soil lime level, the soil fertility level and the soil moisture available. Proper inoculation, whether by nature or by man, of the legumes is one of the basic requirements for maximum nitrogen fixation.

Soil temperature strongly influences the activity of all soil microorganisms, and it is especially important in the transformation of ammonia to nitrate (Figure 1.18). Nitrification is most rapid in well-aerated, moist, warm, fertile soils containing abundant ammonium nitrogen. There is hardly any significant nitrification until the soil temperature reaches about 50°F (10°C). This is the reason reference is often made to this temperature for the application of ammonium nitrogen. Young plants can utilize ammonium nitrogen in the cool 50°F (10°C) weather in the spring, and nature made this so. Later on in the growing season, when the soil warms, plants utilize nitrate more efficiently. Nitrification increases rapidly with an increase in soil temperature above 50°F (10°C), reaching a maximum rate above 90°F (30°C).

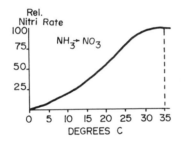

Figure 1.18 Soil temperature greatly influences nitrification rate.

Denitrification, as mentioned previously, is the conversion of nitrate to nitrous oxide and hence to elemental nitrogen that escapes to the atmosphere. In this process, certain bacteria use the oxygen in nitrate (NO_3^-). Poor drainage and the lack of air to provide oxygen favor this process (Figure 1.19). It is especially favored in poorly drained upland soils or flooded rice soils containing nitrate nitrogen. Instead of helping to

> ...The use of oxygen in nitrate Nitrogen (NO_3) by some bacteria in water-logged soils forms nitrous oxide(N_2O) or elemental N that escapes to the atmosphere

Figure 1.19 Denitrification.

provide nitrogen, as was the case with *Rhizobia*, the denitrifiers or anaerobic bacteria cause flooded soils to lose much-needed nitrogen to the atmosphere. Hence, ammonium sources of nitrogen are used on rice, and they are preferred over nitrate or urea forms of nitrogen in other soils under water-saturated conditions.

PLANT NUTRITION

The use of large amounts of CO_2 and H_2O to provide chemical energy in seeds was alluded to earlier. This process of converting radiant energy to food energy in plants containing chlorophyll is photosynthesis. This and other processes associated with plant growth and the development of roots in soils will be discussed as plant-soil relationships.

Photosynthesis. Green plants in the presence of light convert carbon dioxide (CO_2) and water (H_2O) into carbohydrates with water and oxygen as by-products in accordance with the following reaction.

$$CO_2 + 2H_2O \xrightarrow[\text{green plant}]{\text{light}} CH_2O + H_2O + O_2$$
$$\text{air} \quad \text{soil} \qquad\qquad \text{carbohydrate}$$

In this process, carbon in CO_2 is partially released from the group of two oxygen atoms. This uncoupling is done by hydrogen released from water through energy provided by sunlight. Hydrogen then combines with carbon to form carbohydrates, such as sugars and starches. Numerous reactions, including enzymes, activators, co-factors and hydrogen acceptors and donors, are involved in this life-giving process within the green leaf. Any nutrient deficiency can sharply curtail the efficiency of this process and thus reduce sugar and starch production.

Nutrient Uptake by Roots. Life of a plant begins with a seed germinating in soil. This initial step, using energy from starch in the seed, produces a seedling that contains a root and a shoot. As soon as the shoot obtains light, it begins the process of photosynthesis while the root begins absorbing water and nutrients to supply raw materials for photosynthesis (Figure 1.20). This early stage in the plant life cycle and especially the root are strongly influenced by soil properties such as moisture content, structure, nutrient availability and salt concentration.

Roots, the underground operators for plants, make up less than one-third of total plant weight, and they occupy only a small fraction of the soil, about one percent in the case of corn (Figure 1.21). The root emerges first from a germinating seed. Thus, the new plant's life begins as the roots meander between soil particles, bathed by the water of the soil solution and fed by the nutrient elements found there. Anything that

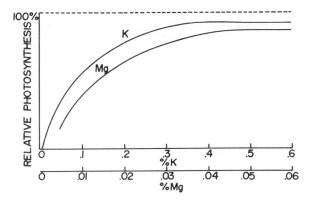

Figure 1.20   Adequate nutrients increase photosynthesis.

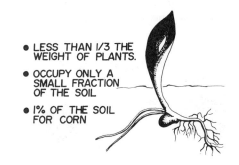

- **LESS THAN 1/3 THE WEIGHT OF PLANTS.**
- **OCCUPY ONLY A SMALL FRACTION OF THE SOIL**
- **1% OF THE SOIL FOR CORN**

Figure 1.21   Underground roots.

restricts root development, and thus water and nutrient absorption, reduces production "upstairs" in the leaves.

Passage of nutrient ions from the soil solution into living roots is basically an ion exchange process like cation exchange on colloid surfaces discussed before. However, nutrients must first get to the surface of roots for adsorption to occur. There are three ways this is accomplished.

1. Root interception. A root runs into some nutrient ions as it grows through soil.
2. Mass flow. The large volume of water entering roots carries soluble substances with it.
3. Diffusion. Ions move to roots along concentration gradients, from higher concentrations to lower ones.

A careful study of these mechanisms reveals the importance of high nutrient levels in soils, and of localized placement in the case of phosphorus, so that inadequate nutrient uptake does not become a limiting factor in plant growth.

Mass flow, especially for nitrogen, sulfur, calcium, magnesium and potassium, and diffusion, especially for phosphorus, provide root surfaces with most of the nutrient supply. Broadcast fertilizer applications, which build up nutrient supply throughout the soil root zone, favor nutrient uptake depending primarily on mass flow. Localized placement for most soils favors the uptake of phosphorus, which is dependent on diffusion and which is very reactive with soil substances that reduce its availability to plants.

The source, which in fertilizer terms refers to the carrier of the nutrient, as well as the method of application influence nutrient availability. In this regard, there is probably more information on phosphorus than on any other nutrient. Percentage of recovery by plants is closely related to its solubility in water. Phosphorus from a source of high solubility diffuses into a much larger volume of soil than the same amount from a source of low solubility. Hence, chances for root contact are better. From the example given in Figure 1.22, it can be calculated that the volume of soil contacted by a highly soluble phosphate fertilizer was eight times that of a low solubility source.

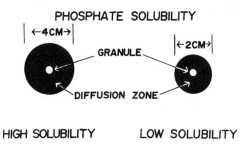

Figure 1.22 Source and method of application affects nutrient to roots.

Water Use and Efficiency. Although a small part of the water taken up by roots is used in photosynthesis as a source of hydrogen, most of it is lost by transpiration from leaf surfaces. This loss is tremendous, amounting to as much as 400 pounds (180 kilograms) for a single corn plant in one season. Figures for this loss in terms of kilograms of water per kilogram of dry matter produced are shown in Figure 1.23. Note that corn and grain sorghum use the smallest amounts of water for a kilogram of dry matter formed. This loss of water per kilogram of plant dry matter produced is called the transpiration ratio.

The two sources of water for plants are precipitation and stored soil moisture. The latter, often called subsoil moisture, is very important in soils of all kinds, but is of particular importance in soils developed in areas of limited precipitation. Roots may extend 4 to 6 feet (120 to 180 cm) deep in soils of good tilth. Such soils are able to hold as much as 2 inches (5 cm) of available water per foot (30 cm) of soil depth. Soil moisture at planting time strongly influences fertilization rates, especially where pre-cipitation is low during the growing season. In some of these areas, soil

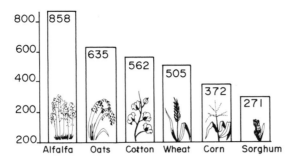

Figure 1.23   Water transpired for each KG of dry matter produced.

moisture tests are made along with soil nutrient tests in order to use more efficiently the soil moisture and nutrients. Adequate fertilization is essential for maximum root development and utilization of soil moisture.

Water-use efficiency is becoming more and more important as higher yields push water supplies to the limit. There are basically two ways to improve water-use efficiency:

1. Reduce evaporation from leaves. Some progress has been made in reducing the transpiration ratios by adding potassium to salty irrigation water. Potassium functions to reduce the size of stomates (holes in the leaf) through which water is lost. Reducing the aperture of the stomates also decreases CO_2 entry and thus may limit photosynthesis; this defeats the purpose of decreasing water loss.
2. Increase the ratio of output of dry matter per unit of water used. This appears to be the most feasible approach at present.

Proper fertilization is one of the most practical methods of increasing crop output or yield per unit of water used. The reason is that low-yielding crops consume just about as much total water as do high-yielding crops. Results of experiments in Nebraska illustrate this. Fertilized corn in this experiment gave a yield increase of 38 bu/A (2554 kg/ha) and produced 2.3 bushels (62.7 kg) of corn more per inch (2.5 cm) of water used.

Plant Nutrition Status.   In addition to soil tests, there are four diagnostic techniques for plant nutrition. These are:

1. Visual symptom of deficiency.
2. Chemical analysis of plants or plant parts.
3. Tissue tests.
4. Measured plant response to added nutrients.

These diagnostic techniques, when combined with soil-test results, can be valuable in identifying and correcting factors that limit yield and quality.

However, in most instances, these techniques are most useful in correcting deficiencies in the next crop because often it is too late to correct damage done to the current crop.

Visual symptoms can be useful for identifying deficiencies. Some symptoms are clearly defined and typical whereas others are not. Also there are times when two or more nutrient shortages produce the same symptoms. Thus, visual diagnoses should be confirmed with one of the other methods when possible.

Total plant analysis, using proper chemical methods, is an exact science. Unfortunately, in the case of trees, shrubs and large crop plants, it is not practical to analyze the total plant in the chemical laboratory, therefore, a plant part, such as leaf, petiole or stem, must be used. Accurate results depend on proper sampling of the appropriate plant part at the appropriate time, and on the essential prior research to calibrate results of the plant part analysis with total plant growth and development. Many laboratories provide these analyses and report deficiencies, sufficiencies and excessive levels of nutrients in most crops.

Tissue tests are made on live plants in the field. Chemical reagents are available in kits for this purpose. They are used to measure nitrate nitrogen, soluble phosphorus and potassium in plant tissue. Tissue tests are used to confirm visual diagnosis and to indicate necessary laboratory tests and analyses. It is one of the essentials of a good field kit. The other is the soil-testing kit. When properly calibrated, some of the same reagents, for use as extracting solutions, can be used for plant tissue and soil tests.

Measured plant response to an added nutrient is still the most conclusive diagnostic tool. The other techniques provide clues or leads in diagnosis, but measured actual response in the growth chamber, glass house or field provides the final answer. Thus, carefully designed field experiments with various sources and rates of nutrients added over a period of several years must become an integral part of each complete diagnostic program aimed at increasing yields from the green leaf.

SUMMARY

Life of a green plant begins with a seed germinating in soil or other appropriate medium containing a means of anchorage, a supply of nutrients and adequate water for growth and development. This initial step in the plant life cycle is strongly influenced by soil moisture, soil structure, soil nutrient status, salt concentration and other factors. From the time the first root begins penetrating pore space in the soil and the first protruding leaf moves above the soil surface to reach radiant energy, the pace at which the plant grows and reproduces is set by a host of soil properties.

How well a grower understands these basic soil properties will largely determine his success in modifying them in order to promote healthy plant growth and to produce high yields at harvest. One soil may be largely organic matter in which fire is a serious hazard, another may be an alkali soil in which lime is a curse, while another may be an acid soil low in organic matter in which lime is a blessing. Modification of the physical, chemical and biological properties of a soil to favor the growth of the green plant is the key to successful soil management.

Modern fertilizers permit growers to add 13 of the essential plant nutrient elements that the green plant requires from the soil to produce profitable yields. By using a fertilizer of the proper kind and amount, plant nutrient deficiency can be eliminated as a factor limiting the growth and development of the green plant.

QUESTIONS

1. Other than oxygen, what are the three most abundant elements found in soil materials?
2. List the sixteen essential elements for plant growth?
3. In what ways do A horizons differ from B horizons?
4. From a physical standpoint, describe sand, silt, clay and organic matter?
5. List the four soil components and give approximate percentages of each?
6. Define ion, cation and anion?
7. Define colloid.
8. The CEC or cation exchange capacity of a soil is a measure of the negative charge of the soil colloids that can be neutralized by easily replaced cations. How is it expressed?
9. What determines the magnitude of the CEC?
10. What is the range of pH values in soils?
11. Are alkaline soils found in regions of low or high rainfall?
12. What is phosphorus fixation?

COMPLETION

13. Microbial plants in soils include _____, _____, _____ and _____. Of these _____ are the most important in soil fertility.
14. Three major roles of bacteria in soil fertility are _____, _____ and _____.
15. Complete the reaction for photosynthesis:

$$CO_2 + 2H_2O \xrightarrow[\text{green plant}]{\text{light}} CH_2O + \underline{\hspace{1cm}} + \underline{\hspace{1cm}}.$$
air   soil   carbohydrate

16. The three ways that nutrients get to the surface of roots so that uptake can occur are _____ , _____ and _____ .

17. A single corn plant can take up as much as _____ of water in a single growing season.

18. Measured plant response to an added nutrient is the most conclusive diagnostic tool. Two other measurements are _____ tests and soil tests.

chapter 2

Nitrogen—The Keystone of Protein

Nitrogen compounds can be used in many ways—as a liquid, gas or solid—to increase protein and the production of food. Here it is added as a liquid by injection with a knife-type applicator. This same equipment can also be used to place nitrogen solution—herbicide mixtures in a band on top of the bed to feed the plant and at the same time act as a pre-emergence weed control measure.

Nitrogen is present in all living matter as a constituent of protoplasm and is, therefore, essential to both plants and animals. The atmosphere is about four-fifths nitrogen gas by volume. Nitrogen is odorless, tasteless, and *worthless* to human beings until combined with other elements. In combination with hydrogen, nitrogen constitutes an integral part of amino acids, which form protein. It also is a part of the chlorophyll molecule and is a constituent of alkaloids, which make certain plants extremely valuable to man. Nitrogen usually ranks fourth, following carbon, oxygen and hydrogen in percentage of total dry matter of plants. Since carbon, oxygen and hydrogen come from air and water, nonlegumes must obtain more nitrogen from the soil than any other essential element, and usually more than all the others, except potassium, combined. It is usually found to be the most deficient essential element in the cultivated soils of the world.

Because soils have little capacity to retain oxidized forms of nitrogen, most nitrogen is associated with inert organic matter and microorganisms in the soil. Nitrogen is not a constituent of most rocks and minerals. Therefore, as the organic reserves become depleted, nitrogen becomes limited for higher plant growth. The most notable exception to this is leguminous plants that receive nitrogen from bacteria of the genus *Rhizobium*. Like an ammonia factory, they combine atmospheric nitrogen into usable compounds. These bacteria invade the tissue of the legume root and exchange nitrogen for carbohydrates of the green plant. Several other microorganisms are known to fix atmospheric nitrogen in the soil, although existing entirely independent of higher plants. These include *Azotobacter* and *Clostridium pastorianum* bacteria and blue-green algae.

Three factors govern the assimilation of nitrogen by plants: the type of plant, the stage of plant maturity and the supply of available nitrogen in the soil. The stage of growth and the degree of leafiness result in a nitrogen content ranging from as little as 0.5 to over 5 percent of the dry weight. The natural soil supply of nitrogen is largely governed by the oxidation rate of the organic matter present. The activity of microorganisms is directly related to the amount of undecomposed organic matter, and the microorganisms compete with higher plants, in most cases, for the nitrogen.

The atmosphere is approximately 78 percent nitrogen and 21 percent oxygen. Calculations show that there are about 36,000 tons (32,760 m tons) of nitrogen over each acre (0.4 ha) of land and water (43,560 ft^2/acre $\times$ 14.7 psi $\times$ 144 in.^{2}/ft^2 $\times$ 0.78 $\div$ 2000 lb/ton = 35,961). However, disregarding the legumes, crops benefit little from this supply. Only about 20 lb/acre (22.4 kg/ha) of nitrogen per year is added by rainfall.

The free-living organisms fix only 15 to 45 lb/acre (16.8 to 50.4 kg/ha) nitrogen per year.[1] Organic matter in the soil decomposes at the rate of 1 to 2 percent a year. This means that 20 to 40 lb/acre (22.4 to 44.8 kg/ha) of nitrogen is made available from this source for plant use. The supply from air, organisms and organic matter is subject to change, usually downward. Thus, all three sources of nitrogen would fall far short of the nitrogen requirements of a 30-ton (27.2 m ton) crop of corn which, if harvested for silage, would remove about 235 lb/acre (263 kg/ha) of nitrogen. Further, the amount fixed in the soil (by a process called azofication) is about the same as the amount lost to the atmosphere by volatilization. For practical purposes, it is not wrong to consider that azofication and volatilization approximately balance each other. Thus, the only practical solution is to convert the nitrogen in the air to a form that will remain *stable* long enough to benefit higher plants.

Anhydrous ammonia, made by compressing hydrogen and air at high temperatures, is the first step in converting nitrogen from the air to a form usable by plants. It is the number one direct-application nitrogen fertilizer in the United States[2] and is widely used in the manufacture of multinutrient fertilizers.

Ammonia and compressed air at high temperatures in the presence of a catalyst and steam will produce nitric acid (HNO_3). Nitric acid, blended with anhydrous ammonia in a stainless steel tank, will produce ammonium nitrate liquor. This liquor, after processing, is used in the manufacture of multinutrient fertilizers and as a liquid for direct application to the soil. When evaporated, it produces solid ammonium nitrate, which is the number two direct-application nitrogen fertilizer in the United States.[3] Nitric acid neutralized with sodium bicarbonate forms nitrate of soda, which is also mined and purified from naturally occurring deposits in Chile. Ammonium nitrate liquor is mixed with finely ground limestone, and the slurry is granulated to produce a mixture containing 20.5 percent nitrogen.

By combining pure ammonia with carbon dioxide gas under high pressure, urea is made. By distilling calcium carbide and pure nitrogen from the air and subjecting them to high temperature, calcium cyanamide is made.

The origin of all these materials is the nitrogen in the air. Ammonia production is the first step in the manufacture of most nitrogen fertilizers. A nitrogen fertilizer production flow chart is shown in Figure 2.1. It was prepared by Southern Nitrogen, Inc., of Savannah, Georgia. For the beginning student of fertilizer technology and use, it is an excellent

[1]A. D. Hall, *Book of the Rothamsted Experiments* (London: E. P. Dutton and Company, Inc., 1905), p. 139.

[2]E. A. Harre, M. N. Goodson, and J. D. Bridges, *Fertilizer Trends,* National Fertilizer Development Center, Tennessee Valley Authority, Muscle Shoals, Alabama, 1976.

[3]*Fertilizer Trends.*

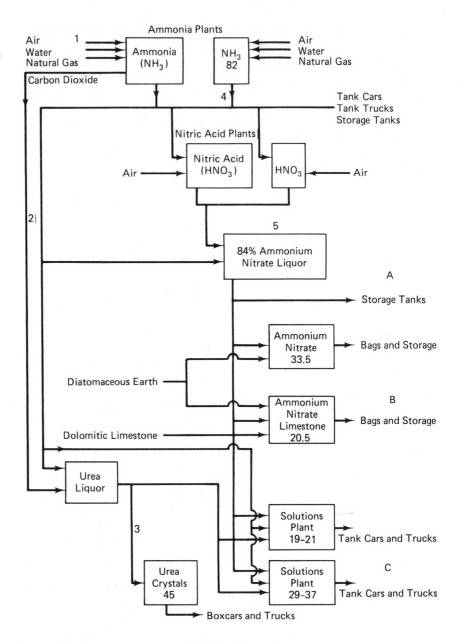

Figure 2.1 Nitrogen fertilizer production flow chart.

description of the manufacture of several nitrogen compounds. A description of the five steps in production follows:

1. Air, water and natural gas are taken into the two ammonia plants where ammonia is produced. A by-product in the production of ammonia is carbon dioxide.

2. The ammonia and carbon dioxide can be pumped together in a reactor at 3,000 psi (204 atm) and 365°F (185°C) to form urea liquor.

3. The urea liquor can be formed into urea crystals that are used for fertilizers and in the manufacture of glues, paints, varnishes, plastics and livestock feed.

4. The ammonia can be sent on to the nitric acid plants where additional air is mixed with it at a ratio of nine parts of air to one part of ammonia. When this mixture is passed over platinum, it gives off oxide of nitrogen. It is then scrubbed with a condensate (distilled water) that gives nitric acid.

5. The nitric acid is then combined with additional ammonia to form 84 percent ammonium nitrate liquor. The ammonium nitrate liquor is sent on to (A) storage tanks, (B) solids plants or (C) solutions plants.

Solids Plants

1. The ammonium nitrate liquor is taken into the ammonium nitrate plant where it is pumped to the top of the prilling tower. The ammonium nitrate is then sprayed out in small droplets; it falls 185 feet (56.1 m), forming small prills. The prills are coated with 3 percent diatomaceous earth. This product contains 33.5 percent nitrogen.

2. Ammonium nitrate liquor is taken into the ammonium nitrate limestone plant where it is sprayed into a rotating bed of dolomitic limestone. The product then goes through a drying process and is screened, cooled, and sent to the warehouse. This material is also coated with diatomaceous earth and contains 20.5 percent nitrogen.

Solutions Plant

The ammonium nitrate liquor is taken into the solutions plant where it can be diluted and made into a straight ammonium nitrate direct-application solution or combined with urea to form an ammonium nitrate–urea solution. Ammonia can also be added to the solutions for producing manufacturing-type solutions.

AMMONIA

In 1913, the Haber-Bosch process of combining nitrogen and hydrogen directly went into commercial operation in Germany. In the United States, the first two successful synthetic ammonia plants were started about 1920 by the Allied Chemical Corporation and by Olin Mathieson at Niagara Falls, New York. World War II brought a need for fixed nitrogen for explosives, and, as a consequence, 11 new plants were constructed. In 1951, tax laws in the United States gave impetus to private industry, and 17 new synthetic ammonia plants were constructed. By 1963, there were 80 plants in the United States with a capacity for producing more than 5 million tons (4.54 million metric tons) of nitrogen annually. In 1973, capacity had increased to 10 million tons (9.09 million metric tons) of nitrogen annually. By 1980 it is estimated that there will be a demand for 17 million tons (15.4 million metric tons).

Along with rock phosphate, potash salts and sulfur, ammonia is a basic raw material of the fertilizer industry. It, or its derivatives, is found in all nitrogen-bearing fertilizers.

Description.   Under normal temperature and pressure, anhydrous ammonia is a pungent and colorless gas with a liquid density of 5 pounds (2.27 kg) per gallon (3.8 liters). Each 5 pounds (2.27 kg) of ammonia contains 4.1 pounds (1.86 kg) of nitrogen. At 50°F (10°C), the gas exerts a pressure of 75 pounds (34.1 kg) per square inch (6.45 cm²), while the pressure at 100°F (38°C) is 197 pounds (89.5 kg) per square inch. It is stored in steel tanks (plastics are sometimes used) that have the capacity for a pressure of 265 pounds (120.4 kg) per square inch. However, a pop-off safety valve is employed to release the gas at 240 pounds (109.1 kg) per square inch in routine operation.

Method of Manufacture.   The synthesis of ammonia is a simple process in contrast to the preparations of the raw materials. All of the nitrogen used in ammonia production comes from the air. Hydrogen for the process is derived from water, naphtha, natural gas, coal or oil. The original Haber-Bosch process is the basis of all the various processes. It consists of subjecting a mixture of nitrogen and hydrogen to high pressures in the presence of an appropriate catalyst.

$$N_2 + 3H_2 \leftrightharpoons 2NH_3$$

The reaction is exothermic, yielding 11.0 kcal per mole NH_3. An iron catalyst is used to increase the yield because the reaction rate is slow. Ammonia synthesis in most U.S. factories is carried out at temperatures of 932°F (500°C) and at an operating pressure of 2,940 to 14,700 psi (200 to 1,000 atm). A complete, typical synthetic ammonia circuit is described by Sharp and Powell.[4] A typical plant has a production capability in excess of 800 tons (728 m tons) per day (Figure 2.2).

[4]J. C. Sharp and R. G. Powell, "Advances in Nitrogen Fixation," in *Fertilizer Technology and Usage,* eds. M. H. McVickar et al. (Madison, Wis.: Soil Science Society of America, 1963).

Figure 2.2   Ammonia plant, dominated by towering reformer furnace, at Columbia Nitrogen Corporation in Augusta, Ga.

As stated previously, nitrogen for the process comes from the air, whereas the hydrogen may come from various sources. It is derived directly and also as a by-product. About three-fourths of the hydrogen for ammonia synthesis in the United States is derived from purified natural gas, which is largely methane (CH_4). Hydrogen is derived from methane by steam reforming and partial oxidation. About 10 percent of the hydrogen in the United States comes from refinery gas. It is handled in the same manner as natural gas.

In countries where the availability of natural gas or refinery gas is limited, water gas is an important source. It is produced by passing steam over red-hot coke in a water-gas generator.

$$H_2O + C \rightarrow CO + H_2$$
$$CO + 2H_2O \rightarrow H_2CO_3 + H_2$$

In 1978, research on this process was revived for U.S. hydrogen production.

For years, ammonia was derived from coke ovens and was captured as ammonium sulfate and ammonia liquor. The nitrogen content of coal is about 1.5 percent and exists in organic combinations consistent with the vegetable matter from which coal was formed. The release and capture of this nitrogen by combustion has never been an entirely efficient process, and the conversion results in the recovery of 15 to 20 percent of the original nitrogen. The use of this process represents less than 2 percent of U.S. ammonia production, but it is still used in Europe where rich coal deposits exist.

Electrolysis accounts for about 6 percent of the hydrogen for ammonia synthesis. It is dependent on a cheap source of electric power or on a situation in which electrolysis is employed for the production of chlorine, and hydrogen is a by-product.

Shipment and Handling. The high pressures exerted by anhydrous ammonia can be dangerous, but they are no more dangerous than the pressures exerted by steam boilers that are in common use in factories and on the farm. Many states have rules and regulations concerning equipment for handling ammonia and steam. Persons handling these gases should become familiar with the regulations. Ammonia is handled as a liquid under pressure and is released as a gas through an ammonia-resistant hose into the soil. Bronze and brass fittings react chemically with ammonia and should not be used.

Incomplete filling of tanks is a safety measure. When the temperature of liquid ammonia in the tank is 30°F (−1°C), the tank can be safely filled to 86 percent of its capacity. At 100°F (38°C), the tank can be safely filled to 95 percent of its volume. If a tank were completely filled with liquid ammonia, a very small increase in temperature would be sufficient to raise the pressure enough to cause an explosion if the pop-off valve failed to work.

Ammonia is shipped from the factory in 10,000-gallon (38,000-liter) tank cars that will hold 25 tons (22.75 m tons) of ammonia. Upon delivery to the railroad siding, it is piped into a 30,000-gallon (114,000-liter) stationary tank that has a capacity of 65 tons (59.15 m tons) of ammonia. This is usually the size tank employed by ammonia distributors. It is sufficient to fertilize 1,000 acres (404.9 ha) with about 100 lb/acre (112 kg/ha) of nitrogen. Some dealers employ twin 6,000-gallon (22,800-liter) tanks that store a 10,000-gallon (38,000-liter) carload of ammonia at a time and permit a 2,000-gallon (7,600-liter) reserve to be maintained. The volume of the dealer's operation may be increased by supplying farmers with 500- or 1,000-gallon (1,900- or 3,800-liter) transport tanks and tractor equipment or by applying the ammonia into the soil for a fee. Transport trucks are used to deliver ammonia from the factory to farms where storage facilities are available.

Farm tractors equipped for ammonia application have 70- to 100-gallon (266- to 380-liter) tanks mounted in the rear of the driver's seat. Trailer ammonia equipment carries much larger tanks (Figures 2.3 and 2.4). A 110-gallon (418-liter) tank holds about 420 pounds (190.9 kg) of

Figure 2.3   This tractor-drawn ammonia rig is used with a chisel plow. Chiselling is an effective method of breaking up hard compacted soil (sometimes called hard pans) that develops at plow-sole depth. The shattering of such pans prevents root constriction. The ammonia provides N for root elongation.

nitrogen, equivalent to about 1.25 tons (1.13 m tons) of 16 percent nitrogen or to about 1,250 pounds (568 kg) of 33 percent nitrogen. Because it is necessary to cover this gas with soil to prevent volatilization, anhydrous ammonia is conducted into the soil from the tank by a hose attached to the rear of applicator knives. A good knife should have a point 3 inches (7.5 cm) long with a sweep 10 inches (25 cm) wide. Such design will permit good sealing of the ammonia even in compact soils and pastures. Disk hillers, shovels or other covering devices may be used to place loose soil against the applicator knives. Ammonia must be placed about 4 inches (10 cm) deep and be well covered with soil to prevent the escape of the gas.

The rate of delivery to the soil is a most important factor because the quantity of ammonia added determines the final crop yield obtained, other factors being equal. Various regulating valves, gauges, flowrators and pumps are employed to control the rate of nitrogen delivery to the soil. The pump is geared to the tractor and maintains a constant rate of ammonia flow to the soil. Orifices, delivery lines and applicator-knife openings should be of uniform size. Each set of applicator equipment should be carefully calibrated. The manufacturer of the various devices should provide calibration data for each machine.

Safety.   Anhydrous ammonia is a safe and economical nitrogen product to handle and use provided that it, like so many other farm chemicals, is

Figure 2.4 Equipment is available to apply anhydrous ammonia separately or in combination with almost any tillage instrument. This trailer-type rig is used to fertilize pasture by injecting ammonia into sod. The 1000 gallon (3800 liter) nurse tank in the background holds enough ammonia to fertilize 41 acres (16.6 ha) with 100 lb/A (112 kg/ha) of N.

used with care by competent workmen. The following safety precautions are suggested:

1. Obtain a copy of the regulations concerning the storage and handling of anhydrous ammonia from the state agency that regulates handling of gases under pressure.

2. Buy tanks, equipment and accessories that are made according to underwriters' codes. Do not use brass or copper fittings because ammonia reacts with copper.

3. Obtain instructions on transferring ammonia from tank to tank. Pick up the line by the hose or pipe, never by the valve (Figure 2.4). Never place your head over the pipe on the tank that houses the relief valve. Wear gloves and goggles.

4. Maintain rubber gloves, goggles, 5 gallons (19 liters) of water and a first-aid kit on all transport equipment.

5. Apply water immediately to any part of the body that has come in contact with ammonia. Use lots of water, but do not apply salves, ointments or greases.

6. Do not put fuel gases, such as butane and propane, in tanks in which ammonia has been stored, unless a competent authority has been consulted. The reaction of these gases and ammonia produces cyanogen (CN_2), a poisonous, colorless gas.

7. Recognize that ammonia can produce injury and handle it accordingly at all times. Park nurse tanks downwind. Before welding make sure that the tank has been purged of ammonia.

In concentrations as low as 0.05 to 0.1 percent by volume, ammonia is extremely irritating to the eyes and to the respiratory system. Concentrations of 0.2 cause excessive coughing and can be fatal at 1.0 percent. It blisters the skin.

Chemical Characteristics.   Ammonia contains 82 percent nitrogen by weight and is 99 percent NH_3. The ignition temperature of an air-ammonia mixture is 1436°F (780°C). Ammonia is soluble in water up to 30 percent NH_3 by weight, resulting in a low-pressure solution containing 24 percent nitrogen that is referred to as "aqua ammonia." Ammonia is a strong alkali and is used as a cleaning material. Ordinary household ammonia is a common cleaning material for floors, walls and metal surfaces.

Because ammonia and aqua ammonia are both basic, the immediate effect of adding them to soils is to raise the pH to above 9 in the zone of application. Ammonia reacts with clay, organic matter and soil moisture. After a few days in warm, aerated soil, microorganisms convert the ammonia to soluble nitrates, which move cations, such as calcium, potassium and magnesium, into growing roots or out of the root zone into percolating waters. Thus, its final effect is to reduce the basicity of the soil and to increase concurrently the soil acidity. The magnitude of the acidity developed is equivalent to 1.8 pounds (0.8 kg) of lime removed from the soil per pound (0.45 kg) of nitrogen added as ammonia.[5] The importance

[5]W. H. Pierre, "The Equivalent Acidity and Basicity of Fertilizers are Determined by a Newly Proposed Method," *J. Assoc. Off. Agri. Chem.* (1934), 101–107.

of determining the lime requirement of soils in humid regions at frequent intervals, for example, every 3 years, and taking care of this need is emphasized. Liming will prevent complications, such as a shortage of calcium and magnesium, and excesses of aluminum and manganese that are usually associated with a low pH level in the soil.

Crop Response. No farm practice has ever spread as rapidly as has the direct application to the soil of anhydrous ammonia. The practice and techniques for its use were first released by Dr. W. B. Andrews of the Mississippi Experiment Station in March, 1947. Today, and for the past several years, more nitrogen is added to the soil in the form of anhydrous ammonia than any other nitrogen material. It is the least expensive nitrogen fertilizer. This gas, when properly used, is as good as any other nitrogen fertilizer. Like other fertilizers, it should be placed several inches (centimeters) away from seed and young roots.

The Cold-Flo system injects liquid ammonia by gravity up to as many as 16 outlets on a primary tillage tool, such as a chisel plow. A nurse tank is pulled behind the tillage implement. Simultaneous tillage and nitrogen application eliminates one trip over the field compared with conventional application of anhydrous ammonia. This procedure is of most benefit to farmers with larger acreages.

NITRIC ACID

The industrial uses of nitric acid are extensive. The product is used, for example, in glass products, explosives, rocket fuels, synthetic rubber, wearing apparel, photographic film and pharmaceuticals. For agricultural purposes, nitric acid is used in the preparation of nitric phosphates and other nitrate fertilizers.

Until the ammonia industry began to move ahead, most nitric acid was prepared by the reaction of sodium nitrate with sulfuric acid as follows:

$$2NaNO_3 + H_2SO_4 \rightarrow Na_2SO_4 + 2HNO_3$$

The availability of inexpensive nitrogen in the form of ammonia provided a more economical process for the production of nitric acid, and this acid is now produced almost entirely by the oxidation of ammonia with air or oxygen in the presence of a platinum catalyst. As pointed out by several writers, there are three processes for making nitric acid from ammonia and air:

1. Ammonia oxidized at atmospheric pressure and the oxides absorbed in water at atmospheric pressure.

2. Ammonia oxidized under pressure and the oxides absorbed in water under pressure (Figure 2.5).

3. Ammonia oxidized at atmospheric pressure and the oxides absorbed in water under pressure.

The cost of plant installation and operation has resulted in most nitric acid being produced by the second of these processes.

Description. Nitric acid is a water-colored to light-yellow liquid weighing 12.6 lb/gal (1.5 kg/l) at normal temperature and pressure. It is completely soluble in water, is extremely reactive with metals (except stainless steel), tends to darken in sunlight and reacts vigorously with plant and animal tissue. Nitric acid is a strong oxidizing agent as well as a strong acid.

Method of Manufacture. The initial essential reactions are:

$$4NH_3 + 5O_2 \rightarrow 4NO + 6H_2O$$
$$2NO + O_2 \rightarrow 2NO_2$$

Figure 2.5 Nitric acid plant, Columbia Nitrogen, Augusta, Ga.

Generally speaking, the reactions are exothermic in nature and, once started, are self-sustaining.

In general, the pressure process used in the United States involves compressing anhydrous ammonia vapor to 180 pounds (81.8 kg) per square inch (6.45 cm²) and mixing with compressed air in the proportion of 9 parts air to 1 part ammonia. This mixture is preheated to about 550°F (288°C). (If the air is furnished by a centrifugal compressor, there is no need to preheat because the air is already hot enough). The mixture is passed through a platinum rhodium catalyst at 1720°F (938°C) that converts 90 to 95 percent of the ammonia to oxides of nitrogen. The oxidation of NO to NO_2 takes place primarily after the gases are cooled in a condenser, whereupon it is combined with steam to form nitric acid according to the reaction:

$$3NO_2 + H_2O \rightarrow 2HNO_3 + NO$$

The acid is 56 to 58 percent strength and is stored in stainless-steel or concrete tanks lined with acid brick. A diagram of the nitric acid manufacturing process is shown in Figure 2.6.

Product Use. When the relative costs of HNO_3 and H_2SO_4 are favorable to HNO_3, the latter can be used to acidulate rock phosphate. This becomes a most economical method of manufacturing mixed fertilizers in areas where sulfur is supplied to crops by sources other than fertilizers.

AMMONIUM NITRATE

Description. Ammonium nitrate produced for the fertilizer market is usually in the form of white granules, or prills, about the size of a number-4 birdshot. The Stengel process produces a flaked product; most other processes produce a spherical-shaped pellet, or granule, in the size limits of 6- to 16-mesh (3.3 to 1.2 mm). The pure salt is a white crystal about the size of fine salt.

Method of Manufacture. Ammonium nitrate is produced by several slightly different processes that utilize the basic product, ammonia, to produce nitric acid, which in turn is neutralized by more ammonia to form ammonium nitrate. The basic reaction is:

$$HNO_3 + NH_3 \rightarrow NH_4NO_3$$

The reaction is exothermic, yielding about 23,000 calories per mole of NH_4NO_3. Nitric acid is blended with anhydrous ammonia in the stainless-steel tank to produce ammonium nitrate liquor (Figure 2.7).

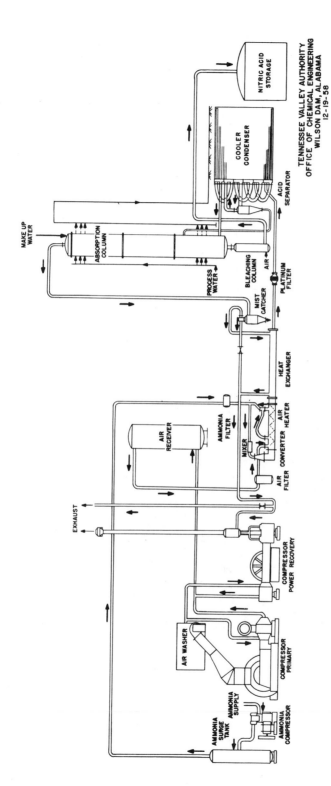

Figure 2.6 Nitric acid.

43

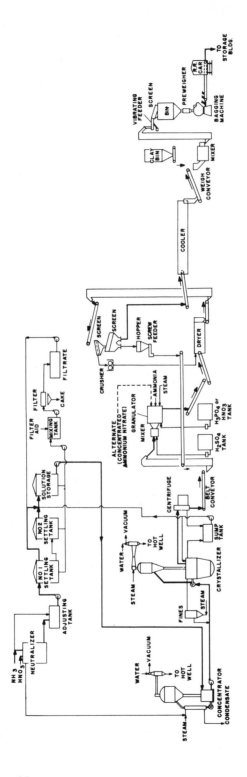

Figure 2.7   Granulated ammonium nitrate.

44

The process is automatically controlled by pH determination equipment that blends the mixture of ammonia and nitric acid in the adjusting tank stepwise to pH 6.4. As a matter of fact, many liquid multinutrient-fertilizer manufacturing processes that utilize ammonia are controlled automatically by pH determination. The liquor contains 82 to 83 percent ammonium nitrate. The 83 percent liquor is used directly in making solutions or is concentrated for use in blending solutions or for prilling. Concentration is accomplished in a battery of evaporators, filters or centrifuges, which produce 93 to 99 percent ammonium nitrate.

It is prilled by spraying into a prilling tower up to 185 feet (56.1 m) high (Figure 2.8) or granulated in a rotary granulator (Figure 2.7). Farmer demand and competition require that solid nitrogen materials be prilled or granulated. The white, shiny prills are sometimes referred to by salesmen as ''pearls of plant food.'' The old-style, fine crystalline material is no longer marketable as fertilizer in the United States.

Figure 2.8 Ammonium nitrate plant, with prilling tower in background, stands alongside railroad siding at Columbia Nitrogen Complex in Augusta, Ga.

Handling Characteristics. Fine crystalline ammonium nitrate is hydroscopic in humid climates and cakes so badly that the ungranulated salt cannot be applied successfully with an ordinary fertilizer drill. This explains the demand for and the manufacture of granular, prilled, or flaked, conditioned products that store well and are conveniently added to the soil with ordinary fertilizer equipment. The commercial product is shipped in plastic or multilayer paper bags that hold 50, 80 or 100 pounds (23, 36 or 45 kg). The prilled, flaked, or granular product can be mixed with most other fertilizer components of approximately the same size and specific gravity. Mixed with urea, ammonium nitrate absorbs water from the air.

A convenient and popular ammonium nitrate water solution for direct application contains 19 to 20 percent nitrogen. Because of its corrosive nature, it will corrode steel tanks, and therefore is stored in aluminum tanks.

Chemical Characteristics. Pure ammonium nitrate contains about 35 percent nitrogen by weight, but commercial fertilizer material is guaranteed at 32 to 33.5 percent. One-half of the nitrogen is in ammonia form, and the other half is in the nitrate form. Much is said by salesmen regarding the two forms of nitrogen in this material—one form (NH_4) feeds the young plant, and the other (NO_3) feeds the older plant to a "high-yielding maturity."

Ammonium nitrate is soluble at 32°F (0°C) to the extent of 0.26 lb (118.2 g) in 0.22 lb (100 g) of water. Because of its high solubility and convenience of application, considerable quantities of ammonium nitrate water solutions are being sold for direct application to the soil.

When exposed to sparks or open flame, ammonium nitrate is a hazardous material. In fact, when treated with fuel oil, it is a substitute for dynamite. In 1964, about 700,000 tons (634,900 m tons) of ammonium nitrate was sold for explosives. However, other nitrogen materials (including nitrate of soda, urea and potassium nitrate) will support combustion, will ignite and will detonate under certain conditions. All nitrogen materials and the bags in which they are shipped should be protected from flames or sparks.

In the liquid form, ammonium nitrate is used as an oxidizing agent. It is also used as a source of anesthetic nitrous oxide.

Ammonium nitrate tends to develop as acid residue in the soil. The magnitude of the acid residue is sufficient to neutralize 1.8 pounds (0.8 kg) of lime ($CaCO_3$) for each pound (0.45 kg) of nitrogen added. Thus, to neutralize the acidity produced by 100 pounds (45 kg) of ammonium nitrate containing 33.5 percent nitrogen requires 33.5×1.8 or 60 pounds (27 kg) of lime. One pound (0.45 kg) of lime at the quarry costs one-tenth of one cent. Therefore, the lime required to neutralize the acidity of 100 pounds (45 kg) of ammonium nitrate costs 6 cents.

Crop Response.   More than 80 percent of the ammonium nitrate manufactured is used for crop production. Ammonium nitrate and urea are the most widely used solid sources of fertilizer nitrogen in the United States. Ammonium nitrate and urea together amount to about one-half of the nitrogen supplied by anhydrous ammonia for crops in the United States and its possessions. Product use and development, rapidly following field crop research, show that nitrogen in ammonium nitrate is as good as other sources of nitrogen plant food. Because of the high concentration of nitrogen in ammonium nitrate (33.5 percent) and urea (45 percent), they are usually the most economical sources of solid nitrogen materials.

To supply a corn crop with 100 lb/acre (112 kg/ha) of nitrogen, 300 lb/acre (336 kg/ha) of ammonium nitrate is required. A two-bale crop of cotton removes 50 lb/acre (56 kg/ha) of nitrogen in the seed alone, and an additional 15 lb/acre (16.8 kg/ha) in the lint. To supply cotton crop with this amount of nitrogen requires nearly 200 lb/acre (224 kg/ha) of ammonium nitrate. A 2,000 lb/acre (2,240 kg/ha) tobacco crop removes 110 lb/acre (123 kg/ha) of nitrogen, and a 5,000 lb/acre (5,600 kg/ha) crop of rice requires 80 lb/acre (90 kg/ha) of nitrogen.

Ammonium nitrate is readily available to plants. It contains nitrogen in both the ammonia and nitrate forms. It is ideally suited for all field and vegetable crops. It may be applied before planting, at planting, or as a side or top dressing. In long-term tests, this material has produced as much crop yield per unit of nitrogen as other nitrogen materials. Per unit of nitrogen, ammonium nitrate is more expensive to manufacture than anhydrous ammonia.

UREA

Description.   Urea is a white, crystalline organic chemical that, in the pure form, contains 47 to 48 percent nitrogen. The material is used as an ingredient in the manufacture of paints, glues, plastics, paper, textiles, feeds, weed-control chemicals and (in the pelleted and conditioned form) fertilizer. The fertilizer grade contains 45 to 46 percent nitrogen. It is rapidly replacing ammonium nitrate as the world's leading solid nitrogen fertilizer because of its economy of production, higher nutrient content and fewer pollution problems during manufacture.

Method of Manufacture.   Urea is a product that is suitable for manufacture by an ammonia plant. The needed raw materials, CO_2 and NH_3, are available in an ammonia plant as may be noted by the following basic reactions that take place in ammonia synthesis:

$$H_2O + C \rightarrow CO + H_2$$
$$CO + 2H_2O \rightarrow H_2CO_3 + H_2$$
$$H_2CO_3 \rightarrow H_2O + \underline{\underline{CO_2}}$$
$$3H_2 + N_2 \rightarrow \underline{\underline{2NH_3}}$$

Ammonium carbamate is formed by combining pure ammonia with pure carbon dioxide gas at 5000 psi (340 atm). It loses a molecule of water and is thus converted into urea.

$$2NH_3 + CO_2 \xrightarrow{\triangle} NH_2COONH_4$$

$$NH_2COONH_4 \rightarrow CO(NH_2)_2 + H_2O$$

Great care must be taken in the final production stage to prevent formation of biuret, which, if present in large enough quantities, is damaging to growing plants. The solution of urea is concentrated in vacuum evaporators and is finally dried by spraying into a tower where it solidifies into pellet or prilled form.

Handling Characteristics. Although urea readily absorbs moisture, the solid fertilizer product on the market is granulated and flows evenly through ordinary fertilizer distributors. It stores as well as ammonium nitrate and other solid nitrogen materials. It has the highest payload of any solid fertilizer nitrogen and is applied extensively as a top-dressing on rice, other small grains and Douglas fir by helicopter and fixed wing aircraft.

On fir trees in the U.S. Northwest, urea is added at a rate of 150 lb/acre (168 kg/ha) of nitrogen every 5 years. The application cost by helicopter was 3 to 6 cents per pound (0.45 kg) of nitrogen. The total cost was about $40.00 per acre (0.40 ha). This was $8.00 per acre (0.40 ha) per year. In 1970, an expenditure of $6.00 per acre for fertilizer was considered profitable on Douglas fir.

A convenient and popular direct-application nitrogen solution contains 30 percent nitrogen (Table 2.1). It is made using 42 percent ammonium nitrate, 33 percent urea, and 25 percent water. It is stored in steel tanks since they are much less expensive than aluminum tanks, which are required for storage of ammonium nitrate solutions. Also, liquid multinutrient solutions containing nitrogen, phosphorus and potassium may be stored in the same kind of steel tanks. Urea is used in nitrogen solutions and multinutrient liquids to permit the manufacture of a product with a higher nitrogen content and lower salting-out temperatures.

Chemical Characteristics. Urea and cyanamide are classified by the American Association of Official Agricultural Chemists as synthetic, nonprotein, organic compounds. Urea is soluble in water to the extent of 50 percent by weight of solution. It is identical to the urea found in animal urine. The formula is $CO(NH_2)_2$. One pound (0.45 kg) of urea and 7 pounds (3.18 kg)

Table 2.1
COMPOSITION AND PHYSICAL PROPERTIES OF NITROGEN SOLUTIONS

Solution number	37	30	32	19	21	29
Ammonium nitrate, %	66.8	42.2	45.1	54.3	60.0	83.0
Anhydrous ammonia, %	16.6					
Urea, %		32.7	34.8			
Water, %	16.6	25.1	20.1	45.7	40.0	17.0
Total nitrogen, %	37.00	30.02	32.00	19.00	21.00	29.05
Nitrate nitrogen, %	11.65	7.39	7.89	9.5	10.50	14.53
Ammonia nitrogen, %	25.35	7.38	7.89	9.5	10.50	14.52
Urea nitrogen, %		15.25	16.22			
Approx. specific gravity at 60°F (15°C)	1.184	1.303	1.326	1.254	1.284	1.379[1]
Weight per gal–pound	9.87	10.86	11.05	10.45	10.70	11.50
Weight per 3.8 liters–kg	4.48	4.94	5.02	4.75	4.86	5.23
Lbs nitrogen per gal	3.65	3.26	3.52	1.99	2.25	3.34
Kg nitrogen per 3.8 liters solution	1.66	1.48	1.60	0.90	1.02	1.52
Approx. vapor pressure, psi gauge at 104°F (40°C)	1 (70 g/cm²)	None	None	None	None	None
Approx. temp. at which crystals begin to form	56°F (13°C)	15°F (−10°C)	32°F (0°C)	34°F (1°C)	51°F (10°C)	154°F (68°C)

[1]At 158°F (70°C).

of crushed corn has a protein content equal to 8 pounds (3.63 kg) of cottonseed meal.[6] Feed-grade urea can also be added to high grain silage at the rate of 10 pounds (4.53 kg) per ton (909 kg). Urea has been manufactured as a feed ingredient for many years. The feed grade is sometimes referred to as 262. This number designation is the product of 42 percent nitrogen multiplied by 6.25, which is the factor used by chemists to convert nitrogen to protein equivalent. Urea is used only in mixed feeds for ruminants. Equivalent protein from a urea source should not exceed one-third of the total protein in the mix. Urea is also used as an ingredient of nitrogen solutions that are employed in mixed-fertilizer manufacture.

Crop Response.   Farmers and agricultural scientists have found that the response of crops to the nitrogen in urea compares favorably with the nitrogen in NH_4NO_3. It is not quite as quick-acting as nitrate nitrogen because the nitrifying bacteria require a few days in warm moist soil to convert ammonia to the nitrate form. Urea converts rapidly to ammonia in the soil by hydrolysis. Urea will move with soil moisture until it changes to ammonia, after which it attaches itself by a chemical reaction to the clay and humus colloids.

It is a chemical nitrogen that can be used as a foliage spray for leaf feeding and is extensively used for this purpose on citrus and avocados. It is compatible with most organic insecticides and can be mixed with forms such as methylparathion, which controls boll weevils and bollworms on cotton.

Nitrogen solutions made from mixtures of urea and ammonium nitrate are further mixed with herbicides and are widely used for nitrogen fertilization of corn and for weed control. When such solutions do double duty, it is necessary to add the liquid on top of a well-prepared seedbed or a recently cultivated row in order to minimize gaseous loss of nitrogen from urea.

There are numerous references in the literature that show that urea added to the surface of organic matter (such as living leaves and plant debris) and left on the surface can be decomposed by the urease enzyme, and that a portion of it can be lost as gaseous nitrogen. It has been shown that urea solutions added directly to well-drained cultivated soil have very slight, if any, losses of this kind. Also, the length of time between the urea addition and the first rain or irrigation that will carry it into the soil is important. Temperature is another factor. The urease enzyme is less active in cold weather than it is at 80° to 85°F (27° to 29°C). Thus, for the most efficient use, urea should be incorporated into the soil or added directly onto a debris-free, cultivated soil.

[6]J. N. Williams, "Urea for Beef Cattle Feeding," Clemson University Livestock Leaflet No. 16, 1972.

Ureaform is a yellowish white pellet of about sand size that is made by combining urea with varying amounts of formaldehyde. Ureaform handles well for a small-sized granule. It can absorb a high percentage of water without changing its physical condition. It is made by mixing urea and formaldehyde in proportions that react to form a material that is about 38 percent nitrogen, 70 to 75 percent of which is cold water insoluble. Fertilizer ureaform is a slowly available source of nitrogen for plants, and, because it is not very soluble, little of it leaches out of the root zone. Like urea, it is acid-forming. Because of its cost, solubility, and slow-acting quality, it is used mostly for golf greens, lawns, turfs and other nonfarm uses.

Calurea is a dry, odorless, white crystal sometimes referred to as calcium nitrate–urea. It contains 34 percent nitrogen and 10 percent calcium. The double salt has an excellent physical condition for spreading in a fertilizer distributor. If it is mixed with superphosphate, the resulting mixture has a poor physical condition.

A large number of chemicals block the conversion of reduced forms of nitrogen fertilizers to oxidized forms by microorganisms or block the action of soil enzymes. Two such materials (Toyo Koatsu AM and Toyo Koatsu ST) are produced commercially in Japan. In the United States, 2-chloro-6-(trichloromethyl)-pyridine (N-Serve®) is made by Dow Chemical; similar materials are available from other manufacturers. The objective of these products is to control nitrification and to keep the nitrogen in the ammonium form. In this form, nitrogen does not readily leach or convert to a gas that escapes from the ground.

Reference has been made to the manufacture and use of nonpressure nitrogen solutions for direct application to the soil surface as well as to the use of anhydrous ammonia for direct application to the soil and the need to cover this gas with the soil to prevent loss by volatilization.

A large array of pressure and nonpressure nitrogen solutions are made (Figure 2.9) for use in manufacturing multinutrient fertilizers or for direct application to the soil by combining various amounts of anhydrous ammonia, ammonium nitrate, urea and water. The direct-application solutions range in total nitrogen content from 19 to 37 percent and range in vapor pressure at 104°F (40°C) from 0 to 1 pound (0.45 kg) per square inch (6.45 cm²). (See Table 2.1 on page 49.)

Anhydrous ammonia may be chemically trapped in the soil or in a factory mixing vat; consequently, it can be used for either direct application to the soil or as a manufacturing ingredient. It contains 82.2 percent nitrogen and has a vapor pressure of 211 pounds (96 kg) per square inch (6.45 cm²) at 104°F (40°C). The economy of using this most basic nitrogen material has been discussed. Where it is not practical to use anhydrous ammonia, the most economical source of nitrogen is the nitrogen solutions.

One advantage of the solutions that is not shared by anhydrous ammonia is that the nonpressure solutions may be, and are, widely employed as diluents for pesticides. They serve as diluents and at the same time supply nitrogen for the crop. These nonpressure solutions, made with ammonium nitrate and water or with ammonium nitrate, urea and water, will be emphasized in this discussion.

Description. Nitrogen solutions are usually colorless and odorless, and they have a salty taste. For identification and merchandising, some

Figure 2.9   Solutions manufacturing plant, Columbia Nitrogen, Augusta, Ga.

manufacturers put in coloring matter. To determine the nitrogen content of ammonium nitrate–water solutions, the hydrometer method, which relates specific gravity to salt content, can be used. If urea is a part of the mixture, the nitrogen content may be determined by the Kjeldahl method, which converts the organic nitrogen in the urea to ammonia.

There are six groups of nitrogen solutions. They are described according to the particular nitrogen compounds present as follows:

1. Ammonia–ammonium nitrate.
2. Ammonia-urea.
3. Ammonia-urea–ammonium nitrate.
4. Ammonium nitrate.
5. Ammonium nitrate–urea.
6. Miscellaneous.

Method of Manufacture. The manufacture of solid ammonium nitrate and urea has been discussed. For example, ammonium nitrate liquor, prior to being pumped to the top of the prilling tower, can be taken to the solutions plant where it may be processed into ammonium nitrate direct-application solution or combined with urea to form ammonium nitrate–urea solutions. Tapping the liquor before the prilling procedure begins eliminates in the processing of nitrogen solutions six steps used in the processing of bagged solid nitrogen materials. These steps are evaporating, prilling, drying, cooling, coating and bagging. Their elimination serves to emphasize the cost advantage of nitrogen solutions over comparable solid prilled materials.

Chemical Characteristics. Ammonium nitrate solutions for direct application contain 19 to 21 percent nitrogen. Ammonium nitrate and urea are made into solutions containing 30 to 32 percent nitrogen. The composition and physical properties of these solutions, along with one made from a mixture of ammonium nitrate and anhydrous ammonia, are given in Table 2.1.

In 1958, a Nitrogen Solution Code was established in order to standardize the description of various nitrogen solutions. For example, solution 37 in Table 2.1 is coded as 370(17-67-0). The 370 indicates 37.00 percent nitrogen; inside the parentheses, the 17 refers to 17 percent free ammonia, and 67 signifies 67 percent ammonium nitrate, and the zero indicates that no urea is used in the mixture. Other chemical characteristics of the individual compounds used as fertilizers are discussed under the headings, "Ammonia," "Ammonium Nitrate" and "Urea."

Handling Characteristics. There are three physical chemical properties of nitrogen solutions that influence their use as fertilizers. These are:

1. Vapor pressure.

2. Temperature at which crystals begin to form.
3. The corrosive nature of the salt solution.

The first property that influences the handling of nitrogen solutions is the vapor pressure. It may be noted in Table 2.1 that solution 37 is the only one that exerts an appreciable vapor pressure at 104°F (40°C). Thus, on a hot summer day, this solution must be incorporated in the soil to minimize losses to the atmosphere. Losses of nitrogen to the atmosphere are also possible with solutions 30 and 32 when they are placed on top of "trashy" land containing plant residues or weeds, both of which are hosts for the urease enzyme which biologically volatilizes urea. Fortunately, these losses are usually no larger than 10 to 15 percent, except under very favorable conditions for the urease enzyme which would prevail in a field of Coastal Bermuda grass or other grass sod when the temperature is 84°F (30°C) or higher. This was discussed briefly under the section on urea.

The second property influencing the handling of solutions is the temperature at which the nitrogen salt begins to crystallize. The temperature below which the 19-percent nitrogen solution may not be used because of salting out is 34°F (1°C). Obviously, on a freezing winter day, it would not be practical to spray the 19-percent solution.

Because of the varying degrees of corrosiveness of the various nonpressure nitrogen solutions, different kinds of tanks are designed for their storage. The 19-percent nitrogen solution made with 54.3 percent ammonium nitrate and 45.7 percent water requires an aluminum tank. The aluminum tank is not designed for the storage of multinutrient mixes. On the other hand, the 30-percent nitrogen solution made with 42.2 percent ammonium nitrate, 32.7 percent urea and 25.1 percent water requires a less expensive, mild steel tank for storage. Furthermore, the mild steel tank can also be used for the storage of liquid multinutrient nitrogen-phosphorus-potassium mixtures. The nozzle tips, through which the solutions are metered to the ground, should be of hardened stainless steel.

Before any sprayed material can be accurately metered, it is necessary to know the rate of spray discharged. Where metering pumps are used, it is only necessary to select the correct nozzle-tip size for the desired rate of application and tractor speed. These pumps are positive displacement pumps with variable stroke, and they are ground-driven. The rate of output is independent of speed and pressure. The nozzle-tip size should be chosen to maintain only enough pressure to give a satisfactory spray pattern. High pressure should be avoided to reduce spray drift.

Where metering pumps are not available, the gallons per acre (l/ha) added depend on:

1. Speed of the tractor or other pulling machine.
2. Orifice size of the nozzle tip.
3. Pressure of the liquid at the orifice tip.

The first step is to select the most desirable operating speed. The second is to select a nozzle-tip size that will meter the desired number of gallons in the speed range selected. Since old nozzle-tip holes tend to become enlarged with use, it is necessary to check the flow of old tips and to select sets that discharge equal amounts. This calibration of a broadcast boom sprayer is easy, if the following steps are observed:

1. Set up a spray boom with a nozzle-tip size to give approximate desired number of gallons per acre (l/ha) at 3 to 5 miles per hour (4.83 to 8.05 km/hr).
   8000—4 to 6 gallons per acre (37.7 to 56.5 l/ha)
   8001—6 to 10 gallons per acre (56.5 to 94.2 l/ha)
   8002—10 to 20 gallons per acre (94.2 to 188.4 l/ha)
   8003—20 to 30 gallons per acre (188.4 to 282.6 l/ha)
   8004—30 to 40 gallons per acre (282.6 to 376.8 l/ha)
   8006—40 to 50 gallons per acre (376.8 to 471.0 l/ha)

2. Tie rags around a tractor tire so that wheel turns can be easily counted. Measure distance traveled in one wheel turn.

3. Set sprayer pressure at 20 to 30 pounds (9.1 to 13.6 kg); then refill tank with water.

4. With tractor at speed to be used in spraying operation and throttle or speedometer setting noted, turn on sprayer and drive 60 to 80 wheel turns. Multiply the number of turns by the wheel circumference to obtain total distance traveled.

5. Cut off sprayer, return, and carefully measure amount needed to refill tank to same level.

6. To figure acre (ha) rate of discharge, use this formula:

$$\frac{43{,}560 \text{ ft}^2/\text{acre } (10{,}127 \text{ m}^2/\text{ha}) \times \text{gallons used } (3.8 \text{ l})}{\text{ft } (0.30 \text{ m}) \text{ traveled} \times \text{boom length in ft } (0.30 \text{ m})} = \text{gal/acre (l/ha)}$$

7. Adjust pressure up or down or select other nozzle-tip size until desired number of gallons per acre (l/ha) is reached.

The gallons (l) of nitrogen solution to be added per acre (ha) will depend on the concentration of nitrogen in the solution and on the amount of nitrogen required by the crop (Table 2.2).

NITROGEN SOLUTION–HERBICIDE MIXTURES FOR CROP PRODUCTION

The use of nonpressure nitrogen solutions mixed with a herbicide has become a standard farm operation (Figure 2.10). These mixtures furnish at the same time nitrogen to make the crop grow and a herbicide to kill

Table 2.2
GALLONS (L) OF NITROGEN SOLUTION TO BE ADDED PER ACRE (HA) WITH VARYING AMOUNTS OF NITROGEN REQUIRED AND WITH VARYING CONCENTRATIONS OF NITROGEN IN THE SOLUTIONS

Nitrogen Required		Ammonium Nitrate Solutions				Ammonium Nitrate–Urea Solutions			
		19 percent		21 percent		30 percent		32 percent	
					Rate of Application				
lb/A	kg/ha	gal/A	l/ha	gal/A	l/ha	gal/A	l/ha	gal/A	l/ha
20	22	10.1	95.9	8.9	83.8	6.2	58.4	5.7	53.7
40	44	20.2	191.9	17.8	167.7	12.3	115.8	11.3	106.7
50	56	25.2	239.4	22.2	209.1	15.4	145.0	14.1	132.8
60	67	30.3	287.8	26.7	251.5	18.5	174.2	16.5	155.4
80	89	40.4	383.8	35.6	335.4	24.6	231.7	22.6	212.9
100	112	50.4	478.8	44.5	419.2	30.8	290.1	28.3	266.5
120	134	60.6	575.7	53.3	502.1	37.0	348.5	33.9	319.3
Pounds nitrogen per gallon		1.98		2.25		3.25		3.54	
Kilograms nitrogen per liter			0.23		0.27		0.39		0.42
Salting-out temperature		34°F	1°C	51°F	10°C	15°F	−10°C	32°F	0°C

Figure 2.10   Row spraying of liquid is depicted here. This equipment is widely used to side-dress with a nitrogen solution-herbicide mixture that will control such weeds as morning-glory seen here and to feed the corn nitrogen at the same time. (Photo courtesy Allied Chemical Corporation, New York, N.Y.)

weeds. Maximum benefits can be derived from this practice with a thorough understanding of both the nitrogen solutions and the herbicides. Some examples of this double duty for cotton, corn, sorghum, small grain and perennial grass pasture are given below. The types of weeds to be controlled by nitrogen-herbicide mixtures are shown in Table 2.3. READ THE LABEL ON THE CONTAINER BEFORE USING ANY PES-TICIDE.

Nitrogen Solution + Fluometuron for Cotton.   Use fluometuron at the rate of 0.4 lb/acre (0.45 kg/ha) active ingredient in a 13- to 14-inch (32- to 35-cm) band over the row. Nitrogen solutions (ammonium nitrate–urea type solutions may be used) should be applied so as to have 20 to 30 lb/acre (22.4 to 33.6 kg/ha) nitrogen in a 13- to 14-inch (32- to 35-cm) band over the row. Adequate agitation in the spray tank is necessary in order to keep

Table 2.3
WEED CONTROL TO BE EXPECTED WITH HERBICIDES USED IN NITROGEN SOLUTIONS[1]

2,4–D	Excellent control of emerged cocklebur *(Xanthium)*, morning-glory *(Ipomoea)*, pigweed *(Amaranthaceae)*, mustard *(Brassicaceae)*, blessed thistle *(Cnicus benedictus)*, etc., with no control of annual grasses. Coffeeweed *(Sesbania exaltata)* is controlled, but coffeeweed *(Cassia obtusifolia)* is not.
Atrazine	Excellent control of annual grasses and broadleaf weeds less than 1½ inches (3.8 cm) tall. Will provide residual control of annual grasses and broadleaf weeds not yet emerged.
Fluometuron	Used as a preemergence treatment in cotton *(Gossipium)* and controls annual grasses and broadleaf weeds such as pigweed *(Amaranthaceae)* and morning-glory *(Ipomoea)* for 6 to 8 weeks.
Ametryne	Excellent control of emerged annual grasses and broadleaf weeds 2 to 6 inches (5–15 cm) high. May provide some residual control of annual grasses and broadleaf weeds not yet emerged.
Simazine	Used as a preemergence treatment in corn *(Zea mays)*. Good control of all annual grasses and broadleaf weeds such as cocklebur *(Xanthium)*, morning-glory *(Ipomoea)*, ragweed *(Ambrosia* spp.) and pigweed *(Amaranthaceae)*.

[1]READ THE LABEL ON THE CONTAINER BEFORE USING ANY PESTICIDE.

the fluometuron uniformly suspended in the nitrogen solution. For pre-emergence control of weeds, the application of nitrogen solution + fluometuron should be made within 3 days after planting cotton. If it is used later on in the cotton-growing season, at the last cultivation, cotton should not be followed by crops susceptible to fluometuron phytotoxicity such as tobacco, small grain, peanuts or soybeans. If cotton is followed by cotton, corn or grain sorghum, fluometuron may be used later in the cotton-growing season.

Where nitrogen solution + fluometuron is used in cotton, a 0-6-12 (0-14-14)[7] or a 0-4-16 (0-10-20) analysis of fertilizer may be used to supply the phosphorus and potassium. It is usually not desirable to apply all the nitrogen needed by the cotton at planting time in areas where excessive rainfall will leach nitrogen out of the growing root zone.

Nitrogen Solution + Atrazine or Simazine for Corn. Use atrazine or simazine at rates of 0.6 lb/acre (0.7 kg/ha) active ingredient in a 13- to 14-inch (32- to 35-cm) band over the row. Nitrogen solutions should be applied at a rate to give 25 to 35 lb/acre (28 to 39 kg/ha) of actual nitrogen in the 13- to 14-inch (32- to 35-cm) band. A 0-6-12 (0-14-14) or a 0-4-16 (0-10-20) analysis of fertilizer may be used to supply the phosphorus and potassium. Application of nitrogen solution + atrazine or simazine should be made within 3 days after planting corn. As with cotton, it is not

[7]Figures in parentheses refer to fertilizer grades in terms of $N–P_2O_5–K_2O$; $P_2O_5 \times 0.44 = P$; $K_2O \times 0.83 = K$.

desirable to apply all the nitrogen needed at planting in sandy areas where excessive rainfall will leach nitrogen to a depth beyond reach of growing roots.

Postemergence use of nitrogen solution–herbicide mixtures has given excellent results. The mixture should be applied to corn when it is 20 to 30 inches (50 to 76 cm) high. Corn should be relatively weed-free when 15 inches (38 cm) high. Use one of the following herbicides in a nitrogen solution derived from ammonium nitrate or an ammonium nitrate–urea source:

1. ⅓ lb/acre (0.37 kg/ha) of acid equivalent of a low-volatile ester of 2,4-D.
2. 1 lb/acre (1.12 kg/ha) of active ingredient of atrazine.
3. 0.75 lb/acre (0.8 kg/ha) of active ingredient of ametryne.

The nitrogen solution–herbicide mixture should be applied directionally in such a manner that the mixture will contact only the lower 4 inches (10 cm) of the corn plant. Adequate agitation is needed in order to keep the herbicides for corn uniformly suspended in the nitrogen-solution mixture. Apply the nitrogen solution with postemergence herbicide at a rate to give the amount of nitrogen needed by corn for growth to maturity. In addition to the 25 to 35 lb/acre (28 to 39 kg/ha) added at planting, side dress with 65 to 75 additional pounds per acre (73 to 84 kg/ha) of nitrogen.

Nitrogen Solution + 2,4-D for Small Grain.   Use one of the following in a nitrogen solution derived from an ammonium nitrate–urea solution:

1. ⅓ lb/acre (0.37 kg/ha) of acid equivalent of a low-volatile ester of 2,4-D.
2. ⅓ lb/acre (0.37 kg/ha) of acid equivalent of 2,4-D acid formulation.

Treat small grain with one of the above mixtures when grain is fully tillered or when 4 to 8 inches (10 to 20 cm) tall, applying with a sprayer, which is equipped for spraying nitrogen solutions uniformly, and with adequate tank agitation. Do not graze or cut for forage until 2 weeks after treatment. Apply the nitrogen solution at rates to give the amount of nitrogen recommended for small grain at this stage of growth, usually 30 to 45 lb/acre (34 to 50 kg/ha).

Nitrogen Solution + 2,4-D for Perennial Grass Pastures.   Apply nitrogen solution + 2,4-D at rates to give 0.5 to 1.0 lb/acre (0.56 to 1.12 kg/ha) of 2,4-D and amount of nitrogen normally applied to grass at this time. Use a low-volatile ester form of 2,4-D or a 2,4-D acid formulation. Apply when weeds are first noticeable in the spring. Do not graze or cut for hay until 45 days after application.

Some burning of the foliage may result from this application on small grain and grass pasture. However, grain yields are not reduced.

On grasses such as Coastal Bermuda, application should be made before growth starts or just after mowing or grazing.

Nitrogen Solution + 2,4-D Acid Formulation for Sorghum. Apply as a directional spray 5 weeks after planting sorghum. Use 2,4-D at the rate of ⅓ lb/acre (0.37 kg/ha). Apply nitrogen solution at rates to give amount of nitrogen recommended for 5-week-old sorghum.

Nitrogen Solution + Trifluralin for Cotton. Urea plus ammonium nitrate solutions (30 to 32) may be used as the total carrier for trifluralin, provided that continuous agitation is maintained. It is further suggested that trifluralin be preemulsified [1 quart (0.95 liter) in 1 gallon (3.8 liters) of water] prior to adding it to the nitrogen solution. With the addition of 6 to 8 fluid ounces (170 to 227 grams) of COMPEX®, a product of KALO Laboratories Inc., the need for agitation is reduced. Ammonium nitrate solutions (17 to 19) should not be used as the total carrier for trifluralin. The present practice is to broadcast the trifluralin and to incorporate it into the soil prior to planting.

Handling Characteristics. Nitrogen solutions are extremely corrosive to normal spray equipment. When nitrogen solutions are being used, sprayers must be equipped with nitrogen-resistant parts in order to avoid corrosive deterioration. Pumps, nozzles, tanks, pressure gauges, etc., that are resistant to the corrosive action of the nitrogen solutions, are commercially available. The basic spray-machine circulatory system for applying herbicides mixed in nitrogen solutions is diagrammed in Figure 2.11.

Herbicides mentioned in this chapter are not soluble in the nitrogen solutions. For this reason, adequate agitation is needed to keep the herbicide suspended in the nitrogen solutions. A by-pass requirement of 2 to 3 gallons (7 to 11 liters) per minute is usually needed in order to maintain a uniform suspension. A jet agitator is available on many sprayers to maintain suspension of the herbicide.

High pressures should not be used in applying the nitrogen solution–herbicide mixtures. This is especially true when using 2,4-D. High pressures create problems by discharging small particles that can drift for considerable distances and can damage susceptible crops, such as cotton, which have dicotyledon-type seed. Lower pressures discharge large particles that hit the ground almost immediately after leaving the nozzle and thereby eliminate drift problems to a great extent. Research had demonstrated that a pressure of 20 to 30 psi (1.36 to 2.04 atm) is desirable for the application of the nitrogen solution–herbicide mixtures.

Many people have encountered difficulty after mixing nitrogen solutions and 2,4-D. Observations indicate that "old" 2,4-D will cause precipitation when the 2,4-D is added to the nitrogen solution. To avoid this,

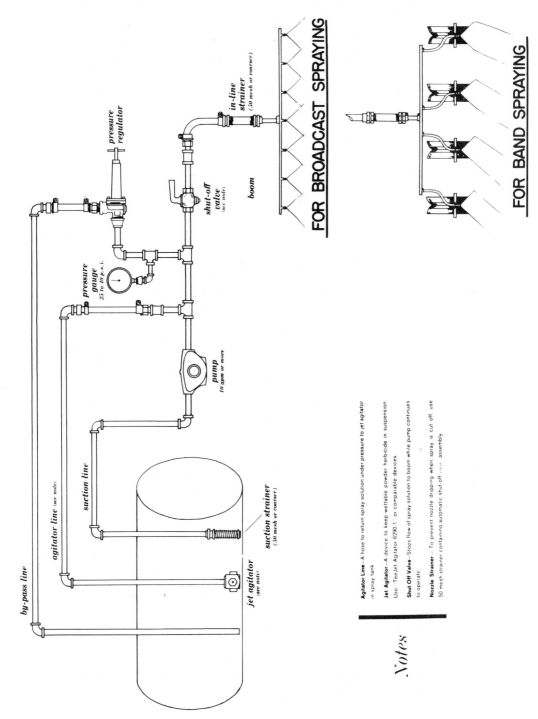

by-pass line

agitator line (see note)

suction line

pressure
gauge
25 to 10 p.x.i.

pressure
regulator

pump
10 gpm or more

shut-off
valve
(see note)

in-line
strainer
(50 mesh or coarser)

boom

jet agitator
(see note)

suction strainer
(50 mesh or coarser)

FOR BROADCAST SPRAYING

FOR BAND SPRAYING

Notes

Agitator Line—A hose to return spray solution under pressure to jet agitator in spray tank

Jet Agitator—A device to keep wettable powder herbicide in suspension
Use 'TeeJet Agitator 6290-1' or comparable devices

Shut Off Valve—Stops flow of spray solution to boom while pump continues to operate

Nozzle Strainer—To prevent nozzle dripping when spray is cut off use 50 mesh strainer containing automatic shut-off valve assembly

Figure 2.11   The basic spray machine circulatory system for applying herbicides mixed in nitrogen solutions.

premix a small amount of the nitrogen solution and the 2,4-D that you plan to use in a small jar. If "lumps" form, then you should not use the 2,4-D.

The amine form of 2,4-D can cause problems when added directly to nitrogen solutions; this form of 2,4-D causes a solid precipitate to form in the spray tank. Use only the low-volatile ester or acid forms of 2,4-D in accordance with recommendations and premix, using a surfactant, before adding it to the spray tank.

Any solution tank to which 2,4-D has been added should be so marked and should not be used to apply nitrogen solution to 2,4-D-susceptible crops until all traces of the 2,4-D have been removed from the tank.

AMMONIUM SULFATE

Fertilizer-grade sulfate of ammonia contains 20 to 21 percent nitrogen and 24 percent sulfur. Today more sulfate of ammonia is manufactured from synthesized ammonia than from coke-oven by-product ammonia. Ammonia nitrogen is retained by the soil in an exchangeable form until it is oxidized to nitrate by nitrifying bacteria. It does not leach as readily as nitrogen in the nitrate form. Much of the nitrogen in mixed fertilizers is in the form of $(NH_4)_2SO_4$.

Because sulfate of ammonia is acid-forming, it is preferred in areas having neutral or slightly alkaline soil. Soils of an acid nature that are limed properly can be, and are, fertilized with sulfate of ammonia to provide the nitrogen needs of the crop.

Description. Ammonium sulfate is a white, crystalline salt that, in pure form, contains 21.2 percent nitrogen and 24.2 percent sulfur. The commercial product most popular for agricultural use is a light gray, free-flowing material made specifically for direct application and for blending in other granular materials. It is sold with a guaranteed analysis of 20 percent nitrogen.

Method of Manufacture. There are two processes used for the manufacture of $(NH_4)_2SO_4$ in the United States. One of them involves the use of ammonia that is produced as a by-product of the destructive distillation of coal to produce coke; "destructive distillation" is a term used where heating is accomplished with the exclusion of air. The other process utilizes ammonia produced synthetically.

Beginning about 1950, the tonnage of ammonium sulfate made from synthesized ammonia surpassed that of the coke-oven industry even though the by-product process has an economic advantage because of low ammonia cost. In each of the processes, the final reaction is closely related to the following:

$$2NH_3 + H_2SO_4 \rightarrow (NH_4)_2SO_4$$

Neutralization takes place in a reactor, and the slurry is transported to a crystallizer where the sulfate of ammonia crystals are taken off the bottom and centrifuged. The mother liquor is recycled to the reactor. Crystal growth is regulated by airflow, time and temperature.

A German process, which involves the use of ammonia, gypsum and carbon dioxide at 5 atmospheres of pressure, is of interest. It is of importance in the event of a sulfur shortage.

$$2NH_4OH + H_2O + CO_2 \leftrightharpoons (NH_4)_2CO_3 + 2H_2O$$

$$(NH_4)_2CO_3 + CaSO_4 \rightarrow (NH_4)_2SO_4 + CaCO_3$$

$CaCO_3$ precipitates and is separated from the liquor by centrifuging. The $(NH_4)_2SO_4$ is then recovered by evaporation of the water.

Handling Characteristics. Most of the sulfate of ammonia for direct application to the soil is granulated in the form of large crystals or prills. The hard free-flowing granules are sized to minimize segregation in blending. They store and handle well for uniform distribution from spreaders. Some of the fine crystalline material is still used to make powdered mixed fertilizers, but it will absorb moisture and will not remain free-flowing for long periods before it is mixed with superphosphate.

Chemical Characteristics. Granular sulfate of ammonia, when tapped until density increases less than 0.5 percent, has a tapped bulk density of 54 pounds per cubic foot (876.6 kg/m³). At this density, it consists of particles approximately in the range of $-10 + 20$ mesh (-1.9 mm $+ 0.8$ mm). It is soluble in water at ordinary temperatures. Its solubility at 32°F (0°C) is 0.167 lb (70.6 g) per 0.22 lb (100 g) water and is 0.228 lb (103.8 g) per 0.22 lb (100 g) water at 212°F (100°C).

Investigators have shown that ammonia may be retained by the soil colloids in an ion-exchangeable form until it is oxidized by nitrifying bacteria. Ammonia does not leach as readily as nitrogen in the nitrate form, such as in $NaNO_3$ and NH_4NO_3.

Sulfate of ammonia has a physiological acidifying effect on the soil equal to about 5.3 pounds (2.3 kg) of lime per pound (0.45 kg) of nitrogen.

Although the practice of adding $(NH_4)_2SO_4$ to superphosphate to make multinutrient fertilizers is not as widespread as it once was, one of the nitrogen compounds found in mixed fertilizers is sulfate of ammonia. The major nitrogen solutions used to manufacture mixed fertilizers today are those containing more or less ammonia and ammonium nitrate. The ammonia reacts with the superphosphate, which is essentially monocalcium phosphate and gypsum ($CaSO_4$). Sulfate of ammonia and dicalcium phosphate are formed according to the following reaction:

ordinary superphosphate   ammonia   4-7-0(4-16-0)

$$CaH_4(PO_4)_2H_2O + CaSO_4 + 2NH_3 \rightarrow Ca_2H_2(PO_4)_2 + (NH_4)_2SO_4 + H_2O$$

monocalcium	gypsum	dicalcium	sulfate of
phosphate		phosphate	ammonia

The end product contains about 4 percent nitrogen and 7 percent phosphorus (4-16-0.)[7] All of the nitrogen in this end product, sometimes referred to as "base goods," is in the ammonium sulfate form.

Crop Response.   Because of its acidifying effect on the soil, sulfate of ammonia is a most popular direct-application nitrogen material in semi-arid regions west of the Mississippi River and particularly in California where the soil reaction is neutral to slightly alkaline. In the Rice Belt where deep well waters containing bases (primarily bicarbonates of calcium) are used for irrigation purposes, the soil reaction is frequently built up to a slightly alkaline condition. In Texas, Arkansas, Louisiana and California, sulfate of ammonia is very popular as a source of nitrogen for rice, and it is used on potatoes in Idaho to help control scab, which is the result of an actinomyces bacteria that thrives in neutral to slightly alkaline mediums.

Where sufficient amounts of limestone are added to the acid soils in humid regions, sulfate of ammonia is as good a source of nitrogen as other materials. In addition, it contains sulfur that is an essential element for plant growth.

AMMONIUM NITRATE SULFATE (30-0-0-5S)

TVA ammonium nitrate sulfate is a granular fertilizer containing 30 percent nitrogen and 5 percent sulfur. Of the 30 percent nitrogen, about 17 percent is ammoniacal and 13 percent is in the nitrate form. The sulfur is present as ammonium sulfate and is uniformly incorporated in each granule. It is produced by reacting ammonia with nitric and sulfuric acids. Granulation is accomplished in a rotary drum or pan granulator. The granules are coated with a diatomaceous earth conditioner and have good storage and handling properties. It is less explosive than ammonium nitrate.

Its physical and chemical properties are:

pH	5.1	
Acid-forming potential, lb		
$CaCO_3$ equivalent per ton	1,425.0	(647 kg/909 kg)
Bulk density—loose pour,	50.9	(23 kg/28 dm²)
lb/ft³, tapped, lb/ft³	54.4	(25 kg/28 dm²)

Its size distribution, percent retained on Tyler screen size indicated, is:

Ammonium nitrate sulfate is frequently used for direct application in place of ammonium nitrate in sulfur-deficient areas and for blending with other granular materials to produce mixed fertilizers.

Crop response to the nitrogen in ammonium nitrate sulfate is comparable with that obtained from ammonium nitrate. The nitrogen-to-sulfur ratio in this fertilizer closely approaches the ratio of these two nutrients in plants. Usually, application of ammonium nitrate sulfate fertilizer at rates needed to provide nitrogen for crops will also supply the sulfur needed.

CALCIUM CYANAMIDE

Description. Calcium cyanamide is a dark-colored, granulated material that contains 20 to 21 percent nitrogen. The dark color results from the fact that it contains calcium carbide. It is considered to be a nonprotein, organic nitrogen.

Method of Manufacture. The raw materials required for its manufacture are coal and limestone. Although calcium cyanamide can be produced by heating a mixture of limestone with coal in an atmosphere of nitrogen, it has proved more economical to carry out the process in four separate steps.

1. The burning of calcitic limestone in a vertical or rotary kiln at about 2012°F (1100°C):

$$CaCO_3 \xrightarrow{\text{heat}} CaO + CO_2$$

2. The production of calcium carbide by heating coke or coal and lime in an electric furnace:

$$CaO + 3C \rightarrow CaC_2 + CO$$

3. The production of pure nitrogen gas from the air by liquefaction and fractional distillation.

4. The treatment of the finely powdered carbide with pure nitrogen at approximately 1832°F (1000°C) to form calcium cyanamide:

$$CaC_2 + N_2 \rightarrow CaCN_2 + C$$

It is estimated that there is presently a manufacturing capacity of several hundred thousand tons of calcium cyanamide in the world. None is being produced in the United States, but considerable quantities are manufactured at Niagara Falls, Ontario, Canada. In 1968, 40,000 tons (36,400 m tons) of calcium cyanamide was imported into the United States to be used for agricultural purposes.

Handling Characteristics. In dry condition, calcium cyanamide is dusty and disagreeable to handle unless it has been oiled. Fortunately, the product on the market today is granulated and in excellent physical condition for use in ordinary fertilizer drills. It is also used as a cotton defoliant. Frequently, the material is applied by airplanes at a rate of 30 lb/ acre (33.4 kg/ha) and must therefore be in good physical condition.

It is poisonous and irritating to the skin and mucous membrane and should be washed off promptly after contact.

Chemical Characteristics. The fertilizer-grade calcium cyanamide contains 21 percent nitrogen, about 11 percent calcium, 11 percent free carbon, 5 percent oil, 2 to 4 percent water, and some aluminum, iron and silicon oxides. It is distinctly alkaline. If moisture and air get to the product in storage, dicyandiamid, a poisonous compound may be formed. It is for this reason that manufacturers recommend that "only calcium cyanamide from newly opened bags be used." In soil of pH 7 or less, this material is converted into urea and lime within a week of addition. It is nonacid and leaves a physiological alkaline residue in the soil equivalent to 2.8 pounds (1.3 kg) of $CaCO_3$ per pound (0.45 kg) of nitrogen applied. It is the most alkaline of all commercial nitrogen fertilizers.

Crop Response. Cyanamide is a combination defoliant, pesticide and fertilizer. Farmers and scientists have found that the response of crops to calcium cyanamide used as a fertilizer compared favorably with $NaNO_3$ and $(NH_4)_2SO_4$ when it is thoroughly mixed with soil. It is not as fast-acting as nitrate sources of nitrogen because it requires a week or so to convert to urea in the soil. For this reason, it is considered more appropriate for preplant, plow-down application than for top-dressing small grains or side-dressing cotton, corn and tobacco.

A special grade of cyanamide may be very effectively used in the control of weeds in tobacco, vegetable and other plant beds when applied at the rate of 1 to 2 lb/yd^2 (0.45 to 0.9 kg/0.84 m^2), 60 to 90 days before seeding. Also, a partial sterilization of the soil always follows an application of calcium cyanamide. This results in the destruction of many disease-producing microorganisms such as bacteria, actinomyces, fungi, molds and nematodes.

A mixture of urea and cyanamide has been used in tobacco plant beds to eliminate some of the alkalinity that can be developed with large amounts, or repeated use, of cyanamide on the same plant bed. However, the use of methylbromide or other gas fumigation is rapidly replacing the use of urea and cyanamide for the control of weeds and diseases in plant beds.

The Chilean government owns one-half interest in the mining and manu-facturing of nitrate of soda in Chile. Probably 80 percent of the nitrate of soda used in the United States is imported from Chile. The other 20 percent is produced in the United States by two chemical companies—Allied Chemical at Hopewell, Virginia and the Chemicals Division of Olin at Lake Charles, Louisiana. In addition to the agricultural markets for sodium nitrate, it is used in the manufacture of glass. Imported nitrate of soda is of the same quality as domestic nitrate of soda in regard to physical condition and crop-producing potential.

Description. It appears on the market today in a pellet or prilled form about the size of number-6 birdshot and is very white and shiny. The domestic product and the imported material can be distinguished only by a very experienced observer.

Method of Manufacture. Two processes are used for the production of nitrate of soda. In Chile, it is extracted from a mined product known as caliche. It is manufactured in the United States from nitric acid and soda ash by the Solvay process, in which salt and oyster shells are used. Nitric acid is then reacted with soda ash to form a neutral solution of sodium nitrate. Carbon dioxide gas is liberated and subsequently trapped to produce more carbonic acid.

$$CaCO_3 + 2HCl \rightarrow CaCl_2 + H_2CO_3$$
$$H_2CO_3 + NH_4OH \leftrightharpoons NH_4HCO_3 + H_2O$$
$$NH_4HCO_3 + NaCl \leftrightharpoons NaHCO_3 + NH_4Cl$$
$$NaHCO_3 + HNO_3 \rightarrow NaNO_3 + H_2CO_3$$

The solution of sodium nitrate is evaporated to remove most of the water. It is then heated to a high temperature, piped to the top of a tower where, similar to the manufacture of shot, it is sprayed out of nozzles. While falling, the raindrop-shaped droplets cool and solidify to form smooth round pellets that land at the bottom of the tower. The pellets, sometimes called prills, are allowed to cool and are later coated with a conditioner and bagged.

Caliche is mined in Chile. It contains (in addition to 8 to 20 percent sodium nitrate) potassium and magnesium salts of borates, sulfates and chlorides. Approximately 10 tons (9.1 m tons) of caliche is processed and purified to form 1 ton (0.91 m ton) of product containing 99 percent sodium nitrate. The process consists of dissolving the caliche in warm water and then cooling it to about 32°F (0°C) at which time the nitrate of soda crystallizes. The crystals are then circulated through a series of three heat interchangers. The circulation keeps the nitrate crystals suspended until they have joined to form pellets that are the proper size for use.

Handling Characteristics. The present-day product has a low moisture content and is usually in excellent physical condition. Most of the Chilean and some of the domestic nitrate is shipped in bulk by water freight and is bagged in ports as near to the market as is practical. Like other flammable nitrogen materials, it must be stored under specified conditions if it is insured.

Chemical Characteristics. All fertilizer sodium nitrates are guaranteed to contain 16 percent nitrogen, 26 percent sodium and traces of impurities. Practically all fertilizers, regardless of their source, contain some impurities, some of which may be essential for plant growth. Nitrate of soda is soluble in water. It has a neutralizing effect on soil acidity equivalent to 600 pounds (273 kg) of limestone per ton (0.91 m ton) of nitrate of soda. Another way of expressing the neutralizing value of nitrate of soda is 1.8 pounds (0.82 kg) of $CaCO_3$ equivalent per pound (0.45 kg) of nitrogen.

$$16\% \times 2,000 \text{ lb } (909 \text{ kg}) = 320 \text{ lb } (145.5 \text{ kg}) \text{ nitrogen}$$
$$320 \times 1.8 \text{ lb} = 576 \text{ lb } (262 \text{ kg}) \text{ of } CaCO_3 \text{ equivalent}$$

High-grade calcitic limestone analyzes 95 percent $CaCO_3$.

$$0.95X = 576$$
$$X = \frac{576}{0.95} = 606 \text{ lb } (273 \text{ kg}) \text{ of calcitic limestone}$$
equivalent per ton (0.91 m ton) of $NaNO_3$

Crop Response. The bulk of nitrogen absorbed by crops growing on well-aerated soils is in the nitrate form. Consequently, the nitrogen in nitrate of soda is as available to plants as in any other nitrogen fertilizer. The nitrogen content of nitrate of soda is less than that of most other inorganic nitrogen fertilizers. Only organic nitrogen materials, such as tankage and cottonseed meal, contain less nitrogen than nitrate of soda, and they are expensive for fertilizer use; they are used mainly to supply protein in feeds.

Agricultural experiment station results from Arkansas, Texas, Louisiana, Mississippi, Georgia and the Carolinas indicate that one brand of nitrate of soda is as good as another for increasing crop yields. Research at these experiment stations also shows that sodium can substitute for about 15 percent of the potassium needs of crops.

Flue-cured tobacco fertilizers used at transplanting time should contain at least 50% of the total nitrogen in the nitrate form. Nitrogen that is side dressed 15 to 20 days after transplanting should be in the nitrate form.

Sodium potassium nitrate is marketed in the form of white pellets. This material is refined from Chilean caliche and is a mixture of sodium and potassium nitrates and is difficult to manufacture into granular form. It contains 15 percent nitrogen and 14 percent K_2O (11.5 percent potassium). It is nonacid-forming. Tests on cotton have shown it to be as good a source of potassium and nitrogen as other forms. Tests on tobacco have shown it to be a superior source of nitrogen and potassium because of the small amount of chlorine it contains; chlorine influences the burning quality of tobacco.

CALCIUM NITRATE

Calcium nitrate is produced in the form of a small, white crystal. Its manufacture is based on the following basic chemical reaction:

$$2HNO_3 + CaCO_3 + 2H_2O \rightarrow Ca(NO_3)_2 \cdot 2H_2O + H_2CO_3$$

It was the first chemical nitrogen fertilizer put on the market. A very small quantity of calcium nitrate is imported into the United States. It has never proved to be entirely satisfactory as a fertilizer, primarily because of its tendency to absorb water. It is offered as a prilled material for bulk mixing and as powdered material for use in sprays. It contains 15 percent nitrogen and 21 percent calcium. Before World War II, it furnished 8 percent of the nitrogen employed by the world's agriculture. It is still an important source of nitrogen in Europe. Although relatively unimportant in the United States, it is still recommended to control blossom-end rot of tomatoes.

AMMONIUM NITRATE LIMESTONE

Ammonium nitrate limestone is a white to gray chalky powder or pellet. The color of this product depends on the color of limestone used in its manufacture. To make this material, ammonium nitrate is combined variously with $CaCO_3$, finely ground chalk, high calcium limestone or dolomite. It has been successfully used in England, the Netherlands, Turkey, Germany, Italy, Norway and the United States. It is usually sold in a granular form. When made with dolomitic limestone, it contains 20 percent nitrogen, 6 percent calcium and 4 percent magnesium. A-N-L®, Cal-Nitro® and Ammonium Nitrate Limestone® are examples of trade names. Cal-Nitro® is produced in the Netherlands and shipped to the

69

United States. It is usually made with calcitic limestone. By using less limestone and more ammonium nitrate a nitrogen content up to 28 percent is possible.

These materials made with calcium salts are excellent suppliers of nitrogen and calcium for all crops and soils. They are nonacid-forming and are, therefore, most desirable for acid soils. In the United States where limestone is plentiful, it will usually be more economical for the farmer to buy lime by the ton and to get nitrogen from the most inexpensive source. The production and use of these mixtures of ammonium nitrate and limestone will depend largely on the cost and historical market acceptance.

AMMONIUM CHLORIDE

Ammonium chloride occurs as small, white crystals. It may be produced by reacting ammonia with hydrochloric acid.

$$NH_3 + HCl \rightarrow NH_4Cl$$

Or it may be produced by a modification of the Solvay process for making sodium carbonate.

$$NaCl + NH_3 + H_2CO_3 \rightarrow NaHCO_3 + NH_4Cl$$

The product is difficult to manufacture into granular form. Ammonium chloride has been tested in the United States but it does not handle well in fertilizer distributors. Ammonium chloride contains 26 percent nitrogen and is acid-forming. Crystal analysis of mixed fertilizers indicates that NH_4Cl is responsible for much of the poor handling characteristics of lumpy fertilizer. Tests on cotton, potatoes and barley have shown ammonium chloride to be a good source of nitrogen. Tests on rice have indicated that it is a superior agronomic material.

FALL- VERSUS SPRING-APPLIED NITROGEN FOR SPRING-SEEDED CROPS

No discussion of nitrogen would be complete without mentioning the practice of adding nitrogen fertilizers to the soil in the fall for use by spring-seeded crops (Figure 2.12 and Table 2.4).

The most useful data obtained in the southeastern United States were summarized in USDA Technical Bulletin 1254, issued in December 1961. In essence, the report shows that regardless of source, fall-applied nitrogen for a 4-year period was 49 percent as efficient as spring-applied

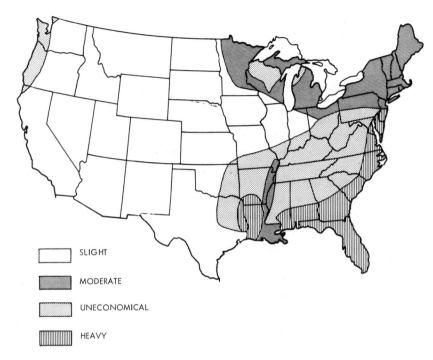

SLIGHT

MODERATE

UNECONOMICAL

HEAVY

Figure 2.12 Winter nitrogen losses.

Table 2.4
NITROGEN CONTENT OF SOME NITROGEN
FERTILIZERS

Material	Percent Nitrogen
Ammonia	82
Aqua ammonia	24
Ammonium nitrate	33.5
Ammonium nitrate–limestone mixtures	20–28
Ammonium nitrate sulfate	26–30
Ammonium sulfate	20.5
Ammonium chloride	26
Urea	45–46
Ureaform	38
Calurea	34
Calcium nitrate	15
Calcium cyanamide	20–21
Nitrogen solutions	19–37
Nitrate of soda	16
Sodium potassium nitrate	15

nitrogen.[8] The soils on which the experiments were run in Georgia, Alabama and Mississippi were similar to North and South Carolina soils. The same general winter temperatures and rainfall patterns prevail over the area.

There is very little delay in the conversion of urea nitrogen to ammonia forms in these soils, and ammonia is converted to nitrate by microorganisms practically the year-round. Thus, no differences among nitrogen fertilizers are found provided they are soil-incorporated.

Other research in Iowa, Minnesota and Wisconsin shows that soil-incorporated fall-applied nitrogen for spring-seeded crops is as efficient as nitrogen added in the spring. Thus, as one travels from the warm, humid southeastern United States to the cooler, less humid regions of the Corn Belt, the losses of fall-applied nitrogen become less and less until they become negligible where soil is in deep-freeze for the winter period that the nitrogen is in the ground and waiting to be used by the spring-seeded crops.[9]

Fertilization is most effective when it complies with the needs of the growing plant. Any divergence from this axiom results in some reduction in plant development.

SUMMARY

1. 35,961 tons (32,724 m tons) of nitrogen over each acre (0.4 ha) of land and water in the world.

2. Nitrogen ranks fourth after carbon, hydrogen and oxygen in the percentage of total dry matter of plants; 0.5 to 5% of dry weight in plants is nitrogen.

3. Nitrogen is usually found to be the most deficient essential element in the cultivated soils of the world.

4. The natural soil supply of nitrogen is governed by oxidation rate of organic matter present.

$$\frac{\text{OM decomposing}}{\text{1 to 2\% annually}} = 20 \text{ to } 40 \text{ lb/acre (22 to 45 kg/ha)}$$

5. 20 lb/acre (22 kg/ha) comes down annually in rainfall. Coal burning supplies some of this.

6. Nitrogen is fixed in the soil by the free-living organisms *Azotobacter, Clostridium* and blue-green algae. These fix 15 to 45 lb/acre (17 to 50 kg/ha) nitrogen annually. About the same amount is lost each year to the atmosphere by volatilization.

[8]R. W. Pearson et al., *Residual Effects of Fall- and Spring-Applied Nitrogen Fertilizers on Crop Yields in the Southeastern United States*, United States Department of Agriculture Technical Bulletin 1254, December 1961.

[9]L. B. Nelson and R. E. Uhland, "Factors That Influence Loss of Fall-Applied Fertilizers and Their Probable Importance in Different Sections of the United States," *Soil Sci. Soc. Amer. Proc.* 19 (1955), 492–496.

7. All three of the above mentioned sources of nitrogen would fall far short of the nitrogen requirements of a 150 bu/acre (10,080 kg/ha) crop of corn, which, if harvested for silage, would remove about 235 lb/acre (263 kg/ha) of nitrogen.

8. The only practical way to obtain more nitrogen is to convert nitrogen from the air into a stable form that is usable by plants.

9. Legumes fix all of their own nitrogen by using *Rhizobium* bacteria. As much as 300 lb/acre (336 kg/ha) is fixed annually. Carbohydrates are traded for nitrogen in a symbiotic relationship between legume and bacteria.

10. Factors governing nitrogen assimilation by the plant are:

    a. Type of plant

    b. Stage of maturity—after transplanting, tobacco uses about 80 percent of its total nitrogen needs during the 3rd through 7th week of growth.

    c. Supply of available nitrogen in the soil.

11. Table 2.4 is a summary of the fertilizers discussed along with their nitrogen percentage.

12. As a general rule, the higher the nitrogen percentage the more economical the carrier is.

QUESTIONS

Directions for Nos. 1–3. Column A lists the nitrogen content of three common nitrogen fertilizer/water solutions. Column B lists the nitrogen compounds found in five nitrogen solutions. For each percentage figure in Column A, write in the answer column the *number* preceding the compounds in Column B from which the solution is made.

Column A	Answer	Column B
1. 19 percent	( )	1. Urea, Ammonium nitrate
2. 30 percent	( )	2. Ammonium nitrate
3. 37 percent	( )	3. Ammonia, Ammonium nitrate
		4. Urea, Ammonia
		5. Ammonia, Urea, Ammonium nitrate

Directions for Nos. 4–8. Each of these questions consists of a statement concerning the use or the properties of certain nitrogen fertilizers. For each statement, select from the list of four compounds in Column A the nitrogen fertilizer with which that statement is most closely associated, and write in the answer column the *number* preceding the compound in the list. (A number may be used more than once.)

	Answer	Column A
4. This compound is a synthetic organic nitrogen and is widely used for aerial application because it is the highest analysis, solid nitrogen fertilizer on the market.	( )	1. Ammonium nitrate 2. Ammonium nitrate-lime-stone
5. This compound is a gas at standard temperature and pressure and is the most concentrated of all nitrogen fertilizers commonly used.	( )	3. Ammonia 4. Urea
6. This nonacid-forming fertilizer contains 20 percent nitrogen.	( )	
7. This compound provides more nitrogen than the others for direct application to the soil and is usually the least expensive per pound (0.45 kg) of nitrogen.	( )	
8. This compound is usually more expensive per pound (0.45 kg) of nitrogen than the others, but it contains calcium, a secondary essential element for plant growth.	( )	

Directions for Multiple Choice Questions No. 9–12. In the answer column write the number(s) corresponding to the answer(s) you select from the list.

	Answer	
9. Acid-forming nitrogen fertilizers are:	( ) ( ) ( ) ( ) ( )	1. Ammonia 2. Sodium nitrate 3. Ammonium nitrate 4. Calcium nitrate 5. Urea 6. Ammonium sulfate 7. Solution 30
10. When the organic matter of virgin soils is depleted by cultivation and cropping, the first essential element to become deficient is usually:	( )	1. Potassium 2. Calcium 3. Nitrogen
11. To supply an acre (0.40 ha) of tomatoes with 120 lb/acre (134 kg/ha) of nitrogen from	( )	1. 34 gal/acre (319 l/ha)

a 32-percent nitrogen solution containing 3.54 lb/gal (0.42 kg/l) of nitrogen requires about:

    2. 40 gal/acre
       (375 l/ha)
    3. 45 gal/acre
       (422 l/ha)

12. The bacteria genus in association with leguminous plant roots that is known to fix atmospheric nitrogen is: (   )

    1. *Azotobacter*
    2. *Clostridium*
    3. *Nitrobacter*
    4. *Nitrosomonas*
    5. *Rhizobium*

True or False Nos. 13–21.

13. ___ For the most profitable corn production (other growth factors not limiting), it is necessary to add 100 to 120 lb/acre (112 to 134 kg/ha) of nitrogen and to have a population of 12,000 to 16,000 plants per acre (30,000 to 40,000 plants per hectare).

14. ___ Organic matter in the soil decomposes at the annual rate of about 1 to 2 percent supplying about 20 to 40 lb/acre (17.8 to 35.7 kg/ha) of nitrogen.

15. ___ Protein-organic sources of nitrogen are superior to inorganic sources for the production of tobacco, ornamentals and vegetables.

16. ___ About the same amount of $CaCO_3$ would be required to neutralize the physiological acidity produced in the soil by 100 pounds (45 kg) of ammonium nitrate containing 33 percent nitrogen, 100 pounds (45 kg) of a 33-percent nitrogen solution and 56 pounds (26.3 kg) of ammonium sulfate containing 20 percent nitrogen. [Ammonia, ammonium nitrate and urea require 1.8 lb (0.81 kg) of lime to neutralize each pound (0.45 kg) of nitrogen added whereas ammonium sulfate requires 5.3 lb (2.4 kg) of lime per pound (0.45 kg) of nitrogen.]

17. ___ About the same amount of $CaCO_3$ would be required to neutralize the physiological acidity produced in the soil from 100 pounds (45 kg) of ammonium nitrate containing 33 percent nitrogen, 100 pounds (45 kg) of a 33-percent nitrogen solution, and 33 pounds (15 kg) of nitrogen from ammonia. (A 33-percent nitrogen solution is made with urea and ammonium nitrate.)

18. ___ For most efficient use, urea should be incorporated into the soil or added directly onto a debris-free, cultivated soil.

19. ___ In a limited capital situation, fertilizer sells for $22 per 200 pounds (90 kg), and the last 200-pound (90-kg) increment of fertilizer brings $117 additional value to the yield of a certain crop. In this case, it would be more profitable to invest in a milking parlor that returns $4 per dollar invested than to add the last unit of fertilizer.

20. ___ The bulk of nitrogen absorbed by plants growing on well-aerated soils is in the nitrate form.

21. ___ The use of nonpressure nitrogen solutions mixed with a herbicide has never become a common farm practice.

chapter 3

Sulfur—Essential For Protein

Sulfur in several forms is used on peanuts; as gypsum to supply the nutrients, calcium and sulfur; and as copper-sulfur dust or sprays to control leaf spot diseases. Here peanuts growing in South Carolina are being treated with copper-sulfur dust to control leaf spot–*Cercospora arachidicola* and *C. personata*.

Sulfur is important in the plant's synthesis of oils. Like nitrogen, sulfur is a constituent of several amino acids, including methionine, cystine and cysteine, which are essential components of plant and animal proteins. About 90 percent of the sulfur in plants is found in these three amino acids. If methionine can be considered as a standard, essential amino acid that is present in plants and animals, the nitrogen-to-sulfur ratio in good quality protein should be around 15:1. Sulfur acts more like nitrogen than any of the other essential plant nutrients; indeed, sulfur deficiency symptoms in plants closely resemble nitrogen deficiency symptoms.

Leaching losses of sulfur from the soil are very much like those of nitrogen caused by percolating waters through light-textured sandy and silty soils. Thus, there is need for annual supplies of sulfur to meet the requirements of growing plants. If this need is not met by decomposition of organic matter, irrigation water, rain or natural soil supplies, it must be added in the form of fertilizer. Jordan and Ensminger[1] predicted that the incidence of sulfur deficiency would increase with time because of the rapid change to fertilizers containing low sulfur content and of the change of factory fuels from coal to gas. The former change has already taken place but the latter one has not. In 1975, there was about 2 times as much sulfur in rainwater as in 1955 (Table 3.1). Their suggestion of a planned program of reexamination of sulfur needs from time to time in order to avoid serious limitations from unknown sulfur deficiencies has considerable merit.

Cotton, corn, rye and wheat yields were increased when 12 lb/acre (13 kg/ha) of sulfur was added to two sandy soils in Georgia. In Alabama, it is recommended that all fertilizers for cotton contain sufficient sulfur to provide 8 lb/acre (9 kg/ha). Similar suggestions appear in South Carolina and Mississippi fertilizer publications based on work of Jordan and Bardsley.[2] Annual requirements for sulfur-sensitive crops are about 15 to 25 lb/acre (17 to 28 kg/ha). Information is being collected on crops and soils in the area to serve as a basis for including sulfur analysis in soil-testing laboratories currently being used to determine other plant nutrient needs.

In rural areas of the southern United States from North Carolina to

[1]H. V. Jordan and L. E. Ensminger, "The Role of Sulfur in Soil Fertility," in *Advances in Agronomy*, ed. A. G. Norman (New York: Academic Press, Inc., 1958).

[2]H. V. Jordan and C. E. Bardsley, "Responses of Crops to Sulfur on Southeastern Soils," *Soil Sci. Soc. Am. Proc.* 22 (1958), 254–256.

Table 3.1
SULFUR COLLECTED IN RAINWATER, 1953–55 *VS.*
1973–75

Collection location	Average annual amount of sulfur collected during the periods	
	1953–55	1973–75
	kg/ha	kg/ha
Clemson, S.C.	8.9[2]	10.1
Columbia, S.C.	6.4	11.8
Charleston, S.C.	5.7	12.1
Average of 3 S.C. towns[3]	6.3[4]	11.3
Watkinsville, Ga.	6.4	
Waynesville, N.C.	6.2	

[1]Data adapted from Jones, "Sulfur Content."
[2]Average for years 1953 and 1955. Data for 1954 not reported.
[3]All South Carolina, Georgia and North Carolina 1953–55 data adapted from Jordan and Bardsley, "Responses of Crops."
[4]The 1953–55 mean is significantly different from the 1973–75 mean at the 1-percent level.

Texas, the amount of sulfur in rain averaged 5.4 lb/acre (6.0 kg/ha) per year.[3] In Indiana, sulfur in rainfall averaged 27 lb/acre (30 kg/ha) per year when the high value of 127 lb/acre (142 kg/ha) for the industrial city of Gary was omitted.[4] In 1923, it was reported that 29.5 lb/acre (33 kg/ha) of sulfur came down in rainfall each year at Ithaca, New York. The sulfur content of water from 19 rivers used for irrigation in the western United States varies from 3 to 1,849 lb/acre (3 to 2,075 kg/ha) per foot (30 cm) deep. Most, but not all, irrigation waters contain sufficient sulfur to prevent deficiencies. Research in South Carolina, twenty years after the data reported above by Jordan and Bardsley, showed that sulfur in rainfall had increased from 5.4 lb/acre (6.0 kg/ha) in 1953–55 to 10.0 lb/acre (11.2 kg/ha) in 1973–1975.[5]

The lithosphere of the earth averages approximately 0.08 percent sulfur. But the amounts in soils vary from less than 100 ppm to several tenths of 1 percent. About 70 to 90 percent of the total surface soil sulfur is found in organic matter. The remainder of the sulfur is present as sulfides and as soluble and insoluble sulfates. Usually only 5 to 10 percent

[3]Jordan and Bardsley, "Responses of Crops."

[4]B. R. Bertramson, M. Fried, and S. T. Tisdale, "Sulfur Studies of Indiana Soils," *Soil Sci.* 70 (1950), 27–41.

[5]U. S. Jones, "Sulfur Content of Rainwater in South Carolina," in *Environmental Chemistry and Cycling Processes*, Department of Energy Symposium Series 45 (Conf.-760429), 1978, 394–402.

of the sulfates are found in the surface furrow slice of humid region, cultivated soils. In humid region soils, most of the sulfates are found 12 to 14 inches (30 to 35 cm) deep and are associated with oxides of iron and aluminum, especially in acid soils having kaolonitic-type clays. Topsoils in the southeastern United States contain less than 446 lb/acre (500 kg/ha) of total sulfur and frequently as little as 89 lb/acre (100 kg/ha). The sulfur content of soils, like the nitrogen content, is closely correlated with the organic matter and clay content.

Prediction of Sulfur Needs.   Soil analysis can be useful in diagnosing sulfur deficiency. Several chemical methods for determining the portion of total soil sulfur that is available to plants have been proposed and evaluated against field results in the southeast.[6] Soils having less than 14 ppm of acetate-extractable or 33 ppm of bicarbonate-extractable sulfur in the top foot (30 cm) were found to be responsive to applied sulfur. Plant analysis also offers reliable diagnosis of sulfur deficiency when the plant part and stage of growth are standardized and the data correlated with yield increases or reductions due to sulfur.

All of the sources of natural supplies of sulfur from soils, water and air being considered, the greatest need for sulfur-containing fertilizers would be expected in upland, leached, nonirrigated soils of low organic-matter content located in areas far from industrial centers. Soils in the Cotton Belt of the Mississippi Valley, where no sulfur-containing fertilizers are used, are sometimes sulfur deficient. On the other hand, amounts of sulfur sufficient to suppress plant growth have been found in industrial areas having an uncommonly large amount of SO_2 released to the atmosphere,[7] and in agricultural areas where large quantities of soluble sulfates have been accumulated from sulfur-rich irrigation waters or from sulfur-rich pesticides. Sulfur-deficient areas have been described on the Pacific coast and in the northwestern, southeastern and midwestern United States, and in the Canadian provinces of Alberta, British Columbia, Manitoba and Saskatchewan.

Many growth factors influence the level of sulfur in crops. Vegetable crops, such as cabbage, turnip and onion, have been rated as having high sulfur requirements. Corn, sorghum, peanuts, tobacco and cotton require large amounts. Small grain crops need less sulfur. Legumes, particularly alfalfa, have intermediate sulfur requirements. Critical levels of sulfur in alfalfa have been reported in Oregon to be 0.22 percent and in Nebraska to be 0.33 percent. Critical level of Coastal Bermuda grass has been reported to be 0.4 percent in Georgia. A nitrogen to sulfur ratio of about 18:1 in plant material gives a rough estimation of the supply of sulfur in

---

[6]C. E. Bardsley and V. J. Kilmer, "Sulfur Supply of Soils and Crop Yields in the Southeastern United States," *Soil Sci. Soc. Am. Proc.* 27 (1963), 197–199.

[7]R. G. Hoeft et. al., "Nitrogen and Sulfur in Precipitation and Sulfur Dioxide in Atmosphere in Wisconsin," *J. Environ. Qual.* 1 (1972), 203–208.

terms of needs for plant nutrition. For ruminant animal nutrition, however, forage having a nitrogen to sulfur ratio of about 10:1 is preferred.

Elemental sulfur and materials containing this element to correct the sulfur deficiencies existing in the cultivated soils of the world are available and reasonably priced. The long-term growth rate in consumption of sulfur is estimated to be 4 percent per year.

A description of these materials and how they are made and used is the subject of the following discussion.

ELEMENTAL SULFUR

Description. It is a light yellow, odorless, nonmetallic element that burns with a blue flame. As a fertilizer-manufacturing material, it can be used as a solid, or heated to form a liquid or molten mass. As a fertilizer for direct application or for mixing, it is used as a finely divided powder or as a spherical, prilled, light yellow pellet of about 10-mesh (0.065-in. or 1.63-mm) size that contains about 5 to 10 percent swelling clay.

Method of Manufacture. There are two principal methods of producing elemental sulfur. One is a mining process named Frasch after Dr. Herman Frasch. The other recovers sulfur as a by-product from sour gas; it will be referred to as *recovered*.[8] Sour gas is a naturally occurring fuel gas that contains hydrogen sulfide (H_2S) as an impurity. Elemental sulfur is produced by one or both of these processes in the United States, Mexico, France, Canada, Turkey, Poland and several other countries.

Sulfur is also produced in petroleum refining and metallurgical operations and as sulfur dioxide (SO_2) by the roasting of pyrites. Pyrites are the source of about 10 percent of the free world's sulfur supply.

The Texas-Louisiana sulfur deposits are said to be the largest in the world. These deposits occur between the barren cap rock and the barren anhydrite salt domes at depths ranging from 500 to 1,500 feet (150 to 450 m).

Frasch. The sulfur-mining process described here is that used on the coast of the Gulf of Mexico in the states of Louisiana and Texas. There are four basic operations:

1. Drilling the well to a sulfur formation.
2. Heating large amounts of water in a power plant.
3. Pumping superheated water into the deposit to melt the crystalline sulfur.
4. Raising melted sulfur to the surface.

[8]Texas Gulf Sulfur Company, *Golden Sulphur* (New York, 1964).

The well drilling is done with a rotary rig for sinking an 8- to 10-inch (20- to 25-cm) pipe to the top of the cap rock and inserting a 6-inch (15-cm) pipe through the large pipe down to the anhydrite formation below the sulfur deposit. A 3-inch (7.5-cm) pipe equipped with a collar that seals the space between the 3-inch (7.5-cm) and 6-inch (15-cm) pipes is inserted down almost to the bottom of the sulfur-bearing rock. Next into the 3-inch (7.5-cm) pipe is placed a 1-inch (2.5-cm) air line that penetrates almost to the bottom of the 3-inch (7.5-cm) pipe.

The first step in the process consists of pumping superheated water down into the sulfur deposit. The amount of water needed to melt a ton of sulfur varies from 1,000 to 12,000 gallons (13,800 to 45,600 l) depending on the character of the formation and the skill of the operator. Before the water can be used, it must be softened by running it through a sodium-exchange material to replace the scale-producing calcium and magnesium salts with sodium. Because of the quantity and quality of water needed for the operation, sulfur companies set up rather elaborate water reservoirs and purification systems similar to those which supply water to large cities. From the water-treating facility, the water moves to high-pressure boilers at the power plant, where part of it is converted to steam. The remainder goes to heaters where the temperature is raised to 320°F (154°C) by direct contact with steam. Insulated pipes then carry the water to tanks near the wells from which it is fed into the hot-water pipe.

Steaming the well is accomplished by metering the hot water down the space between the 10-inch (25-cm) and 6-inch (15-cm) pipes (Figure 3.1). There are two sets of holes near the bottom of the 10-inch (25-cm) pipe, one set above the collar of the 6-inch (15-cm) pipe and one set below it. The water flows through the top set of holes and melts the sulfur in the region through which it percolates. The process is a continuous one; if the flow of hot water to the well is interrupted, the sulfur freezes in the pipes and may cause the loss of the well.

Since sulfur is heavier than water, the melted liquid sulfur drops to the bottom of the hole and forms a pool at the foot of the 6-inch (15-cm) pipe. Here it goes through the lower perforations of the 6-inch (15-cm) pipe and is forced into the bottom of the 3-inch (7.5-cm) pipe that is set slightly above the bottom of the 6-inch (15-cm) pipe. The sulfur rises perhaps one-half to two-thirds of the distance to the surface. Compressed air is then inserted to the bottom through the 1-inch (2.5-cm) pipe and forces the sulfur all the way to the top of the 3-inch (7.5-cm) pipe.

Meanwhile, the hot water cools in the lower reaches of the well and is removed by draining through underground crevices. The cooled water, drained from the formation, is piped to another plant where it is treated to remove soluble sulfides and other impurities.

Recovered.   Much of the production of elemental sulfur in the world comes from Frasch mines on the Gulf of Mexico. However, sulfur recovered from sour natural gas and petrol refinery gas accounts for a larger share of

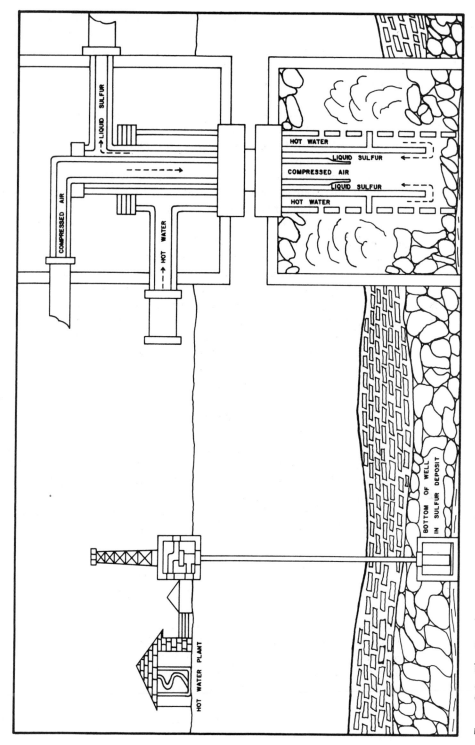

Figure 3.1   The Frasch process—putting hot water into the well, and getting liquid sulfur out.

83

total production in North America each year. Output in the United States of recovered sulfur, i.e., that chemically combined with other elements, increased 8 percent in 1977, amounting to a 35 percent share of total brimstone output in the United States. At the same time, consumption of elemental sulfur increased 4 percent, but total production fell as a result of the decline in the production of Frasch sulfur.[9]

Development of sulfur recovery from sour gas, stimulated by the demands of World War II for greater quantities of this element, began in the early 1940s with the operation of a pilot plant in Arkansas by Texas Gulf Sulphur Company. By the end of the decade, six recovery plants were operating, and four others were being built. Discovery and development of vast sour gas fields in Canada and France gave impetus to the industry, and by 1965 there were more than 150 sour gas recovery plants in operation around the world. In Canada alone, the sulfur contained in gas reserves has been estimated at 260 million tons, enough to supply the free world's total needs, at present consumption rates, for 14 years.

Before sour natural gas can be sold, it must be sweetened to remove the rotten-egg odor of hydrogen sulfide. The rate of recovered sulfur production depends, to a large extent, on the level of demand for sweet natural fuel gas. As fuel gas demand goes up, sulfur production goes up. In some wells, sulfur is a by-product, in some it is a co-product, and in others it is the main product, depending on the amount of hydrogen sulfide in the gas. Natural gas deposits have been found that are so rich in hydrogen sulfide, up to 87 percent, that the contents of the deposit is virtually a sulfur well.

The first step in processing sour gas consists of separating the liquid hydrocarbon condensate and selling it to refineries (Figure 3.2). The sour gas is then dried and sent to a cleaning plant where the hydrogen sulfide is removed by an absorbent solution. The sweetened natural gas is piped to market, while the absorbent moves to a reactivator that removes the hydrogen sulfide, now called "acid gas" by people in the industry. The sulfur plant acid gas is burned with a calculated amount of air in a furnace to produce liquid sulfur and sulfurous gases according to the following reactions:

$$H_2S + {}^3/_2 O_2 \rightarrow SO_2 + H_2O$$
$$2H_2S + SO_2 \rightarrow {}^1/_2 S_6 + 2H_2O$$

A coalescer extracts more liquid sulfur from the gases, which are then passed to the catalyst chamber where the bauxite catalyst converts any remaining hydrogen sulfide into sulfur vapors. Passing into the second coalescer, small droplets of sulfur combine into larger particles and condense as additional liquid sulfur. Waste gases from the last coalescer

[9]United States Brimstone. Production Declines. *Sulphur* No. 130; 1977. The Sulphur Institute, Washington, D.C.

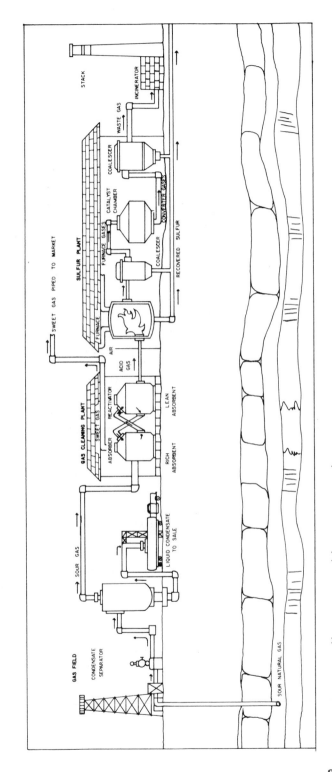

Figure 3.2 How sulfur is recovered from sour natural gas.

are burned in an incinerator. Still in liquid form, the sulfur is either stored in steam-heated tanks or piped to outdoor vats, where it is sprayed in thin layers over a large area to cool and solidify.

Chemical Characteristics.   Sulfur is an element, but it exists in various forms with a complex molecular structure that influences its physical properties and handling characteristics. It has no odor nor is it soluble in water. It oxidizes slowly on contact with soil, air or the atmosphere. It is in the same group in the periodic table with oxygen and selenium. It is therefore a congener of these elements, which are sometimes called chalcogens or chalk formers. Sulfur is also near nitrogen and phosphorus in the periodic table and exhibits some of their chemical properties.

Sulfur is a poor conductor of heat and electricity, but friction on sulfur produces a negative charge. Under certain conditions of friction or grinding, sulfur dust and air form a potentially explosive mixture. The explosive mixture of potassium nitrate, charcoal and sulfur has been used as gunpowder since it was invented in China around A.D. 1200.

Sulfur ignites at a temperature between 478° and 501°F (250° to 260°C). When burned, it produces sulfur dioxide, a pungent gas, which is widely used in the manufacture of sulfurous and sulfuric acid. Indeed 75 to 85 percent of all elemental sulfur is used for this purpose.

Handling and Shipping.   More than 85 percent of the elemental sulfur delivered in the United States in 1964 was in liquid form, and it was rapidly replacing solid sulfur shipments in Europe.[10] Shipment in liquid form requires that the sulfur be heated by steam coils in the bottom of an insulated tank if transport is by railway tank car, truck or barge. It is necessary to maintain a temperature of 270° to 290°F (132° to 143°C). At the destination, the molten sulfur is pumped out of the transport tank into storage tanks where it is kept in liquid state until used.[11] The advantages of shipping in liquid form are many.[12] There is the saving on labor for handling as well as the saving due to elimination of wind loss and prevention of potential dust explosions. Further, the cost of operating and maintaining melting equipment at the destination can be eliminated if shipping schedules of the liquid sulfur can be geared to production schedules of sulfuric acid. Most new sulfuric acid plants are being designed to utilize the liquid product.

The other older method of storing elemental sulfur is the so-called vat. The sulfur vat is a huge, outdoor, rectangular pile of solidified sulfur. Some of the vats are large enough for four elevated football fields to fit end to end on top of the pile. Such a vat contains 400,000 to 500,000 tons

[10]Texas Gulf Sulphur Company, *Golden Sulfur*.

[11]Freeport Sulphur Company, *Liquid Sulphur by Rail* (New York, 1961).

[12]Freeport Sulphur Company, *Freeport Sulphur Handbook* (New York, 1959).

(364,000 to 455,000 m tons) of pure sulfur.[13] The pile is built up during months of production by spraying thin layers of liquid sulfur as it comes from the mines or recovery plant. The vat is built with these thin layers because sulfur does not lose heat rapidly and therefore must be spread thinly over a large area to insure uniform cooling and solidification. To prepare for shipping, the solid sulfur is blasted or bulldozed over the side of the pile, crushed, and loaded with power-driven, clamshell buckets onto cars or trucks. For loading into barges or ships, the crushed solid sulfur is loaded onto conveyor belts that carry it to the hold of the vessel.

Up to this point, the discussion of handling characteristics has been concerned with the processing and delivery of elemental sulfur from mines or recovery plants to factories where the product is eventually used for further manufacture. The remainder of this discussion is devoted to the handling of an elemental sulfur product designed for blending with nitrogen-phosphorus-potassium mixtures. One product that has been tested is prilled sulfur, which is 88 to 90 percent sulfur and 10 to 12 percent swelling clay.[14] Without prilling with clay, the size of individual sulfur particles is 60-mesh (0.098-in. or 0.25 mm). Prills are manufactured from waste petroleum refinery gas. The handling advantages of prilled sulfur reported by the manufacturer are:

1. Resistance to breakage and problems of fines.
2. Safe to handle because problems of fines and dust are minimized, thus eliminating explosion hazard.
3. Mesh size selected (Table 3.2) to match prilled and granular fertilizers offered for sale.

Table 3.2
TYPICAL PRILLED SULFUR MESH SIZE

ASTM Screen No.	Tyler Screen No.	Size Opening (inches)	(mm)	Percent by Weight
Greater than > 6	6	0.129	3.3	0.6
8	8	0.091	2.3	10.6
12	10	0.065	1.6	70.8
16	14	0.046	1.2	18.0
20	20	0.032	0.8	0.4
Smaller				0.2
than <	20			100.0

[13]Texas Gulf Sulphur Company, *Golden Sulfur*.

[14]G. Rehm, "Sulfur Response on Irrigated Corn in Nebraska," *Sulphur Inst. J.* 12 (1976), 13–14.

The purpose of producing such a product is to facilitate the manufacture of a homogeneous blend of prilled sulfur and other granular fertilizers from the point of blending or mixing for application in the field. The material can be used where needed as an ideal form of elemental sulfur for spreading on the land without prior mixing with other fertilizer materials. It is shipped bulk or bagged in railway cars and trucks.

Crop and Soil Response.   Elemental sulfur, as well as certain other forms that will be discussed later, is added to soils for the purpose of making food and fiber crops grow better. It is usually referred to as agricultural sulfur and contains from 50 to over 90 percent pure sulfur. It exerts its influence on crop growth in one of two ways. In the first place, it is an essential nutrient without which plant and animal life on this planet could not exist. Plants take up sulfur almost exclusively as the sulfate form. Alfalfa recovered 21.7 lb/acre (24.3 kg/ha) of sulfur in 3 years from gypsum ($CaSO_4:2H_2O$) added at the rate of 30 lb/acre (33.6 kg/ha) of sulfur. From the same amount of sulfur added as elemental sulfur, alfalfa recovers 9.2 lb/acre (10.3 kg/ha). Elemental sulfur should be added to the soil about 6 months before planting and should have a fineness of 60-mesh (0.0098-in. or 0.25 mm) or smaller.

In the second place and perhaps most important in regard to elemental sulfur, is its capacity to reduce soil pH and thereby to improve the production of excessively alkaline soils or those made alkaline by excessive additions of lime. After oxidation and hydrolysis in soils, 32 parts of elemental sulfur will convert to sulfuric acid and neutralize the equivalent of 100 parts of calcium carbonate. The neutralization of alkaline salts in the soil frequently results in increased availability of other plant nutrients, such as phosphorus, iron, manganese and zinc, along with subsequent improvement of plant growth.

Elemental sulfur is oxidized to sulfuric acid in a warm, moist, aerated soil by bacterial action. The sulfur-oxidizing bacteria, *Thiobacillus*, are found in most soils throughout the world. These soil bacteria oxidize elemental sulfur to sulfuric acid in the following manner:

$$S + O_2 \rightarrow SO_2 + energy$$

$$2SO_2 + O_2 \rightarrow 2SO_3 + energy$$

$$SO_3 + H_2O \rightarrow H_2SO_4$$

Other things being equal, the rate of the reaction depends on the fineness of the sulfur. With finely ground material, e.g., 60 to 100 mesh (0.25 to 0.14 mm), the large surface area speeds up oxidation of sulfur. Prilled sulfur that contains a swelling clay like bentonite will oxidize more slowly than ground sulfur; in some circumstances, the slow oxidation rate may be desirable because the reaction may occur throughout the growing season and for several years. Usually, about three years' supply of this product is suggested for the first year and annual rates thereafter. The amount of sulfur required to lower soil pH is shown in Table 3.3. Lowering the pH of alkaline soils with sulfur, sulfuric acid or aluminum sulfate will help to flocculate the finely divided soil particles, thus reducing the incidence of puddling and generally improving the soil's physical condition. Elemental sulfur is one of the most economical sources of this nutrient, and world agricultural demand for it is likely to grow more rapidly in the future than in the past.

Table 3.3
FINELY GROUND SULFUR NEEDED TO LOWER SOIL pH TO ABOUT 6.5

Soil pH Found by Measurement	Kilograms Per Hectare[1]			
	Broadcast Application to Whole Soil Mass		Band or Furrow Application	
	Sandy Soils	Clay Soils	Sandy Soils	Clay Soils
7.5	400–600	800–1000	200–250	300–500
8.0	1000–1500	1500–2000	300–500	600–800
8.5	1500–2000	1500–2000	600–800	800–up
9.0	2000–3000		800–up	

[1]Divided by 0.9 = pounds per acre

SULFURIC ACID

The growing importance of fertilizers throughout the world and particularly in underdeveloped and Communist block countries will result in an increased demand for sulfur because the major single use for sulfur is the production of fertilizer using sulfuric acid. In 1977, sulfur consumption in the Western World for fertilizer use is expected to total 16.7 million tons (15.2 million m tons).[15]

Description.   Sulfuric acid is a heavy, colorless, oily, and very strong acid, sometimes referred to as oil of vitriol. The sulfur it contains has a valence of 6—H_2SO_4. It has a specific gravity of 1.8.

Method of Manufacture.   Sulfuric acid is made by two processes. One is known as the chamber process, and the modern one is called the contact process (Figure 3.3). Very few of the older chamber acid plants are still in

[15]"Western World Sulfur Demand," *Sulphur*, No. 131 (Washington, D.C.: The Sulfur Institute, 1977), 5–8.

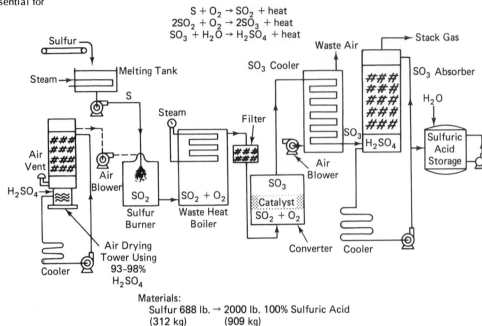

Figure 3.3   Flow diagram of contact process for sulfuric acid.

operation; one or two were operating in the 1970s in the older fertilizer-consuming areas of the southeastern United States. Both methods involve burning sulfur or sulfur ores, such as pyrites, to sulfur dioxide; the use of a catalyst to convert sulfur dioxide to the trioxide; and the absorption of the latter in dilute sulfuric acid. The major differences between the two methods are the types of catalysts, the design of the systems and more particularly the strength of the sulfuric acid manufactured.

The *contact* process begins by burning elemental sulfur in purified air according to the following reaction:

$$S + O_2 \rightarrow SO_2 + \text{heat}$$

The rapid conversion of sulfur dioxide to the trioxide in presence of the catalyst is effected by maintaining the temperature in the converter at about 1022°F (550°C).

$$2SO_2 + O_2 \rightarrow 2SO_3 + \text{heat}$$

Since this reaction is reversible, it is forced to the right by keeping a 150 percent excess of oxygen in the converter during the contact period. The reaction speed is dependent on the temperature, the activity of the catalyst and the time of contact with the catalyst. The procedure gets the name contact process from the fact that sulfur dioxide is absorbed on the surface of the vanadium or similar catalyst before its reaction with oxygen begins. Upon completion of the reaction, the sulfur trioxide is rapidly piped to the cooler to prevent reversing the reaction. The sulfur trioxide is absorbed in 98 percent sulfuric acid.

$$H_2SO_4 + H_2O + SO_3 \rightarrow 2H_2SO_4$$

The absorbing acid is continuously recycled from the bottom to the top of the tower, and the cooled sulfur trioxide is piped to the bottom. Finally, the acid is collected in tanks, and the concentration is adjusted by adding weaker acid or water. The efficiency of converting elemental sulfur to sulfuric acid by the contact process is approximately 95 percent.[16]

The *chamber* process for production of sulfuric acid utilizes the same reactions employed by the contact process except that the catalyst is nitrogen oxide. The nitrogen oxide catalyst is located in the sulfur dioxide gas pipes and mixes with the sulfur dioxide before it enters the air-cooled, box-shaped lead chambers, from which this process gets its name. It is in the lead chambers that sulfur trioxide is formed and subsequently hydrated to sulfuric acid. The catalyst, nitrogen oxide, is reused after it is scrubbed of the residual gases with 77 percent sulfuric acid in towers. The lead chambers produce sulfuric acid in concentrations on the order of 65 percent, and the concentration of that from the tower is about 77 percent. Concentration to 93 percent sulfuric acid is possible by using heat to evaporate some of the water.

Handling Characteristics. Particular care must be taken to prevent corrosion of containers and transport by sulfuric acid because it is a very strong acid. Materials for containing it include acid-proof brick, steel, and lead-lined tanks and lead pipe. The lead containers are used to handle only sulfuric acid of less than 98 percent concentration. In the fertilizer industry, it had been quite common to find the sulfuric acid plant coexisting with the fertilizer plant, especially where superphosphates, ammonium phosphates and ammonium sulfates were made. The acid was simply piped to the spot where it was to be used for acidulating rock phosphate or neutralizing ammonia. Steel-tank railway, truck and barge transports are now available for shipping the acid long distances.

Product Use. In addition to its use for fertilizer, sulfuric acid has many important applications including the manufacture of chemicals, explosives, film, iron, matches, varnish, paint, pesticides, pharmaceuticals, pigments,

---

[16]M. F. Gribbens, "Conversion of Ammonia to Fertilizer Materials," in *Fertilizer Technology and Resources*, ed. K. I. Jacob (New York: Academic Press Inc., 1953), pp.73–75.

plastics, paper, pulp, rubber, textiles, vegetable oils and petroleum. An interesting arrangement between the petroleum and fertilizer industries involves the use of relatively pure sulfuric acid for petroleum refining and the use of the spent acid in the manufacture of various phosphate fertilizers. The spent acid from petroleum refining is referred to as alkylation acid. Waste sulfuric acid from other industrial processes including the manufacture of explosives, pigments, certain metals and several organic synthesis, is also used in fertilizer phosphate production. The use of spent acid for fertilizer manufacture, especially from certain processes, should be carefully evaluated to avoid concentration levels of toxic materials that might hinder vegetable growth.[17,18]

Crop and Soil Response.   The use of sulfuric acid as such for direct addition to the soil has received little attention in the past except for metering into irrigation waters in the southwestern United States. In some instances, this acid has been added directly to soils by spraying or by chiseling it in with a special applicator. Treatment of saline-alkali soils with sulfuric acid and irrigation waters in arid regions increases water infiltration rates, which are so very important to the efficient use of water in these areas. With the advent of mild steel liquid fertilizer application equipment, sulfuric acid, as such, may become more important as a direct-application fertilizer material in certain areas of the world. See Figure 3.4.

GYPSUM

Gypsum, $CaSO_4 \cdot 2H_2O$, was first recorded as an inorganic soil amendment in 1768 when it was found to be beneficial to clover in a Swiss experiment. It was introduced to colonial USA agriculture by Benjamin Franklin who reportedly added it to a meadow near Washington, D.C. to spell in greener grass, "This has been plastered." A product still sold in the United States has a trademark consisting of a picture of Franklin and the registered trade name BEN FRANKLIN® agricultural gypsum. The value of gypsum is most pronounced with legumes on sandy soils. Gypsum is also widely used in the treatment of soils and irrigation waters in arid agricultural regions. Eighteenth and nineteenth century experiments provided no proof that sulfur in gypsum was the factor responsible for improved growth even though Sachs had shown in 1860 that sulfur was an essential element for plant growth. Lawes (1843) began the superphosphate industry that pro-

[17]K. D. Jacob and W. H. Armiger, "Field Experiments with Alkylation-acid Superphosphate," *J. Am. Soc. Agron.* 36 (1944) 281–286.

[18]L. F. Rader, Jr., D. S. Reynolds, and K. D. Jacob, "Effect of Picric Acid in Superphosphate on Tomatoes and Beans as Indicated by Greenhouse Experiments," *J. Am. Soc. Agron.* 36 (1944), 545–551.

Figure 3.4 Sulfur increased yields in 7 of 12 experiments with clover or clover-grass pastures as reported by Jordan in the Southern United States. Here a scientist is indicating lush fescue rows in a clover-grass pasture. Fescue seed and a mixed fertilizer containing gypsum were added in bands with a grassland drill at planting. Then clover seed was broadcast. (*Sulfur as a plant nutrient in the Southern United States,* H.V. Jordan, U.S. Dept. Agr. Tech. Bull. 1297, 1964.)

vided a readily available material containing in addition to monocalcium phosphate about 60 percent $CaSO_4$. Because of the widespread use of $(NH_4)_2SO_4$ and superphosphate as fertilizers, there was very little interest in sulfur plant nutrition until manufacturers and the Tennessee Valley Authority pointed the way to methods of fertilizer manufacture that included little or no sulfur.

Description.　Although gypsum is usually in the form of a white, powdery material having an analysis of $CaSO_4 \cdot 2H_2O$, it is sometimes colored with iron or other impurities. Gypsum rock is soft and can easily be scratched with a knife. A similar product having an analysis of $CaSO_4 \cdot \frac{1}{2}H_2O$ is referred to as plaster of Paris or land plaster and is used in the manufacture of portland cement. Agricultural gypsum is a finely divided material with the color and consistency of cement. Anhydrous calcium sulfate is sometimes referred to as anhydrite.

Method of Manufacture. Gypsum is a mineral and occurs as a very soft rock or in the form of sand. The dunes of the White Sands National Monument in New Mexico are gypsum sand. The rock is processed by grinding mills to the desired fineness. It is the same gypsum and/or anhydrite as that formed in the manufacture of superphosphate by adding sulfuric acid to ground phosphate rock.

Handling Characteristics. Gypsum handles very much like powdered, 20 percent superphosphate that is 60 percent gypsum. It can be distributed in small quantities with ordinary fertilizer distributors. Also many planters report successful use of the planter hopper method—using approximately four handfuls of gypsum per hopper of seed. In large quantities, it can be added to the soil with lime spreaders (Figure 3.5).

Chemical Characteristics. Pure gypsum has the formula $CaSO_4 \cdot 2H_2O$. Agricultural gypsum contains 50 to 95 percent $CaSO_4$. Pure $CaSO_4$ contains 23.5 percent sulfur and 29.4 percent calcium. A dozen samples of the material used as a fertilizer on peanuts in Virginia and North Carolina contained 21 percent sulfur and 28 percent calcium. It has a solubility of 0.00044 lb (0.2 g) per 0.26 gal (100 ml) of water at 32° to 212°F (0° to

Figure 3.5 Sulfur is added in many ways and in many forms, as gypsum, fine soil sulfur, and sulfur pellets to increase pasture, forage, and rangeland yields and quality. Here it is being added as gypsum with a conventional spreader to pasture land in Texas.

100°C), and will leach out of very sandy soils. It is the most soluble mineral form of calcium, being much more soluble than limestone, and is a neutral salt. Although it contains sulfur, it will not counteract soil alkalinity as will sulfate of ammonia and ammonium phosphate sulfate. Elemental sulfur lowers soil pH only in oxidation from sulfur to SO_4^{--}. The calcium in gypsum will not counteract soil acidity as will calcium carbonate and calcium silicate.

Crop Response. For many years, experiments in the southeastern United States,[19] comparing superphosphates with other phosphorus fertilizers that do not contain sulfur, have shown that superphosphates produce more corn, cotton, oats, tobacco, wheat, alfalfa, red clover and vegetable crops, especially of the cruciferae family. The sulfur removed from the soil by commonly grown crops is equal to or greater than the phosphorus removed. In fact, the sulfur content of many legumes and the cruciferae exceeds that of phosphorus. Sulfur increased yields in 7 of 12 experiments with clover-grass pastures (Figure 3.4).[20]

In parts of the southeastern United States, 600 to 800 lb/acre (670 to 896 kg/ha) of gypsum, as a source of calcium, is added to peanuts at the early bloom stage. In the same area, 300 to 500 lb/acre (336 to 560 kg/ha) of gypsum is added to Coastal Bermuda grass.

Of a number of field and vegetable crops, only corn responded to sulfur additions, such as gypsum or kieserite, on Norfolk loamy sand in the 1973–78 period.[21]

Sulfur deficiency symptoms are characterized by a yellowing of leaves that resembles nitrogen starvation. Both sulfur and nitrogen are essential for protein formation, and research indicates that the nitrogen-to-sulfur-ratio may vary from 15:1 to 18:1 depending on the plant.[22] This ratio makes it possible to distinguish between plant material produced at an adequate and at a distinctly deficient supply of sulfur for plant nutrition.

For ruminant animals' grazing forage, if dietary sulfur levels are adequate (0.20 to 0.25 percent sulfur), a nitrogen-to-sulfur ratio of 10:1 to 12:1 is preferred. Experiments have shown that sheep, cattle, goats and dairy cows utilize sulfate sulfur in the synthesis of methionine, cystine and cysteine. Work in Australia suggests that sulfur fertilization may be superior to dietary supplementation but that it is easy to remedy an insufficient sulfur content in the fodder or silage of ruminants by adding

[19]U. S. Jones, "Phosphate Fertilizers," Mississippi Agricultural Experiment Station Bulletin 503, 1953.

[20]H. V. Jordan, *Sulfur as a Plant Nutrient in the Southern United States,* United States Department of Agriculture Technical Bulletin 1297, 1964.

[21]U. S. Jones, "Atmospheric Desposition of Sulfur as Related to Soil Fertility in South Carolina," *Soil Sci. Soc. Am. J.* in press.

[22]B. A. Stewart and C. J. Whitfield, "A Ratio—Nitrogen:Sulfur," *Agric. Res.* 14 (August 1965), p. 5.

sulfate to the diet when necessary.[23] Silage should be supplemented with one part of sulfur for every ten parts of nitrogen added as urea.

Much sulfur is added to soils in rainwater particularly near cities where immense quantities of hydrogen sulfide and sulfur dioxide are increasingly being discharged into the atmosphere (Figure 3.6).[24] However, cultivated, coarse sandy soils in rural areas may lose, through drainage and cropping, more sulfur than is added in precipitation. Sulfur is also added to soils via many mixed fertilizers that are made from materials containing varying amounts of sulfur (Table 3.4).

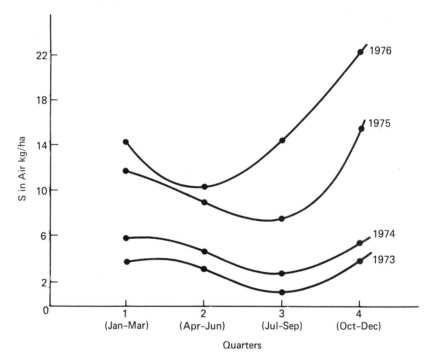

Figure 3.6   Sulfur collection in air averaged by quarters and years for 15 locations in South Carolina. 1973–1976. (U.S. Jones, *op. cit.*)

In arid regions of the world, gypsum is used in large quantities to reduce the incidence of puddling and to improve the permeability of saline-sodic soils so that plants can grow.[25] Calcium sulfate reacts with sodium carbonate or potassium carbonate to form corresponding sulfates and less soluble calcium carbonate according to the following reaction:

[23]S. L. Tisdale, "Sulfur in Forage Quality and Ruminant Nutrition," Sulphur Institute Technical Bulletin 23, 1977.

[24]U. S. Jones, "Atmospheric Desposition."

[25]C. A. Bower, *Chemical Amendments for Improving Sodium Soils*, Agriculture Information Bulletin 195 (Washington, D.C.: United States Department of Agriculture, 1959).

Table 3.4
SULFUR CONTENT OF VARIOUS FERTILIZERS

Material	Sulfur Percent
Aluminum sulfate (alum)	14.4
Ammonium phosphate	2.2
Ammonium phosphate sulfate	15.4
Ammonium polysulfide	40–45
Ammonium sulfate	23.7
Ammonium sulfate nitrate	15.1
Ammoniated superphosphate	11–13
Copper sulfate	12.8
Calcium sulfate (gypsum)	15–18
Iron sulfate (copperas)	11.5
Lime-sulfur	23–24
Magnesium sulfate (kieserite or Epsom salts)	18.2
Manganese sulfate	14–17
Potassium-magnesium sulfate	22.7
Potassium sulfate	17–18
Sodium sulfate (salt cake)	22
Soil sulfur (elemental)	50–99
Sulfur dioxide	50
Sulfuric acid (100% H_2SO_4)	32.7
Superphosphate, normal	12–14
Superphosphate, concentrated	1–3
Zinc sulfate	13–18

$$CaSO_4 + Na_2CO_3 \rightarrow CaCO_3 + Na_2SO_4$$

Irrigation water will then wash out the soluble sulfates and increase penetration of water by 30 to 170 percent depending on conditions. One acre-foot (hectare 30 cm deep) of water dissolves about 250 pounds (280 kg) of gypsum.

Other sulfur compounds such as ammonium polysulfide, sulfuric acid, elemental sulfur and ammonium thiosulfate are just as effective and require much less material. However, they are also more expensive.

ANHYDROUS AMMONIA-SULFUR

The Tennessee Valley Authority at Wilson Dam, Alabama, Shell Chemical, and several other fertilizer manufacturers has been working on a process for adding elemental sulfur to anhydrous ammonia. The product usually contains about 74 percent nitrogen and 10 percent sulfur. Several carloads have been produced by TVA for use in research, education or soil and water conservation.

The chemistry of the process is not well known although it is known that sulfur is soluble in anhydrous ammonia and that the reaction of sulfur with ammonia forms various compounds. When the anhydrous ammonia-sulfur is released in the soil, most of the sulfur separates as elemental sulfur; it is suspected that the system is altered sufficiently to produce some oxidized sulfur compounds. It is known that biological oxidation of elemental sulfur as well as of nitrogen is temporarily stopped when anhydrous ammonia is added to soil because of the sterilization effect of the ammonia in the soil. The ammonia-sulfur-soil reaction products are easily hydrolyzed, especially in alkaline soil conditions. The chemical transformation in the soil could be valuable in making some of the sulfur available to plants until a biological system becomes reestablished.

ALUMINUM SULFATE

This salt, known as "alum," contains 14.4 percent sulfur and has many industrial and pharmaceutical uses. In agriculture, it is used to acidify soils that are naturally too alkaline or that have been made so by injudicious use of liming materials. Aluminum sulfate and iron sulfate are both used to correct iron deficiencies of ornamental shrubs. Alum does not provide iron as a nutrient, but, after hydrolysis in the soil, it makes the iron in the soil more available by increasing the acidity according to the following reaction:

$$Al_2(SO_4)_3 + 6H_2O \rightarrow 2Al(OH)_3 + 3H_2SO_4$$

Alum is also used in municipal water supplies to clarify water by removing the clay. In the reaction shown above, the product $Al(OH)_3$, reacts with the negatively charged colloidal clay or colloidal organic matter, making it insoluble and thus causing it to settle to the bottom of the reservoir and clear the muddy water. Superphosphates will also clear muddy ponds and add phosphorus; like soap powder and other detergents, superphosphates will, with continued use, bring on eutrophication. Alum will not cause eutrophication.

SULFUR-CONCENTRATED SUPERPHOSPHATE

The Tennessee Valley Authority is conducting research designed to produce a material containing 17 percent phosphorus (40 percent P_2O_5) and 20 percent sulfur. Molten sulfur is incorporated into the granule by adding it to the slurry in manufacture of concentrated superphosphate. Such a material is needed for direct application to sulfur-deficient soils and will be particularly useful to fertilizer bulk blenders doing business in sulfur-deficient areas using high-analysis, low-sulfur raw materials.

AMMONIUM POLYSULFIDE

Few, if any, of the raw materials used in the manufacture of clear liquid nitrogen-phosphorus-potassium fertilizers contain appreciable quantities of sulfur. In sulfur-deficient areas where liquid fertilizers are used, there is need for a material that is soluble enough to use in the liquids. This material, ammonium polysulfide solution, contains 20 percent nitrogen and 40 percent sulfur.

$$2NH_3 + H_2S \rightarrow 2(NH_4)S_x$$

It can be added directly to the soil, metered into irrigation water or mixed with anhydrous ammonia solutions for subsequent addition to the soil. Ammonium polysulfide, however, is not completely compatible with some types of liquid fertilizer because of high salt content or high acidity. Samples of this material should be trial mixed prior to full scale batch operation. Suppliers of this product include several chemical companies in the United States and elsewhere.

Calcium polysulfide has been used as a solid in some arid areas. When it or ammonium polysulfide is added to the soil, the particular material decomposes and releases elemental sulfur, which is subsequently oxidized and hydrolyzed to sulfuric acid. Both of the polysulfides are used to improve the physical condition of alkaline soils in the arid and subhumid regions of the western and southwestern United States.

AMMONIUM THIOSULFATE

The solid material contains 19 percent nitrogen and 43 percent sulfur and is being used in irrigation water for fertilizer purposes. A liquid product containing 12 percent nitrogen and 26 percent sulfur is offered by several manufacturers including General Chemical Company, Kerly Chemical Company and Elcor Corporation. It is compatible with aqua ammonia, nitrogen solutions, ammonium phosphate and nitrogen-phosphorus-potassium fertilizer liquids. Alkaline materials such as anhydrous ammonia or acid solutions such as phosphoric acid will decompose ammonium thiosulfate.

These materials are thoroughly discussed in other sections and are mentioned here only because they contain sulfur. Potassium sulfate contains 41 percent potassium (49 to 50 percent K_2O) and 18 percent sulfur. Potassium-magnesium sulfate contains 22.7 percent sulfur, 11.2 percent magnesium and 18.4 percent potassium (22 percent K_2O). The use of these materials in tobacco fertilizers has eliminated most of the chlorine that would be present if potassium chloride were used as the source of potassium.

AMMONIUM SULFATE

This material has been discussed in Chapter 2 because it is normally priced on the basis of its nitrogen content; the sulfur is in effect an incidental ingredient but no less important as a plant nutrient (Figure 3.7). It is mentioned here because it contains 24 percent sulfur in addition to 21 percent nitrogen and, of the group of nitrogen-sulfur salts suitable for use in liquid fertilizers, its cost per unit of sulfur is by far the lowest, based on its historic prices. Its drawback is its relatively low solubility in aqueous solution, compared to some of the other nitrogen-sulfur salts. The low solubility limits the amount of sulfur that may be added to a fertilizer solution as ammonium sulfate.

Trials made in 3-ton (2.73-m ton) batch clear liquid plants indicate that 1 to 1.5 percent sulfur can be added as sulfate of ammonia to N–P_2O_5–K_2O grades such as 12-3-6, 4-8-12 and 5-10-10 without difficulty of salting out. The raw materials used in formulating these mixtures were liquid ammonium polyphosphate (10 percent nitrogen, 15 percent phosphorus), urea–ammonium nitrate solution (32 percent nitrogen) and white muriate of potash (50 percent potassium).

SUMMARY

1. The sulfur content of many plants is equal to or greater than the phosphorus content.

2. The sources of sulfur for plants are air, rainwater, soil and fertilizer. Sulfur in air and rainwater is increasing.

3. Tobacco needs some sulfur, but too much is harmful; 2 to 3 percent sulfur in fertilizer, but not more than 4 percent, is suggested for bright tobacco. Peaches need some sulfur, but they can get too much when large quantities of sulfur are added as fungicide for brown rot.

Figure 3.7   Sulfur increased yields in 7 of 9 experiments with cotton in the Southern United States as reported by Jordan. Eight kg/ha of sulfur, band/placed with the fertilizer, was usually adequate for maximum yields.

4. In rural areas of the southeastern United States in 1953–55, the amount of sulfur in rain averaged 5.4 lb/acre (6.0 kg/ha) per year. Research conducted in 1953–55 on cotton and clover in the southeastern United States also indicated that mixed fertilizers should contain sufficient sulfur to provide 8 lb/acre (9.9 kg/ha) per year. Research in 1973–75 showed that sulfur in rain averaged 10 lb/acre (11 kg/ha) per year in South Carolina. During 1970–71, sulfur in rain averaged 14, 38 and 150 lb/acre (16, 42 and 168 kg/ha) per year in rural, urban and industrial locations in Wisconsin, respectively.

5. Because of the incidental application of sulfur with materials such as superphosphate, ammonium sulfate, and potassium sulfate, plant deficiencies of this element have not been widespread. With the advent of liquid and solid fertilizers containing no sulfur, the sulfur status of soils, air and rainwater needs reevaluating.

6. Good soil management dictates that the sulfur status of soils be known.

7. How much sulfur after oxidation and hydrolysis is required to neutralize 2 tons (1.82 m tons) of limestone that is 95 percent $CaCO_3$ equivalent? The atomic weights are S = 32, H = 1, O = 16, Ca = 40 and C = 12.

$$\frac{x}{32} = \frac{0.95\ (4000)}{100}$$

$$x = 1{,}216\ lb = 552.7\ kg$$

8. The two principal methods of producing elemental sulfur are Frasch and recovered from sour gas.

9. Sour gas is a term applied to natural gas containing various quantities of hydrogen sulfide.

10. Under most conditions, gypsum is as effective or more effective than other sources of sulfur as a plant nutrient.

11. Sulfur performs a triple role in soil fertility and plant nutrition: it is an essential plant nutrient; it lowers pH (increases acidity); and it helps to reclaim saline-sodic soils.

12. As a rule, forage plants make adequate growth when the nitrogen-to-sulfur ratio is between 15:1 to 18:1. However, ruminants seem to perform best when the ratio of nitrogen-to-sulfur in their feed is between 10:1 and 12:1.

13. Soils having less than 14 ppm of acetate-extractable or 33 ppm of bicarbonate-extractable sulfur in the top foot (30 cm) were found to be responsive to applied sulfur.

14. Critical levels of sulfur in vegetative tissue of many plants vary from 0.2 to 0.4 percent sulfur if other growth factors are normal.

QUESTIONS

Agricultural sulfur is used as a soil amendment to reduce excessive alkalinity in soils. Select the three statements below that are applicable to sulfur, and insert the statement number in the answer column.

Answer

1. (   )   1. It increases soil pH.
2. (   )   2. It decreases soil pH.
3. (   )   3. It is not soluble enough to be used in
                 clear liquid fertilizers.
            4. It is a neutral salt.
            5. About 640 pounds (288 kg) is suggested
                 to neutralize 1 ton (0.91 m ton) of limestone.

True or False:

4. _____Agricultural gypsum added to the soil in humid regions increases soil pH.

5. _____About 600 lb/acre (672 kg/ha) of agricultural gypsum are suggested for peanut production in the southeastern United States.

6. _____Sulfur performs a triple role in soil fertility. It is an essential plant nutrient; it lowers soil pH; and with irrigation, it can improve sodium soils.

7. _____The bulk of the sulfur taken up by plants is in the sulfate form.

8. _____Under most conditions, agricultural gypsum is one of the most effective sources of sulfur as a plant nutrient and for reclamation of sodium soils.

9. _____To reduce pH, elemental sulfur should be added to the soil about 6 months before planting and should be no larger than 60-mesh (0.25-mm) size.

10. _____The most important soil bacteria for converting elemental sulfur to sulfate is *Thiobacillus*.

11. _____Sour gas is a term applied to natural gas containing various quantities of SO_2.

12. _____The two principal methods of producing elemental sulfur are Frasch and recovered.

13. _____After oxidation and hydrolysis in soils, 32 pounds (14.4 kg) of elemental sulfur will convert to sulfuric acid and as such will neutralize the equivalent of 2,000 pounds (900 kg) of calcium carbonate.

14. _____The sulfur content of many plants is equal to or greater than the nitrogen content.

15. _____The sources of sulfur for plants are air, soil, water, fertilizers and pesticides.

chapter 4

Phosphorus—The Key to Life

No Phosphorus   Phosphorus

Barren plot between vegetated areas is a phosphorus deficient subsoil exposed by a roadcut near Cartersville, Georgia. An agronomist and a highway engineer are studying erosion control by Andropogon virginicus (Broom sedge). The eroded bare strip, like the vegetated areas on both sides, was seeded by spreading Broom sedge straw from an adjoining field over the entire area. The only difference was that the vegetated area had been fertilized at seeding and the barren area had not.

Phosphorus is required by all living organisms and every living cell. Plant and animal life cannot exist without it. Many early scientists were concerned with a study of this element. In 1769, Gahn, a Swedish chemist, found phosphorus to be an essential constituent of the bones of men and animals. In 1804, Theodore de Saussure reported that phosphorus is an ingredient of the ash of plants and it is removed from the soil by growing plants.

Elie de Beaumont in 1856 advocated the addition of finely ground mineral phosphates to supply the needs of growing plants. The following year, Von Liebig reported that rock phosphate, as well as bones, could be dissolved with sulfuric acid and thus rendered more soluble and efficient for the use of growing plants. In 1859, Thomas Green Clemson, founder of Clemson University wrote, "There can be no civilization without population, no population without food, and no food without phosphoric acid." Since that time, a plethora of literature has been published on the use and function of phosphorus in plants and animals.

Animal Needs. The average adult human has approximately 2 pounds (0.9 kg) of phosphorus in his body, most of which is in the bones and teeth. The body of a fat steer sold for slaughter contains about 10 pounds (4.5 kg) of phosphorus. Dairy cows have a particularly high requirement for the element since milk has a high phosphorus content. Animals on phosphorus-deficient feeds develop unusual appetites, and phosphorus is frequently dissolved from their own bones and teeth to satisfy the needs of other body functions. Fortunately, the phosphorus needs of animals, including man, can be fulfilled by forage and food grown on soils that are well supplied with soil or fertilizer phosphorus. To insure against a phosphorus deficiency, dairy animal rations are often supplemented with bone meal, calcium phosphate or defluorinated phosphate rock, all of which are also used as fertilizer.

Phosphorus Sediments. Phosphorus sediments are widely distributed in the lithosphere of the earth. The earth's crust contains approximately 0.12 percent phosphorus. It occurs most commonly as calcium phosphate, though a multitude of other forms and compounds exist in the many different soils of the earth's mantle. The larger deposits in the world are either directly or indirectly sedimentary. They have been and are now being laid down in beds on the ocean floor by marine organisms. These beds will later be elevated to the land masses. Some of these sediments may then be redeposited, interbedded and cemented with limestone to form hard rock. Upon weathering, the cementing material, which is usually calcium carbonate, is leached away leaving some deposits in the form of loose pellets or pebbles.

Estimates of world reserves of phosphates that can be mined economically range up to 54.9 billion tons (50 billion m tons). The principal known reserves are in Africa, North America, the Soviet Union and Europe, in that order. They are ample for several hundred years.

Life Functions. Phosphorus has a vital role in the breakdown of the carbohydrates and other foods produced by photosynthesis in the plant. Photosynthesis is the process that enables plants to utilize the sun's energy to produce food and fiber and thus to provide man with food, clothing and shelter. An absence of phosphorus would prohibit photosynthesis and limit the plant's capability to produce carbohydrates such as sugar, starch and cellulose. A lack of phosphorus would also limit the formation of amino acids and proteins that are the building blocks of new plant cells.

Perhaps as vital as the process of photosynthesis is the role phosphorus plays in the reproductive and inheritance process of plants and animals. Phosphorus is an essential ingredient of the nucleoproteins, which are located in the nucleus of cells. Cell division, which is responsible for the growth of higher plants and animals, is not possible without phosphorus. Furthermore, deoxyribonucleic acid (DNA) molecules, which are the vehicles of transfer of inherited characteristics from generation to generation, are compounds that could not exist without phosphorus. Phosphorus has been attributed a role in the following plant growth processes:

1. Carbohydrate breakdown for energy release.
2. Cell division.
3. Transfer of inherited characteristics.
4. Stimulation of early root growth and development.
5. Hastening maturity of plant.
6. Fruiting and seed production.
7. Energy transformation.

Phosphorus Nutrition. A plant's uptake of phosphorus is largely governed by three major factors. These are the type of plant, the stage of plant maturity and the competition between the plant roots and soil chemicals for soil and fertilizer phosphorus. The $H_2PO_4^-$ form of phosphorus is the form most available to plant roots.

Plants vary widely in their ability to obtain sufficient phosphorus from soils testing low in available phosphorus. For example, buckwheat and alfalfa are considered "strong feeders" of phosphorus from soils having a low available supply and from those having been fertilized with rock phosphate. The legumes that are capable of utilizing phosphorus from relatively insoluble compounds also have a high calcium requirement. Oat hay contains about 0.18 percent phosphorus and alfalfa hay 0.3 percent when they are both on soils containing ample supplies of available phosphorus. Crops grown on soils well supplied with available phosphorus and producing high yields can be expected to be of high quality in terms of phosphorus content. High-yielding crops require ample amounts of available phosphorus throughout the growing season.

Young plants assimilate phosphorus very rapidly and accordingly need water-soluble forms that migrate in soil water to the meager root system. Later, the root develops into a more extensive and efficient system that is in contact with more soil volume and is therefore able to contact physically and subsequently to absorb sufficient phosphorus from less soluble forms. However, by the time plants have produced about 25 percent of their total dry weight, they have accumulated as much as 75 percent of their phosphorus needs. Winter wheat, for example, absorbs 70 percent of its phosphorus between tillering and flowering. This emphasizes the importance of having an adequate supply of soluble phosphorus in the soil for plants that have a short growing season and for those with a limited root system. Plants with more extensive root systems absorb phosphorus from a larger volume of soil and are therefore better able to obtain sufficient phosphorus from the soil than are those plants with limited root systems.

Root and Soil Compete. Perhaps the most important factor influencing the assimilation of phosphorus, particularly by young plants, is the phosphorus-fixing capacity of the soil in which the seed is planted. As previously mentioned, one of the functions of phosphorus in plant nutrition is its role in the breakdown of carbohydrates for energy release. When a seed in the presence of moisture begins to germinate, the starch grains that surround the endosperm begin to break down and are converted to sugar to provide quick energy. During this conversion from starch to sugar, phosphorus, in the form of insoluble organic compounds in the seed, is converted into water-soluble forms. These forms may move with the soil moisture away from the young seedling. It is also obvious that the soluble phosphorus compounds may be precipitated into less soluble forms of iron, aluminum or other phosphorus-fixing metallic compounds which are present in soluble forms in the soil immediately surrounding the young plant. Thus, there is competition between the young seedling and the soil for the phosphorus that nature supplies in the seed to get the young seedling off to a quick and vigorous start. This further emphasizes the importance of an ample supply of water-soluble phosphorus in the soil zone of the germinating seeds.

Recovery Efficiency. From the discussion above, it is evident that the recovery efficiency of available soil and fertilizer phosphorus by plants is quite low. In fact, the recovery efficiency of fertilizer phosphorus amounts to only 10 to 30 percent of that added immediately prior to planting the crop. However, since the loss of phosphorus in percolating waters is very small, the 70 to 90 percent that is not absorbed by the plant remains immobile in the soil unless it is lost by erosion. Thus, the content of phosphorus in soils varies widely depending on the parent material of the soil and the history of phosphorus fertilization.

Soil Phosphorus. Total phosphorus in an acre (0.4 ha) of furrow slice averages about 0.07 percent. A few soils may contain as much as 0.15 percent, but many contain less than 0.03 percent. However, the amount of total phosphorus in the soil is not necessarily a good measure of the amount that is available for growth because a large proportion of the phosphorus may exist in forms that are difficult to utilize. In 1916, a chemical method for determining the available phosphorus of soils was proposed by Truog. Since that time, considerable research effort has produced soil chemical methods that are widely used and that are adapted to nearly every soil in the United States and in many other countries as well. Based on data obtained by U.S. soil-testing laboratories, a high percentage of the agricultural soils of the world test from very low to medium in available phosphorus and must therefore be fertilized with regular additions of phosphorus in order to maintain profitable crop yields. Next to nitrogen, phosphorus is the most deficient element for plant growth in the cultivated soils of the world.

Deficiency Symptoms. Plant deficiency of phosphorus is not as clearly defined as are the deficiency symptoms of some other essential plant nutrients. The symptoms usually are expressed in restricted early growth. In cool weather, the leaves of young corn, small grains and grasses turn purple if phosphorus is limiting growth. Genetic characters may also be responsible for purple color in certain species and varieties.

Soil pH. A discussion on phosphorus deficiency and availability would not be complete without including the importance of soil pH. When basic materials such as limestone are added to a soil with a neutral pH, the availability of calcium phosphates will be decreased. On the other hand, additions of limestone to acid soils, which contain iron and aluminum phosphates, will increase the phosphorus availability by the formation of more soluble calcium phosphates. In general, phosphorus is more available at a pH of 6.0 to 6.5 than at higher or lower pH values (Figure 7.3—see page 220). At a soil pH above about 6.5, fertilizer phosphorus is precipitated (fixed) by calcium and magnesium. At a soil pH of 5.0, fertilizer phosphorus is precipitated by aluminum and iron.

An acid soil, which was limed to near neutrality, was found to be optimum for cotton production. At a pH between 6.0 to 7.0, 10 lb/acre (11.2 kg/ha) of fertilizer phosphorus was as efficient for cotton production as 20 lb/acre (22.4 kg/ha) at pH 5.0. A soil containing 15 lb/acre (16.8 kg/ha) of available phosphorus is considered to be well supplied with this essential plant nutrient.

Recently, phosphate nodules in what may be considered recoverable quantities have been discovered on the ocean floor off the coast of southern California. Geologists, who have studied the larger sedimentary deposits of calcium phosphates, consider them to be of marine organic origin.[1] It is not surprising then that rock phosphate was formerly sold on the basis of its bone phosphate of lime content, usually referred to as BPL. In chemical language, BPL is tricalcium phosphate $Ca_3(PO_4)_2$. The predominant mineral form of phosphate in the larger deposits of the world is francolite, which is calcium carbonate–fluorapatite, represented by the formula $Ca_{10} F_2(PO_4)_6 \cdot xCaCO_3$. During World War II, rock phosphate from North Africa was used as ballast in the holes of cargo ships bringing troops home to the United States from the Mediterranean theater. Commercial deposits are also located in South Africa, the Soviet Union, Australia, India and the Spanish Sahara. It is estimated that world consumption in 1980 will be 105.5 million tons (96 million m tons). Uranium may be recovered as a by-product in phosphate processing.[2]

Deposits. Domestically, there are four major phosphate rock-producing areas, Florida and North Carolina, the western states, and Tennessee. Of these four areas Florida and North Carolina produce the greatest quantity of rock, over 80 percent of the total marketable production in the United States each year since 1968. In 1975, total production was 49 million tons (44.6 million m tons), of which Florida and North Carolina contributed 83 percent.[3] The western states have the largest reserves, but most of the mining activity is in the southeast.

Description. Four kinds of phosphate rock are recognized: hard-rock phosphate, soft phosphate, land-pebble phosphate, and river-pebble phosphate.

The hard rock is far the best grade produced in this country, but the land-pebble deposits are the most extensively mined, comprising over 95 percent of Florida phosphate production. The land pebble occurs as light gray to black pebbles mixed with clay and sand. A typical matrix has an average composition of 6.6 percent phosphorus. The North Carolina deposits are the latest to be exploited, starting in the mid 1960s. North Carolina deposits lie beneath strata ranging in thickness from 45 to 250

[1] J. R. Van Wazer, "Occurrence and Mining," in his *Phosphorus and Its Compounds* Vol II (New York: Interscience Publishers, 1961) pp. 955–967.

[2] J. A. Barr, Jr., J. W. Ruch, and R. F. Borlik, "Recovering Uranium as a By-Product in Phosphate Processing," *Rock Prod.* 58(10) (1955), 96–102.

[3] D. A. Paul et al., "*The Changing Fertilizer Industry*," Economic Research Service, United States Department of Agriculture, Agric. Econ. Rep. No. 378, 1977.

feet (13.6 to 75.8 m). The phosphate column consists of sand, clay and interbedded shell limestones. The P_2O_5 content varies from 2 to 21 percent and is apparently proportionate to the allophane content of the area. Reserves in the Beaufort County area are among the largest in the United States, comprising 700 square miles (1,813 km²) of phosphate deposits. These phosphate grains are brown in color, smooth and glossy, and spheroidal to ovate in shape.

The phosphate deposits of Tennessee occur as a very hard and compact apatite-bearing rock, in contrast to the loose structure of the Florida deposits. They may be classified as blue, brown and white deposits. Operations on the blue deposits were suspended after the discovery of brown phosphorus. White phosphorus is of more recent geologic origin than the brown or blue and resembles very much the hard-rock phosphate of Florida. The white deposits, like Florida hard-rock deposits, are high grade and carry from 72 to 90 percent of BPL or tricalcium phosphate. However, they occur irregularly and are not as accessible as the brown phosphorus. The brown deposits are the most extensively mined of the Tennessee deposits. They are ground to a texture of brown sugar and are probably more extensively marketed for direct application to the soil than any other rock phosphate in the United States. The phosphate deposits of the western United States are located largely in Idaho, Wyoming, Utah and Montana. Idaho has maintained its position as the leading phosphate-rock producer of the western states. These deposits appear to be precipitated, resulting from a reaction between phosphoric acid and calcium carbonate in the waters of an ancient sea. Some of the rock is soft, and some is very hard. They vary in color from light gray to jet black. After World War II and the Korean Conflict, phosphate rock from these deposits was purchased by the U.S. Army and shipped to occupied areas in the Pacific Ocean.

Method of Manufacture.   Almost all of the mining in the United States is by strip-mining methods. In Florida, overburdens averaging about 15 feet (4.5 m) thick must be removed to expose the matrix, which is about 20 feet (6.0 m) deep.

When Florida land-pebble or hard-rock phosphates are prepared for market, the soft phosphates which are present in the matrix are washed into waste ponds where they settle out with clay and other impurities. The pebbles remaining after washing are ground or crushed before they are marketed. Some superphosphate and phosphoric acid manufacturers have Raymond mills and do their own grinding. Others buy the rock phosphates already ground. For years, the waste ponds containing the soft phosphates, clay and other impurities were not recovered. Today, some of the waste material is marketed under the trade names Colloidal Phosphates®, Mineral Colloids®, Calphos® and Phos-cal-oids®. These products are known in the industry as waste-pond phosphates. Because of the fineness of the waste-pond phosphate, claims have been made regarding its availability to crops.

The availability to crops of the unacidulated waste-pond phosphate will be discussed later.

North Carolina mining is a pumping and dredging operation because the deposits are far below the water table.

Tennessee brown deposits occur in beds of varying thicknesses overlying phosphatic limestone. They are mined much the same as in Florida, i.e., from open surface pits using a dragline, power shovel or scrapers.

Western phosphate rock is mined by underground and open-pit methods. The Florida and Tennessee phosphate require washing before it is marketed; some of the phosphate deposits of the western United States do not need this washing. Lower grade ore is the bulk of the production there and must be processed like Florida and North Carolina deposits; this ore must be calcined in order to make it suitable for wet-process H_3PO_4 manufacture.

Chemical Characteristics. Florida hard-rock phosphate contains 78 to 80 percent BPL. It has a low content of iron, aluminum and fluoride. Florida soft phosphate usually contains quantities of calcium carbonate and clay, and for this reason the BPL or tricalcium phosphate content ranges from 40 to 60 percent. Florida land-pebble phosphate has been formed by wave action on phosphate-bearing formations. Land-pebble phosphate contains 68 to 75 percent BPL and frequently carries 5 to 6 percent iron and aluminum oxides. These are the most extensively mined deposits in the United States and comprise over 95 percent of the Florida phosphate rock production. The Florida river-pebble phosphate deposits represent stream deposits of mixed phosphate pebble, sand and clay. They were among the first deposits to be mined. Since they yield a lower grade phosphate ore, mining of this type of material ceased in 1908. Analysis of representative samples of Florida phosphate rock by Matson[4] is given in Table 4.1.

Table 4.1
ANALYSIS OF REPRESENTATIVE SAMPLES OF FLORIDA PHOSPHATE ROCK

	Hard-rock Phosphate (percent)	Land-pebble Phosphate (percent)	River-pebble Phosphate (percent)
Phosphoric acid (P_2O_5)	35.52	34.43	28.47
Tricalcium phosphate ($Ca_3(PO_4)_2)_3$	77.50	74.85	61.77
Phosphorus (P)	15.50	14.02	10.82
Insoluble (SiO_2)	2.03	4.78	
Ferric oxide (Fe_2O_3)	0.10	4.15	0.83
Aluminum oxide (Al_2O_3)	2.30		1.37
Calcium oxide (CaO)	51.45		45.69
Carbon dioxide (CO_2)	2.45		6.37

[4]G. C. Matson, "The Phosphate Deposits of Florida," U.S. Geol. Survey Bul. 604, 1915.

Average chemical composition of beneficiated North Carolina phosphate rock, uncalcined and calcined, is given in Table 4.2. Particle size of unground and uncalcined North Carolina phosphate is given in Table 4.3.

Tennessee deposits before beneficiation are estimated to contain from 9 to 55 percent tricalcium phosphate. Beginning about 1949, all quotations of Tennessee phosphate rock are given on a P_2O_5 basis instead of on the BPL or $Ca_3(PO_4)_2$ content formerly used. The finely ground brown rock phosphate is sold for direct application to the soil and is guaranteed to contain 30 to 33 percent P_2O_5 (13 to 14 percent phosphorus), 3 to 4 percent fluorine, and in addition, 7 to 10 percent neutralizing value in terms of $CaCO_3$ equivalent.

The phosphate content of the workable beds of western deposits varies from 65 to 80 percent tricalcium phosphate. Analyses of representative samples of phosphate rock from Tennessee, Idaho, Florida, South Carolina, Curacao, Morocco and Tunis are given in Table 4.4.

Table 4.2
AVERAGE CHEMICAL COMPOSITION OF
BENEFICIATED NORTH CAROLINA PHOSPHATE
ROCK,[1] UNCALCINED AND CALCINED

Component	Uncalcined Percent	Calcined Percent
P_2O_5	30.64	33.55
CaO	48.13	53.11
MgO	0.78	0.79
Al_2O_3	0.58	0.78
Fe_2O_3	0.61	0.89
Na_2O	1.02	1.12
K_2O	0.14	0.12
SiO_2	3.40	2.26
SO_3	1.77	1.80
F	3.60	3.91
CO_2	5.72	2.70
H_2O	1.27	0.02
Organic C	1.30	0.29
Total	98.96	101.34
$\dfrac{CaO}{P_2O_5}$	1.57	1.58

[1]Data supplied by Texas Gulf Sulphur. Samples were collected near Aurora, North Carolina.

Crop Response and Product Use. Over 90 percent of the rock-phosphate production is used in acidulating plants to produce superphosphate and

Table 4.3

PARTICLE SIZE OF UNGROUND AND UNCALCINED NORTH CAROLINA PHOSPHATE[1]

ASTM Screen No.	Size Opening (mm)	Percentage at Different Mesh Sizes		
		Test 1	Test 2	Test 3
+28 mesh	0.61			0.3
+30 mesh	0.59		3.12	
+35 mesh				6.9
+48 mesh				31.4
+60 mesh	0.25	54.0		
+65 mesh				78.0
30 to 100 mesh			92.73	
60 to 100 mesh		43.0		
+100 mesh	0.15			98.3
100 to 200 mesh		2.0	4.10	
+150 mesh	0.10			99.7
less than 200 mesh		1.0	0.05	
+ 200 mesh	0.07			99.9
		100.0	100.0	

[1]Data supplied by Texas Gulf Sulphur. Samples were collected near Aurora, North Carolina.

phosphoric acid. Less than 8 percent is used for direct application to the soil. About 2 percent is used for stock and poultry feed.

There are two products, one from the Tennessee deposits and one from the Florida mines, that are sold for direct application to the soil without acidulation. These are waste-pond phosphates from Florida, sold under various trade names, and Tennessee brown phosphate. The waste-pond phosphate contains 18 to 23 percent P_2O_5 (7 to 9 percent phosphorus), of which 1 to 2 percent may be classed as available, according to the AOAC method.[5] The AOAC method for "available" phosphorus determines the solubility in neutral $1N$ ammonium citrate. This does not mean that the phosphorus material is water soluble. The Tennessee brown phosphate contains 30 to 40 percent P_2O_5 (13 to 17 percent phosphorus), of which 2 to 3 percent may be classed as available. These materials have been carefully evaluated and compared with superphosphate as a source of phosphorus for crops.

The following summary statements were drafted after most of the

[5]Association of Official Analytical Chemists, *Methods of Analysis* 12th ed., (Washington, D. C.: AOAC, 1975).

Table 4.4
CHEMICAL AND PHYSICAL CHARACTERIZATION DATA FOR SOURCES OF ROCK PHOSPHATE USED IN GREENHOUSE STUDIES.[1]

Source of rock phosphate	Total P_2O_5	Fluorine Content	Specific surface	Fineness		Fraction of total phosphorus dissolved by various solutions				
				100 mesh (0.14mm)	200 mesh (0.07mm)	Sequestrene solution[2]			1% Lactic acid	Neutral ammonium citrate
						pH 4.3	pH 5.8	pH 6.7		
	pct.	pct.	m²/g	pct.	pct.	pct.	pct.	pct.	pct.	pct.
Tennessee	34.4	3.80	7.8	100	61	17.5	18.9	8.4	6.7	4.7
Idaho	32.4	3.27	5.7	100	60	17.1	22.2	13.9	7.2	6.8
Florida	33.5	3.82	11.0	100	68	21.5	32.8	23.6	7.4	9.0
South Carolina	26.9	3.57	6.6	100	65	35.1	44.2	42.5	8.6	17.2
Curacao Island	38.6	0.74	6.4	100	69	34.4	49.0	46.7	15.2	14.0
Morocco	34.1	4.27	22.4	100	84	36.3	60.9	62.9	13.4	15.7
Tunis	29.0	3.60	37.2	100	64	40.1	53.6	58.0	11.2	17.0

[1]L. E. Ensminger, R. W. Pearson, and W. H. Arminger, *Effectiveness of Rock Phosphate As a Source of Phosphorus for Plants*, ARS 41–125, USDA, Jan. 1967.

[2]Extractions were made by shaking 0.5 g of material for two hours with 100 ml of sequestrene (ethylenedinitrilo tetraacetic acid) to one liter of water. The different pH values were obtained by adding 6.85 g of NaOH for pH 4.3, 8.56 g of NaOH for pH 5.8, and 10.28 g of NaOH for pH 6.7.

115

experimental data available were evaluated.[6] Relative yield values for these materials are based on the increased yield from superphosphate as 100:

1. Rock phosphate added at a double rate of phosphorus (2 times as much as superphosphates) produced relative yield increases of 54, 61 and 70 for cotton, pastures and cowpeas, respectively.

2. Colloidal waste-pond phosphate added at a double rate of phosphorus produced relative yield increases of 35, 50, 58 and 58 for oats, corn, cotton and pasture, respectively.

3. An average of three ten-year tests with cotton showed the residual efficiency of rock phosphate to be one-half that of superphosphate. For the residual studies, both superphosphate and rock phosphate were added at the same rate of phosphorus.

4. A seven-year test with cotton, corn and cowpeas showed that rock phosphate added at a quadruple rate of phosphorus produced as much cowpea hay as superphosphate but produced 3 bushels (81.8 kg) less corn and 51 pounds (23.2 kg) less seed cotton per acre (0.4 ha.)

Considerable research has been conducted using uptake of radioactive phosphorus, ^{32}P, to measure the availability of various kinds of phosphate fertilizer. Using superphosphate tagged with ^{32}P as a standard, Ensminger et al.[7] demonstrated that 640 lb/acre (717 kg/ha) of P_2O_5 from rock phosphate added to either unlimed Cecil clay loam or Eutaw clay had an availability equivalent of about 140 lb/acre (157 kg/ha) of P_2O_5 from superphosphate. It appears that 4 to 5 lb/acre (4.5 to 5.5 kg/ha) of phosphorus from rock phosphate are required to give the same yield increase as 1 lb/acre (1.12 kg/ha) of phosphorus from superphosphate. In the same report, the data indicated that the residual effect of rock phosphate was less than the residual effect of equivalent amounts of phosphorus from superphosphate.

The foreign sources of rock phosphate from Curacao, Morocco and Tunis were superior to all domestic sources tested except those from South Carolina (Tables 4.5 and 4.6). Rock phosphate alone, regardless of the source, did not produce maximum yields when applied at high rates. Thus, it is abundantly clear that rock phosphate must be further processed to produce a product that gives maximum yields.

The fundamental processes that are employed to make the phospho-

[6]U. S. Jones, "Phosphate Fertilizers," Mississippi Agricultural Experiment Station Bulletin 503, 1953.

[7]L. E. Ensminger, R. W. Pearson, and W. H. Arminger, *Effectiveness of Rock Phosphate As a Source of Phosphorus for Plants,* ARS 41–125, USDA, Jan. 1967.

Table 4.5
EFFECT OF SOURCES OF ROCK PHOSPHATE AND RATES OF SUPERPHOSPHATE APPLICATION WITH AND WITHOUT LIME ON YIELD AND PHOSPHORUS UPTAKE OF SUDANGRASS AND LADINO CLOVER GROWN ON CECIL CLAY LOAM IN THE GREENHOUSE.[1]

Source	Phosphorus applied		Dry matter per pot							
			1 Harvest of Sudan Grass				3 Harvests of Ladino Clover			
			Yield per pot		P_2O_5 uptake per pot		Yield per pot		P_2O_5 uptake per pot	
		P_2O_5	Unlimed	Limed	Unlimed	Limed	Unlimed	Limed	Unlimed	Limed
	lb/acre	kg/ha	g	g	mg	mg	g	g	mg	mg
None	0	0	1.00	2.67	2.10	9.61	6.56	9.85	21.36	31.92
Conc. superphos.	60	67	1.55	3.21	6.36	14.12	7.31	12.11	27.37	40.59
'' ''	120	134	1.58	3.08	6.95	14.78	8.11	12.77	30.75	48.69
'' ''	240	269	2.23	3.85	11.82	18.87	7.86	14.89	34.70	59.71
Curacao Island	120	134	1.61	3.08	6.44	12.01	8.90	13.35	34.65	46.00
Morocco	120	134	1.58	3.11	6.79	12.75	7.04	13.54	30.90	46.92
Tennessee	120	134	1.42	2.99	5.25	10.17	8.18	11.23	30.50	36.88
Florida	120	134	1.19	3.00	4.76	9.90	7.76	11.22	31.62	38.27
Idaho	120	134	1.20	2.64	4.92	9.77	7.47	9.15	30.96	31.76
Tunis	120	134	1.90	2.85	9.12	11.40	9.42	12.43	36.74	43.35
South Carolina	120	134	1.99	3.06	8.96	11.93	8.21	13.33	31.61	43.44
L.S.D. 0.05			0.38		3.15		0.9		6.10	
0.01			0.51		4.17		1.2		8.07	

[1]L. E. Ensminger, R. W. Pearson, and W. H. Arminger, *Effectiveness of Rock Phosphate As a Source of Phosphorus for Plants*, ARS 41–125, USDA, Jan. 1967.

Table 4.6
EFFECT OF SOURCES OF ROCK PHOSPHATE AND RATES OF SUPERPHOSPHATE APPLICATION WITH AND WITHOUT LIME ON YIELDS OF SUDANGRASS AND LADINO CLOVER GROWN ON EUTAW CLAY IN THE GREENHOUSE[1]

Phosphorus applied			Dry matter per pot							
			1 Harvest of Sudangrass				3 Harvests of Ladino Clover			
			Yield per pot		P_2O_5 uptake per pot		Yield per pot		P_2O_5 uptake per pot	
Source	lb/acre	kg/ha	Unlimed	Limed	Unlimed	Limed	Unlimed	Limed	Unlimed	Limed
	P_2O_5	P_2O_5	g	g	mg	mg	g	g	mg	mg
None	0	0	1.89	2.03	5.67	5.68	7.30	7.29	23.34	25.29
Conc. superphos.	60	67		3.80		14.82	9.32	10.72	29.63	35.49
"	120	134		4.79		20.60	12.34	12.81	43.24	51.27
"	240	269		4.88		22.97	14.26	13.54	56.22	57.75
Curacao Island	120	134	3.63	3.70	15.97	15.54	10.45	10.29	42.11	35.39
Morocco	120	134	3.42	3.46	15.39	14.19	9.43	10.86	40.42	39.69
Tennessee	120	134	2.62	2.86	11.53	10.01	9.55	10.32	38.60	36.10
Florida	120	134	3.64	2.98	16.74	12.52	9.99	10.67	37.51	37.48
Idaho	120	134	2.65	3.21	12.19	12.20	9.72	11.37	36.37	38.99
Tunis	120	134	3.16	3.78	14.85	17.01	9.92	12.12	40.92	46.44
South Carolina	120	134	3.15	4.05	14.49	15.80	9.84	10.79	39.84	42.57
L.S.D. 0.05			0.50		1.67		0.9		5.59	
0.01			0.67		2.21		1.2		7.48	

[1]L. E. Ensminger, R. W. Pearson, and W. H. Arminger, Effectiveness of Rock Phosphate As a Source of Phosphorus for Plants, ARS 41–125, USDA, Jan. 1967.

rus in phosphate rock more available are heat treatment and acid treatment. Both processes destroy the insoluble, crystalline apatite structure resulting in a product that contains phosphorus that is more soluble and more readily available to plants. Heat from gas or oil will do this, but electrical energy is required to produce elemental phosphorus. Acid treatment is effected with sulfuric, phosphoric and nitric acids. The various processes involved and the products from rock phosphate and heat or acid will be discussed in the following sections.

ORDINARY SUPERPHOSPHATE

Superphosphate is a term used in reference to phosphates, the phosphorus of which is in a form readily available to plants. Ordinary or normal superphosphate refers to those that contain up to and including 24 percent P_2O_5 (10.5 percent phosphorus). Since 1930, almost the entire production has been with Florida land-pebble, Tennessee brown or Idaho and Montana phosphate rock.

Description. It is marketed as a gray, brown or almost white material in either powdered or granular form. It has an acid odor.

Method of Manufacture. Superphosphate is a product resulting from the mixing of approximately equal quantities of 60 to 70 percent sulfuric acid (93 to 98 percent acid, diluted with water) with phosphate rock ground to pass a 100-mesh (0.14-mm) sieve and containing 72 percent BPL, $Ca_3(PO_4)_2$. There are two manufacturing processes commonly used. They are the Sturtevant process, rarely used in the United States, and the Broadfield den continuous process. The sulfuric acid and ground phosphate rock are mixed together in a simple cone mixer that uses a swirling action for mixing. A popular manufacturing method in the fertilizer industry is the use of a TVA cone mixer and a Broadfield den that allows the continuous mixing of rock and acid, producing superphosphate at the rate of 40 to 50 tons (36 to 46 m tons) per hour (Figure 4.1). The resulting material, which is dried at a temperature exceeding 212°F (100°C) developed in the reaction, is later cured and ground, or granulated by using a tilted rotating drum. Another process in use is the Sturtevant process or batch mixing process (Figure 4.2). The reaction for both may be expressed as follows:

$$Ca_{10}(PO_4)_6F_2 + 7H_2SO_4 + 3H_2O \rightarrow 3CaH_4(PO_4)_2 \cdot H_2O + 7CaSO_4 + 2HF$$

Handling Characteristics. Powdered, 20 percent superphosphate will usually set up in the bags after a few months storage, even in a dry place. Further, the free acid in the material will usually rot fiber and result in

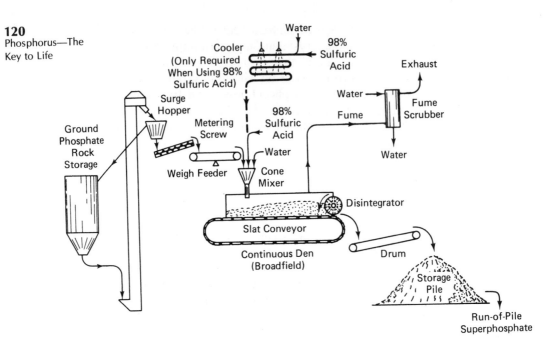

Figure 4.1 Continuous process for the manufacture of normal superphosphate.

broken bags if allowed to remain in storage very long. The granulated material will usually store well and will not bridge or stop up the fertilizer distributor when it is added to the soil. Fortunately, the cost of making granulated superphosphate is no more than the cost of making the finished,

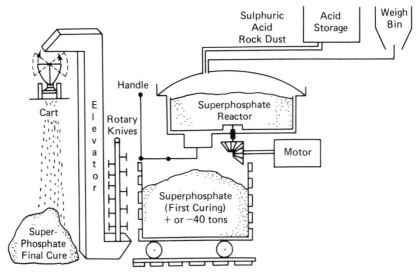

Figure 4.2 Sturtevant process for manufacturing superphosphate.

pile-cured, powdered material because granulation eliminates some of the pile curing.

Chemical Characteristics. Ordinary superphosphate is two-fifths monocalcium phosphate and three-fifths gypsum by weight. It has a pH of about 3. It analyzes about 20 percent P_2O_5 (9 percent phosphorus), 19 to 22 percent calcium and about 10 to 12 percent sulfur. The mean available P_2O_5 content of all grades sold to consumers in the United States in 1960 was 20.23 percent and contained 1.67 percent moisture. Most of its phosphorus, about 85 percent, is soluble in water. A representative sample of ordinary or normal superphosphate also contains traces of magnesium, iron, aluminum, copper, manganese, zinc and chlorine. Because it is of low analysis, its importance for supplying fertilizer phosphorus in the United States is declining.

Crop Response. Because of phosphate fixation by iron, aluminum, calcium and the clay of soils, less than one-fourth of the phosphorus added to the soil in the form of fertilizer is actually removed in the harvested portion of crops during the season following the addition to the soil. Ordinary superphosphate is usually regarded as the standard for phosphorus fertilizers. Although ordinary superphosphate is low in analysis, it is an excellent source of phosphate because it contains calcium and sulfur and because most of its phosphorus is water soluble.

There are three distinct forms of phosphorus that are added to the soil when superphosphate is used: $H_2PO_4^-$, HPO_4^{--} and PO_4^{-3}. Ordinary, unmanipulated superphosphate contains at least 85 percent of its phosphorus in the water-soluble $H_2PO_4^-$ form as monocalcium phosphate. The remainder of the phosphorus is in the HPO_4^{--}, dicalcium, or PO_4^{-3}, tricalcium phosphate, forms. These are not soluble in water. Availability of phosphorus to plants is best correlated with the $H_2PO_4^-$ water-soluble form.

As mentioned above, ordinary superphosphate is a product obtained when rock phosphate is treated with sulfuric acid. This converts the bulk of PO_4^{-3}, tricalcium phosphate, to the $H_2PO_4^-$, monocalcium phosphate, form that is soluble in water and available to growing plants. When superphosphate is ammoniated, about one-half of the $H_2PO_4^-$ form reverts to the PO_4^{-3} form. Ammoniated superphosphate will be discussed in a subsequent section.

PHOSPHORIC ACID

A very small amount of phosphoric acid is added as such to the soil via irrigation water, but for the most part it is considered as a raw material for further processing. Three kinds of phosphoric acid are in various stages of use and product development. These are classified according to the way they are made. The first two are referred to as wet-process and electric-furnace acids. The third is made from acid produced by either one or both of the first two methods but is upgraded to contain up to 76 percent P_2O_5 (33 percent phosphorus) by dehydrating very pure H_3PO_4; it is called superphosphoric acid. Some is used to produce detergents, food stuffs and many other products, but most of it is used in the production of fertilizer.

WET-PROCESS PHOSPHORIC ACID

The same ingredients that are employed in producing superphosphate are used to manufacture wet-process acid. The main difference is that an excess of sulfuric acid is used. A liquid is produced when an excess of sulfuric acid is mixed with finely ground rock phosphate. The principle reaction for producing H_3PO_4 by the wet process is:

$$Ca_{10}F_2(PO_4)_6 + 10H_2SO_4 + 20H_2O \rightarrow 10CaSO_4 \cdot 2H_2O + 2HF + 6H_3PO_4$$

The $CaSO_4$ is removed by decantation and filtration. It requires about 10 to 15 percent more sulfuric acid to produce phosphoric acid by the wet method than to produce ordinary 20 percent superphosphate. It is usually referred to as green acid because of impurities, primarily vanadium. It contains 28 to 30 percent P_2O_5 (12 to 13 percent phosphorus).

Method of Manufacture. The method of manufacture may be best understood by following the arrows in Figure 4.3, starting at the upper left corner of the sketch. Pebble rock is unloaded from a barge truck or a freight car onto a conveyor belt that takes it to grinding mills where the phosphate rock is ground to the fineness of face powder. The finely ground rock phosphate and sulfuric acid are introduced into the reactors. Here, phosphoric acid is formed in a liquid state with gypsum suspended in the liquid. The mixture of phosphoric acid and gypsum is drawn out of the reactor onto a filter. The acid is sucked through a plastic vacuum filter; the gypsum precipitate is washed free of acid with water; and the filtrate is returned to the reactors. The gypsum is drawn off to the disposal pond, and the acid goes to an evaporator where it is concentrated and then to storage tanks.

It requires 3.25 tons (2.96 m tons) of phosphate rock and 2.75 tons (2.50 m tons) of sulfuric acid, a total of 6 tons (5.46 m tons) of material, to produce 1 ton (0.9 m ton) of phosphoric acid. Five tons (4.55 m tons) of gypsum are produced as a by-product of this process.

In ordinary superphosphate manufacture, gypsum is not filtered and therefore must be shipped and handled as a part of low-analysis fertilizers. Thus, high-analysis fertilizers should be less expensive per unit of nutrient

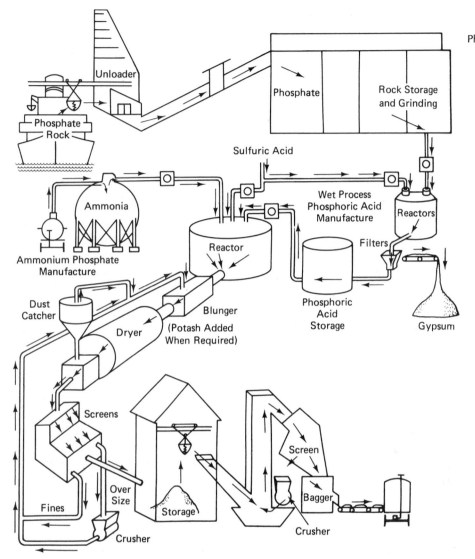

Figure 4.3 Ammonium phosphate manufacturing process.

because shipping costs and handling charges for 1 ton (0.9 m ton) of ordinary superphosphate containing 10 percent phosphorus are the same as 1 ton (0.9 m ton) of high-analysis superphosphate containing 20 percent phosphorus.

Appreciable quantities of fluorine are liberated by the reaction of phosphate rock and sulfuric acid. The fumes are collected and passed through a scrubber to eliminate the corrosive substance from the air. Fluorine may be used as a by-product to produce sodium silicofluoride for use in the fluoridation of city water supplies for tooth enamel protection. It has several other commercial uses.

Chemical Characteristics. The phosphoric acid produced in this way usually contains only 28 to 30 percent H_3PO_4 and is considered unsuitable for use in the manufacture of food additives or detergents unless it is further purified. However, it is eminently satisfactory for the manufacture of high-analysis solid mixed fertilizers. Furthermore, $CaSO_4 \cdot 2H_2O$ which can be used as a fertilizer containing calcium and sulfur, as a soil amendment or which can be processed into sheet rock or wall board, is produced as a by-product.

In 1975, over 85 percent of the fertilizer phosphorus used in the United States and its possessions was derived from the wet-process phosphoric acid method of manufacture because there were 45 to 50 plants producing this acid in the United States at that time. Because market analysis of fertilizer trends point to its increased use for the production of concentrated phosphate fertilizers, this process will undoubtedly account for a greater proportion of the fertilizer phosphorus consumption in the future. It is the least expensive way to manufacture phosphoric acid.

Product Use. Wet-process liquid phosphoric acid has not been generally used for direct application to the soil because its impurities (which include compounds of iron, aluminum, calcium, fluorine, sulfur and solid matter) make it difficult to transport and handle. However, advanced technology permits fertilizer manufacturers to produce a purer form of H_3PO_4 by the wet process. Crude phosphoric acid (28 to 30 percent P_2O_5) can be concentrated to 40 percent P_2O_5 for the production of ammonium phosphate, to 54 percent for the usual merchant grade that is transported for trade, or to 72 percent to produce superphosphoric acid (Figure 4.4). Shipment by rail tank car and ocean liner is well established.

FURNACE-PROCESS PHOSPHORIC ACID

Method of Manufacture. Elemental phosphorus is produced by mixing sand, coke and phosphate rock in an electric furnace at approximately 2900°F (1600°C) (Figure 4.5). Carbon in the form of coke is required for the reduction of fluorapatite, $Ca_{10}F_2(PO_4)_6$, to phosphorus. The silica in the form of sand reacts with the lime (CaO) formed during the reduction to produce calcium silicate slag. At the temperature existing in the furnace, phosphorus as well as some of the fluorine and carbon monoxide are liberated as gases and are drawn off through a duct in the top of a furnace. The calcium silicate slag is a molten, viscous liquid and is periodically tapped from the furnace. Iron oxide, originally a contaminant in the phosphate ore, is reduced during the process to a metallic iron, and it

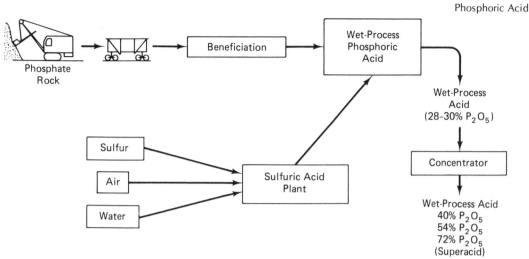

Figure 4.4   Wet-process phosphoric acid production.

combines with some of the phosphorus to form ferro-phosphorus, which at the furnace temperature is a heavy, molten metal alloy that settles to the bottom of the furnace below the molten slag. If the reactions occurring among the contaminants are neglected, the reduction of phosphate rock may be expressed by the following reaction:

$$Ca_3(PO_4)_2 + 3SiO_2 + 5C \xrightarrow[1593°C]{2900°F} 3(CaO\cdot SiO_2) + P_2 + 5CO$$

The furnace gases containing the phosphorus vapor are passed from the furnace into a dust catcher—an electrostatic Cottrell precipitator unit (Figure 4.5) where the entrained dust from coke, silica and phosphate ash is removed. The gases are kept at a temperature of approximately 550°F (290°C) to prevent the condensation of phosphorus. From the dust catcher–electrostatic precipitator unit, the gases containing the phosphorus, fluorine compounds and carbon monoxide are passed into a condenser where a spray of water flows counter to the flow of gases and reduces the temperature enough to condense the phosphorus to a liquid. The carbon monoxide passes through the unit as a gas and is piped to other parts of the plant where it is used as a fuel. The liquid phosphorus is drained into storage tanks where it must be kept under water to prevent its burning. Present phosphorus production capability at a Tennessee Valley Authority plant near Florence, Alabama is about 100 tons (91 m tons) per day although, at present, it is not in operation.

After the phosphorus is condensed, it is kept warm enough to remain

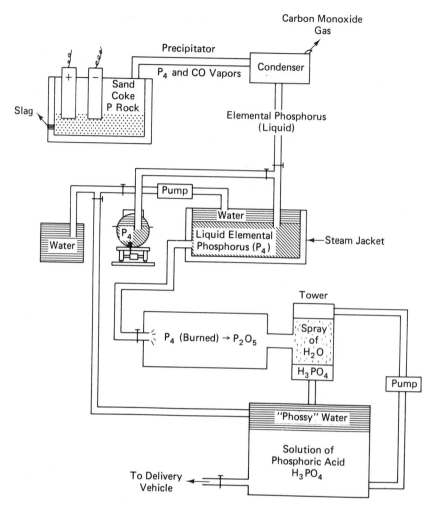

Figure 4.5 Furnace process of manufacturing phosphoric acid.

in liquid form. It solidifies at 112°F (44°C). To produce ordinary phosphoric acid, liquid phosphorus is pumped through a pipe to a burner where it is mixed with air and burned to phosphorus pentoxide (P_2O_5). The P_2O_5 is then sent to a hydrator where water is sprayed into the P_2O_5 fume. The P_2O_5 reacts chemically with water to form phosphoric acid according to the reaction:

$$P_2O_5 + 3H_2O \rightarrow 2H_3PO_4$$

The concentration of this acid is about 80 to 85 percent H_3PO_4. It contains 52 to 60 percent P_2O_5 (24 to 26 percent phosphorus).

Crop Response. Phosphoric acid manufactured by the furnace process has been sucessfully used in irrigation water and added directly to the soil as a source of phosphorus for plant growth. It is much easier to transport and handle after delivery than the unpurified, wet-process acid. In 1975, a unit [20 pounds (9.09 kg)] of P_2O_5 as furnace-process acid was about double the cost of a unit of P_2O_5 from wet-process acid.

SUPERPHOSPHORIC ACID

During 1957, TVA began the production of phosphoric acid up to 76 percent P_2O_5 (33 percent phosphorus). It was an anhydrous phosphoric acid produced from orthophosphoric furnace acid. Presently, a similar product containing up to 72 percent P_2O_5 is made from wet-process acid.

$$2H_3PO_4 \overset{\triangle}{\Rightarrow} H_4P_2O_7 + H_2O \uparrow$$

A considerable portion of this product is pyrophosphoric acid. Further concentration can be effected by producing other polyphosphoric acids in the following manner:

$$H_3PO_4 + H_4P_2O_7 \overset{\triangle}{\Rightarrow} H_5P_3O_8 + H_2O \uparrow$$

Polyphosphoric acid may be defined as any of a series of acids containing more than one atom of phosphorus, such as $H_4P_2O_7$, pyrophosphoric acid.

Description. Superphosphoric acid is a mixture of orthophosphoric acid and polyphosphoric acid. The form of phosphoric acid available in 1963 to fertilizer producers was either 75 percent orthophosphoric acid (54 percent P_2O_5) manufactured by the electric furnace process (white acid) or 72 percent orthophosphoric acid manufactured by the wet process (green acid). Superphosphoric acid made from white acid contains about 74 to 76 percent P_2O_5 (equivalent to about 105 percent orthophosphoric acid); it contains about equal proportions of orthophosphoric and polyphosphoric acids, the latter being mostly pyrophosphoric acid. Superphosphoric acid made from green acid contains 70 to 72 percent P_2O_5 (31 to 32 percent phosphorus), about one-half of which is polyphosphoric acid.

Method of Manufacture. Nearly all of the superphosphoric acid production in 1976 was from wet-process green acid produced by a dehydrate-type process.

When electrical energy is less expensive than acidulation, elemental phosphorus may be produced by the reduction of phosphate rock in an electric furnace. Phosphorus is evolved in the form of a gas and is condensed by spraying with water. The condensed solid is burned in the presence of air to form P_2O_5 gas that reacts with water in a spray tower to form phosphoric acid (Figure 4.5). Either orthophosphoric acid or superphosphoric acid is produced by controlling conditions within the tower.

Physical and Chemical Characteristics. This acid contains about 40 percent more P_2O_5 by weight and 70 percent more by volume than ordinary phosphoric acid. It may be used in conjunction with wet-process phosphoric acid where it serves to sequester the impurities in the acid when it is ammoniated. In addition, the more concentrated acid may be used to produce liquid fertilizers of higher analysis than those obtained from ordinary phosphoric acid. If phosphoric acid is used, the highest grade of 1:3:0 ratio that will not salt out at 32°F (0°C) is 8-11-0 (8-24-0).[8] Using white superphosphoric acid, some firms produce an 11-16-0 (11-37-0) grade; using green acid, several companies are making a 10-15-0 (10-34-0) base solution. This is important to the fertilizer industry because of savings in freight and storage costs. Also, certain micronutrient elements have higher solubility in liquid fertilizers produced from this acid.

Physical and chemical properties of superphosphoric acid are compared with those of furnace-grade orthophosphoric acid in Table 4.7. Information was supplied by Tennessee Valley Authority, Muscle Shoals, Alabama, the National Fertilizer Development Center.

Superphosphoric acid usually contains minute suspended solids such as carbon and silica carried over in the production of phosphorus. The amount is quite small and should give no trouble in fertilizer production.

Table 4.7
PROPERTIES OF ORTHOPHOSPHORIC ACID AND SUPERPHOSPHORIC ACID

Property	Superphosphoric Acid	Orthophosphoric Acid
Specific gravity at 80°F (27°C)	1.92	1.574
Density at 76°F (24°C)	16 lb/gal (1.9 kg/l)	13 lb/gal (1.6 kg/l)
H_3PO_4 content, %	105	75
P_2O_5 content, %	76	54.3
P_2O_5 content at 76°F (24°C)	12 lb/gal (1.4 kg/l)	7.1 lb/gal (0.8 kg/l)
Orthophosphate content, % of total P_2O_5	49	100
Viscosity at 80°F (27°C), centipoises	780	17
Freezing temperature (approx.)	60°F[1] (16°C)	0°F (−18°C)
Boiling point	600°F (315°C)	275°F (135°C)

[1]Possesses supercooling properties. Usually will remain free of crystals for 30 days or more at 32°F (0°C).

[8]Figures in parentheses refer to fertilizer grades in terms of N-P_2O_5-K_2O.

The viscosity of the acid increases as the temperature of the acid is decreased. The efficiency of centrifugal pumps is very low with superphosphoric acid at temperatures below about 100°F (38°C). Rotary, gear or other types of positive displacement pumps should be used at lower temperatures. Meters that are essentially independent of viscosity, such as constant displacement meters, rotodips, metering pumps or magnetic flow metering pumps, should be used.

Superphosphoric acid is generally less corrosive than orthophosphoric acid. At temperatures below 180°F (82°C), several stainless steels (e.g., A.I.S.I. types 201, 304, 316, 430) and other metals, including Hastelloy C, Durimet 20 and Duriron, show satisfactory resistance to corrosion. A number of rubber and plastic lining materials and applied coatings also show satisfactory resistance to corrosion at room temperatures.

Uses in Manufacturing. The principal use of this acid has been for the manufacture of higher analysis, liquid mixed fertilizers that have lower salting-out temperatures than those produced with other phosphoric acids.[9] For example, an 11-15-0 (11-33-0) liquid made with superphosphoric acid has a salting-out temperature of about 0°F (−18°C).

Superphosphoric acid is added in amounts up to 20 percent by volume to liquid fertilizers made from wet-process orthophosphoric acid. Used together with wet-process orthophosphoric acid, the superphosphoric acid acts as a sequestrant for the production of liquid fertilizers in which the impurities of green acid are held in suspension. For more information, see the section on ammonium polyphosphate liquid fertilizer (page 151).

Likewise, a higher analysis solid superphosphate (anhydrous monocalcium phosphate) that contains 54 to 55 percent P_2O_5 (23 to 24 percent phosphorus) has been manufactured with superphosphoric acid. It has good ammoniation characteristics. Bulk handling and storage properties of the higher analysis superphosphate appear to be satisfactory, but the bagged material has shown a tendency to deteriorate the bags.

Crop response. Greenhouse and field tests with superphosphoric acid have shown it to be equal to concentrated superphosphate as a source of phosphorus for plant growth.[10]

Shipping. The same rail freight rate applies to superphosphoric acid as to orthophosphoric acid in the Western Trunk Line, Southwestern, Southeastern, and Illinois Freight Association territories. Shipments are also

[9]J. A. Wilbanks and M. C. Nason, "Liquid Fertilizers from Wet-Process Phosphoric Acid and Superphosphoric Acid," *J. Agric. Food Chem.* 9 (1961), 176–178.

[10]J. L. Anthony, "Phosphate Applied in Fall Effective in Black Belt," *Miss. Farm Res.,* September 1962.

made by commercial truck in stainless-steel trailers. An advantage of 11-16-0 (11-37-0) and 10-15-0 (10-34-0) made from superphosphoric acid compared with 8-11-0 (8-24-0) made from orthophosphoric acid is the savings in shipping and storage costs (Table 4.8).

Table 4.8

WEIGHT AND STORAGE REQUIREMENTS OF SEVERAL LIQUID BASES USED IN LIQUID MIXED FERTILIZER MANUFACTURE

Nominal Grade N–P_2O_5	Pounds of Liquid to Supply 1 ton of N + P_2O_5	Cubic ft Storage Space Required per Ton of N + P_2O_5
8-24-0	5250 (3122 kg/m ton)	80 (2.48 kg/m ton)
10-34-0	4545 (2270 kg/m ton)	52 (1.62 kg/m ton)
11-37-0	4167 (2081 kg/m ton)	48 (1.49 kg/m ton)

TRIPLE SUPERPHOSPHATE

Description.   Triple superphosphate is marketed as a gray, brown or almost white material. It is in either powdered or pelleted form. It has an acid odor. The prefix triple was first used when ordinary superphosphate contained 16 percent P_2O_5 and triple superphosphate contained three times as much, 48 percent P_2O_5.

Method of Manufacture.   Much of the triple and concentrated superphosphate is made by a process on which the Tennessee Valley Authority holds United States patents. It is a continuous mixer steel cone into which phosphate rock and phosphoric acid are fed simultaneously; a slurry is discharged from the cone onto a moving conveyor belt similar to the Broadfield den (Figure 4.1) continuous mixer used for ordinary superphosphate. The slurry solidifies on the belt and is further processed to powder form for use in mixed fertilizer production or to pellet form for direct application to the soil. The reaction between the tricalcium phosphate in the ore and phosphoric acid may be represented simply as follows:

$$Ca_{10}(PO_4)_6F_2 + 14H_3PO_4 + 10H_2O \rightarrow 10CaH_4(PO_4) \cdot H_2O + 2HF$$

Handling Characteristics.   Triple superphosphate of earlier days was difficult to spread satisfactorily because it lumped and bridged in the fertilizer distributor. Also, there was a large amount of free phosphoric acid that made it difficult to cure, and it eventually rotted the bags. However, progress, primarily related to granulating processes, has produced superior products that handle most satisfactorily in the factory, the bag and the fertilizer distributor.

Chemical Characteristics. Triple superphosphate is essentially ordinary superphosphate with most of the gypsum removed. It is largely monocalcium phosphate monohydrate. It analyzes 45 to 50 percent P_2O_5 (20 to 22 percent phosphorus), 12 to 16 percent calcium and 1 to 2 percent sulfur. Most of its phosphorus is in water-soluble form. It is used in compounding high-analysis mixed fertilizers. It supplies 30 to 35 percent of the fertilizer phosphorus used in the United States and its possessions.

Crop Response. As a source of phosphorus for crops, it is as effective as ordinary superphosphate. Unless it is used in quantities large enough to supply 16 lb/acre (18 kg/ha) of sulfur, it does not supply sufficient sulfur for the best growth of some crops in some locations. For this reason, and unless sulfur is present in the soil or is supplied through rainfall or some other means, triple superphosphate in the rural, southeastern United States will usually produce around 90 to 95 percent as much increased yield per acre (0.4 ha) as ordinary superphosphate.[11] Coarse sandy soils in humid regions far removed from industrial and urban air pollution are the most likely soils to be sulfur deficient.

CONCENTRATED SUPERPHOSPHATE

Description. Tennessee Valley Authority concentrated superphosphate contains about 54 percent available P_2O_5 (24 percent phosphorus), most of which is monocalcium phosphate, $Ca(H_2PO_4)_2$. The P_2O_5 in triple superphosphate (45 to 50 percent P_2O_5) is chiefly monocalcium phosphate monohydrate, $Ca(H_2PO_4)_2 \cdot H_2O$. The loss of a molecule of water, small amounts of fluorine and other volatiles in the production of concentrated superphosphate accounts for its higher analysis. The free moisture content is about 1 percent or less.

Method of Manufacture. Tennessee Valley Authority concentrated superphosphate is produced by the acidulation of phosphate rock with superphosphoric acid diluted to 73 to 74 percent P_2O_5. Continuous acidulation and partial granulation are accomplished in a rotating drum. After it is denned and cured, the superphosphate is screened to minus 6-mesh (3.3 mm) (Figure 4.6).

Physical and Chemical Characteristics. Concentrated superphosphate is available as minus 6-mesh (3.3 mm), run-of-pile material. It has a bulk density of about 70 pounds (32 kg) per cubic foot (0.028 m^3). It has good

[11]U. S. Jones, "Sulfur Content of Rainwater in South Carolina " in *Environmental Chemistry and Cycling Processes,* Department of Energy Symposium Series (Conf. 760429), 1978.

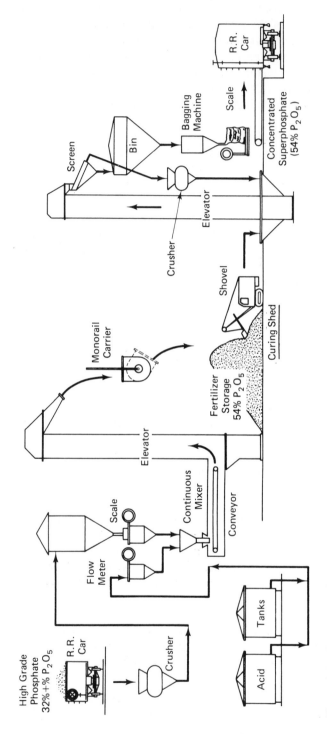

Figure 4.6 Concentrated superphosphate—how it is made.

132

physical properties and should remain relatively free flowing in storage. From 80 to 90 percent of the available P_2O_5 in the concentrated superphosphate is water soluble. The free-acid content, expressed as P_2O_5, is 3.5 percent or less. Tests have indicated that the material will absorb 2.5 to 3.0 pounds (1.1 to 1.4 kg) of ammonia per unit (20 lb or 9.1 kg) of P_2O_5 without excessive loss of ammonia.

Uses in Manufacturing. This material is used in ammoniation-granulation plants or in batch ammoniator plants to produce such high-analysis grades as 13-17-0 (13-39-0), 12-10-10 (12-24-12), 7-12-12 (7-28-14), 10-9-16 (10-20-20), 6-10-20 (6-24-24) and 13-6-10 (13-13-13). In many grades, its use permits consumption of larger proportions of relatively low-cost normal superphosphate, containing about 12 percent sulfur in addition to phosphorus.

Crop Response. Concentrated superphosphate is produced by acidulating phosphate rock with superphosphoric acid, as contrasted to the acidulation of phosphate rock with ordinary phosphoric acid in the production of triple superphosphate. Except for the higher phosphorus content, concentrated superphosphate is essentially the same as triple superphosphate, and crop response to the two materials should be the same.

CALCIUM METAPHOSPHATE

Description. A finely divided glassy-like substance, calcium metaphosphate is usually almost white with a very faint, pale aqua cast. It has a texture similar to fine sand.

Method of Manufacture. Furnace gases are burned to produce elemental phosphorus as shown in Figure 4.5 and as explained in the section on furnace-process phosphoric acid. Molten calcium metaphosphate is produced by burning elemental phosphorus with air and absorbing the resultant P_2O_5 fume in finely ground phosphate rock. The hot phosphorus vapors and the phosphate rock unite to form liquid calcium metaphosphate that, upon cooling, is solidified into a glass. The glassy material is shattered into small pieces that are ground fine enough so that 100 percent will pass through a 10-mesh (2-mm) screen.

$$2P_2 + 5O_2 \xrightarrow{\triangle} 2P_2O_5$$

$$Ca_{10}(PO_4)_6F_2 + 7P_2O_5 + H_2O \rightarrow 10Ca(PO_3)_2 + 2HF$$

Production capability of the Tennessee Valley Authority plant is over 250 tons (228 m tons) per day. Calcium metaphosphate is not likely to be produced commercially in large quantities because it requires elemental phosphorus refined with electrical energy and cannot compete with products based on wet-process phosphoric acid.

Handling Characteristics. Ground limestone is added to help keep the material from forming lumps, and the product is bagged and shipped. It is easy to spread and flows freely through ordinary distributors. It does not bridge in the drill or cake in storage because it is a stable chemical that does not absorb moisture rapidly. Lime spreaders can distribute it uniformly; but, since it is the highest-analysis straight-solid phosphate material on the market, it requires a relatively small amount per acre (0.4 ha) compared with ordinary or concentrated superphosphates. As of this writing, it has not yet been granulated. Granular, high-analysis, multinutrient fertilizers have been made with this material.

Chemical Characteristics. A typical sample made from Tennessee brown rock phosphate assays 65 percent P_2O_5 (28 percent phosphorus), 17 percent calcium, 2 percent silicon, 3 percent iron and aluminum and 0.2 percent fluorine. Its chemical formula is $Ca(PO_3)_2$. This high concentration results in a significant saving in freight charges (per unit of P_2O_5) when the fertilizer is shipped considerable distances. It is the highest-analysis straight-solid phosphate material produced.

Crop Response. On acid soils, calcium metaphosphate compares favorably with concentrated superphosphate as a source of phosphorus for all crops except those for which early, short-season growth is important, such as vegetables. Crop response to calcium metaphosphate on alkaline soils has been variable. Results from field tests conducted with crops on alkaline soils in the western states are conflicting, with the exception that it is usually an excellent source of phosphorus for alfalfa. It is not soluble in water; however, after hydrolysis of the material in soil, it becomes more water soluble. As a source of phosphorus for crops in the humid region of the United States, it is almost as effective as concentrated superphosphate. The small difference in response in favor of concentrated superphosphate is due to the fact that calcium metaphosphate is not as soluble in water. It contains no sulfur and ranks 80 percent as effective as ordinary superphosphate for increasing crop yields.[12] It is probably the most economical and highest-grade phosphorus fertilizer that can be produced by the electric-furnace method.

GRANULAR HIGH-ANALYSIS PHOSPHATE

Method of Manufacture. Granular high-analysis phosphate is made by feeding equal parts of pulverized calcium metaphosphate and run-of-pile concentrated superphosphate to a rotary granulator.

Steam is introduced into the rolling bed and hot water is sprayed

[12]Jones, "Phosphate Fertilizers."

over the bed in order to promote granulation. The product of the granulator is dried, cooled and screened.

Handling Characteristics. Physical conditions are similar to those of calcium metaphosphate.

Chemical Characteristics. The available P_2O_5 content is 57 percent, of which more than 50 percent is water soluble. The water solubility results largely from the phosphorus content of concentrated superphosphate since only a small percentage of the calcium metaphosphate is hydrolyzed during granulation. The free-acid content expressed as P_2O_5 is about 4 percent. The manufacturing and processing characteristics are similar to those of calcium metaphosphate.

Crop Response. This granular high-analysis phosphate, because of its relatively high water solubility, gives agronomic results similar to those of concentrated superphosphate. However, best comparative results may be obtained on acid soils.

FUSED TRICALCIUM PHOSPHATE

Description. Fused tricalcium phosphate is a dark-colored material that looks like sand covered with charcoal.

Method of Manufacture. The presence of fluorine in the apatite crystal (molecule) is in large part responsible for the low availability of rock phosphate to growing plants. A substantial portion of the fluorine is removed at a temperature of 2507° to 2597°F (1375° to 1425°C) in the presence of water vapor followed by quenching of the melt. The reaction is as follows:

$$Ca_{10}(PO_4)_6F_2 + H_2O \xrightarrow[1400°C]{2552°F} [Ca_3(PO_4)_2)_3] \cdot CaO + 2HF$$

Handling Characteristics. Fused tricalcium phosphate has an excellent physical condition, especially when it is screened to 10-mesh (2-mm) size. It will not bridge in the drill or cake in storage because it is a rather stable chemical that does not absorb moisture rapidly. It may be distributed by lime spreaders. No curing is necessary. Quenching of the melt produces a material that is about 90 percent finer than 10-mesh (2 mm) and about 50 percent of which passes a 40-mesh (0.4-mm) screen. The availability of the phosphorus to crops is related to particle size of the product. The 40-mesh (0.4-mm) size is more available, but the larger 10-mesh (2-mm) material gives satisfactory results.

Chemical Characteristics.   The material contains 28 to 35 percent available P_2O_5 (12.4 to 15.2 percent phosphorus), 50 percent calcium, with only 4 percent or less fluorine. Its chemical formula is expressed as $Ca_3(PO_4)_2)_3 \cdot CaO$. The larger-scale operations of TVA indicate that fused tricalcium phosphate is the least expensive heat-treatment process for phosphate rock. It will be noted that the method of manufacture does not require sulfuric acid, coke, elemental phosphorus, or large amounts of electricity, and it can use relatively low grades of phosphate rock. Albeit, after many years of testing and demonstration, the material has not yet gained wide acceptance as a fertilizer probably because its availability to crops decreases with an increase in the lime level of soil. It is not soluble in water.

Crop Response.   As a source of phosphorus for crops in the humid region of the United States, it is about 75 to 80 percent as effective as ordinary superphosphate.[13] Tests in the western United States indicate that this material is not a promising source of phosphorus, even for alfalfa.

OPEN-HEARTH BASIC SLAG

Description.   Slag is heavy, dark brown powder ground so that 70 percent will pass through a 100-mesh (0.1-mm) sieve and 100 percent through a 10-mesh (2-mm) sieve.

Method of Manufacture.   It is produced by the open-hearth method in the production of pig iron. Phosphorus is removed from iron ore because its presence weakens the cast iron produced. Iron ore and lime are heated in an open heater. Additional quantities of iron ore and lime along with an occasional batch of fluorspar—75 pounds (34 kg) per 10 tons (9.1 m tons) of lime—may be added to reduce the viscosity of the slag. When the mixture reaches a temperature at which it becomes red hot, the lime melts and unites with the acid impurities including phosphoric acid. The resulting compounds containing calcium, phosphorus, silicon, magnesium and manganese are lighter than iron, and they rise to the surface of the molten mass and are poured off as slag. After the slag is cooled to a hard, black massive cinder, it is crushed and ground to a finer powder.

Handling Characteristics.   Basic slag weighs more per unit volume than any other fertilizer material. It is so finely ground that it acts like dust when it is spread. Some growers mix slag with pulverized fertilizer and thereby improve its handling characteristics. Truck fertilizer spreaders

[13]Jones, "Phosphate Fertilizers."

with leather fenders on the boom have been used successfully as have tractor-drawn distributors.

Chemical Characteristics. It contains 8 to 12 percent P_2O_5, 3 to 5 percent calcium, and 6 to 8 percent $CaSiO_3$. It is alkaline to litmus and has a neutralizing effect equal to 70 percent $CaCO_3$ equivalent. It is very likely to solidify if stored in a damp place. Some chemists consider it a silicate and a phosphate of lime, [$(CaO)_5 \cdot P_2O_5SiO_2$].

Crop Response. Basic slag is an excellent source of phosphorus and lime. In addition, it contains magnesium, iron and manganese, plus traces of boron. For pastures in the southeastern United States, particularly on sandy soils of the Coastal Plains where no provision has been made for lime and the rarer essential elements, slag has proven to be very helpful. Results on a corn, soybean, wheat and clover rotation in Indiana[14] show the relative effectiveness of basic slag and several other phosphate materials (Table 4.9).

Table 4.9
THE RELATIVE EFFECTIVENESS OF DIFFERENT
PHOSPHATES—1930

Source of Phosphorus[1]	Cost of Fertilizer	Net Returns per Acre per Rotation
Superphosphate	$3.36	$13.53
Basic slag	3.36	13.13
Steamed bone meal	3.36	9.20
Phosphate rock	4.80	7.10

[1]All phosphorus sources except rock phosphate were added at the rate of 48 pounds of P_2O_5 per acre (54 kg/ha). Rock phosphate was added at the rate of 192 pounds of P_2O_5 per acre (215 kg/ha).

BONE PHOSPHATE

Description. Bone phosphate is white to grayish material ground to a powder that is light in weight per volume. It is sometimes referred to as steamed bone meal.

Method of Manufacture. Green bones are boiled and steamed at high pressure for the purpose of removing fats. The removed fatty material is

[14]H. F. Reed and H. G. Hall, "Report of Moses Fill Annex Farm," Indiana Agricultural Experiment Station Circular 172, 1930.

used for the manufacture of gelatin and glue. Bones are then ground for easy handling and distribution. The principal source of supply is the abattoir although small amounts are secured from the manufacture of bone buttons and knife handles.

Handling Characteristics. It is light in weight per volume. When it is broadcast on a windy day, it should be mixed with fine, damp soil or sawdust.

Chemical Characteristics. Steamed bone meal contains 1 to 2 percent nitrogen and 22 to 30 percent P_2O_5 (10 to 13 percent phosphorus). It is similar chemically to $Ca_3(PO_4)_2$ and is commonly referred to as BPL, bone phosphate of lime. Like rock phosphate, it was formerly sold on the basis of its BPL content rather than its phosphorus percentage. As a general rule, young bones contain less phosphorus and more nitrogen than older bones.

Crop Response. It is superior to rock phosphate as a source of available phosphorus. It is slightly lower in availability to crops than basic slag.[15] Liming the soil has a depressing influence on solubility of phosphorus in steamed bone meal. It is rarely encountered today in the field fertilizer trade. Bone fertilizers have been held in high esteem probably because fertilizers of plant and animal origin were, for a long time, considered less likely to burn the plants than an inorganic source. Home gardeners continue to ask for bone meal to use on their flowers, shrubs and lawns because it can be used rather indiscriminately without fear of salt damage (burning).

PREDICTION OF PHOSPHORUS NEEDS AND RESPONSES

Soil Tests. Soil tests for phosphorus have received considerable attention in most areas of the United States. Correlation of chemical methods with greenhouse-grown plants has become standard procedure. A number of studies have been made using bioassay with plants and microorganisms to predict yield of crops in the field. Some tests have been made that attempted to correlate chemical methods with field response. This latter comparison meets the goal of prediction but is the most difficult to define and understand. The present state of the chemical assessment of soil phosphorus is generally limited to predicting response to applied phosphorus in terms as broad as low, medium, high and possibly very high "available" phosphorus. All laboratory or greenhouse-type methods are based on having soil samples that adequately represent a field. As a

[15]Reed and Hall, "Report of Moses Fill."

consequence, the proper sampling procedure deserves a high priority for validating the soil test. In the final analysis, the most accurate assessment of crop response to phosphorus, or any other nutrient, can only be made by growing the specific crop in a specific environment.

In contrast to biological soil tests, chemical procedures are quite rapid and do not require special facilities for plant or organism growth.

Chemical soil tests for phosphorus involve two important steps:

1. Extraction of some fraction of the phosphorus in the soil.
2. Quantitative determination of the phosphorus in the soil extract.

There is a great diversity in extraction techniques for available phosphorus among soil-testing laboratories. In 13 state laboratories in the southern region of the United States, there are 7 different extracting solutions employed.[16] In contrast, the most commonly used extractant in the arid region, particularly where calcareous soils are found, is $0.5M$ $NaHCO_3$ buffered at pH 8.5. However, Alabama and Mississippi use AlF_3 buffered at pH 4.0 for calcareous soils.

Thus, the choice of procedure for determination of phosphorus will depend on the nature of the soils under study.

The soil-test work group of the National Soil Research Committee published one of the most comprehensive studies on phosphorus soil-test methods and related crop response. Soil samples from 74 phosphorus experiments were tested for phosphorus by 55 state and commercial laboratories. These results showed that correlation of soil tests, with greenhouse values was 0.703 and correlation with the field values was only 0.361 since the latter are influenced by rainfall, subsoil phosphorus and many other factors.[17]

Some states indicate a likelihood of response to added phosphorus if the soil-test level is below a certain value. In California, correlation of $NaHCO_3$-extractable phosphorus with cotton indicated a certain likelihood of response when the value is below 1 ppm.

Fertilization Rates. An example of a well-planned set of guides for phosphorus fertilization rates in the Coastal Plain of Virginia involves grouping soils with certain behavioral similarities and making recommendations based on low, medium, high and very high phosphorus levels within the groups. A shortened version of this system is shown in Table 4.10.

[16]W. E. Sabbe et al., "Procedures Used in State Soil-Testing Laboratories in the Southern Region of the United States," So. Coop. Series Bulletin 190, 1974.

[17]Soil-Test Work Group of the National Soil Research Committee, "Soil Tests Compared with Fields, Greenhouse and Laboratory Results," North Carolina Agricultural Experiment Station Technical Bulletin 121, 1956.

Table 4.10

P$_2$O$_5$ RECOMMENDED FOR COTTON IN VIRGINIA COASTAL
PLAIN AS RELATED TO GROUPING OF SOILS BY
IDENTIFIABLE CHARACTERISTICS AND SOIL-TEST CLASSES[1]

Soil-Test Class	P$_2$O$_5$ (lb/acre)		
	Group I[2]	Group II[3]	Group III[4]
L	80–100	60–80	40–60
M	60–80	40–60	30–40
H	40–60	30–40	20–30
VH	20–40	10–30	10–20

[1]Adapted from G. R. Epperson and G. W. Hawkins, *Virginia Agricultural Extension Service Bulletin* 297, 1966.

Author's Appraisal:

[2]Little or no slope; good internal drainage, depth, water relations, etc.

[3]Generally steeper slopes; some less desirable soil characteristics.

[3]Texture, drainage, topography or some other known condition constitutes a production hazard.

Plant Analyses. Plant phosphorus concentrations in the various types and parts of plants usually range from about 0.1 to 1.5 percent on a dry basis.

Two types of plant analysis are used in evaluation of the nutrient status of a crop. The most valuable involves analyzing a sample that is representative either of the entire plant (tops and roots) or, more frequently, of the above-ground plant. Uptake values are used to assess the soil and fertilizer contributions to the plants.

Diagnosis of nutrient adequacy or inadequacy is often performed by sampling some specific plant part and relating the concentration of nutrient found to levels established by previous experimentation. Most of the levels of adequacy have been established in nutrient-culture experiments similar to those outlined by Ulrich.[18]

Critical Levels. It is commonly believed that the best plant tissue to use for measuring the phosphorus status of the cotton plant is the petiole of the most recently matured leaf. There is a slight deviation from this for different growth stages of cotton and for different species of deciduous plants. Soil and plant scientists warn against using immature leaves at the top of the plant. Some specify the use of the petiole of the basal leaf until early bloom and thereafter of the petiole of the first mature leaf.

[18]A. Ulrich, "Plant Analysis—Methods and Interpretation of Results," in *Diagnostic Techniques for Soils and Crops,* ed. H. B. Kitchen (Washington, D. C.: The American Potash Institute, 1948) pp. 199–230.

A search for a critical level of phosphate in leaves or petioles indicates that there is a scarcity of data that allow a tissue test to be evaluated in quantitative terms.

Phosphate-phosphorus in the petiole of cotton is usually between 1,000 and 2,000 ppm and tends to decrease during the season. Evidence indicates that the level should be maintained above 1,000 ppm until after peak bloom. Any deficiency that delays growth, such as lack of nitrogen, causes a decided increase in the phosphate in the petiole.

Petiole phosphate values have been found to drop during periods of moisture stress and to rise following moisture application.[19] This illustrates the fact that any growth factor that is out of balance tends to influence the other growth factors.[20]

Plant Analyses As a Guide To Fertilization. Plant analysis for phosphorus is often most useful as a guide for next year's fertilization program. However, if samples from a long-season crop are obtained early in the season, when side dressing of a soluble phosphorus source may still be effective, tissue analysis may be used to assess the situation effectively.[21]

SUMMARY

1. Phosphorus is one of the key elements in life: it is found in all nuclei of cells and is a component of DNA and RNA, the vehicles of heredity.

2. Phosphorus is not readily leached from the soil. It moves very little and has the tendency to become "fixed." Neutral to slightly acid soils permit the most efficient use of phosphate fertilizers, but basic and acid soils "fix" phosphates in forms unavailable to plant roots.

3. Phosphate reserves are located in four major areas: Africa, North America, the Soviet Union and Europe. There are several kinds of phosphate rock, each with its own characteristics and properties.

4. Calcium phosphates are the most abundant forms on earth but must be further processed by heat or acid in order to produce a product that gives maximum yields.

5. Ordinary superphosphate, made from a mixture of rock phosphate and sulfuric acid, frequently gives crop yields superior to equal quantities of phosphorus from other phosphate materials because it contains more calcium and sulfur and because most of its phosphate is water soluble.

[19]W. W. Masters, C. W. Wendt, C. Harvey, and A. W. Young, "Nutrient Assimilation by Cotton as Measured by Petiole Analysis," *Agronomy Abstracts,* 1962, p. 38.

[20]U. S. Jones and C. E. Bardsley, *Phosphorus Nutrition of Cotton* (Ames, Iowa: Iowa State University Press, 1968) Ch. VII.

[21]J. B. Jones, "Plant Analysis Handbook for Georgia," University of Georgia Agricultural Extension Service Bulletin 735, 1974.

6. Phosphoric acid as such has not been used for direct application to the soil because it is corrosive and therefore difficult to handle on the farm. Uses for wet-process phosphoric acid have been increasing. Its role in the manufacture of liquid and solid ammonium phosphate has been growing. Furnace-process phosphoric acid has not been able to compete with wet-process acid because the cost of electricity for the furnace is prohibitive.

7. Superphosphoric acid has been used in the manufacture of liquid and higher analysis solid fertilizers. Experiments have shown it to be equal to concentrated superphosphate as a source of phosphate for plant growth.

8. Concentrated superphosphate and triple superphosphate are essentially the same, being largely composed of monocalcium phosphate. Crop response should be the same.

9. Calcium metaphosphate compares well with concentrated superphosphate on acid soils; crop response to it on alkaline soils has been variable.

10. Fused tricalcium phosphate is about 75 to 80 percent as effective as ordinary superphosphate, indicating that it is not a promising source of phosphorus.

11. Some basic slags are excellent sources of phosphorus and lime. They also contain iron, magnesium and manganese.

12. Bone phosphate is superior to rock phosphate as a source of available phosphorus and is slightly lower in availability to crops than basic slag.

13. It is the pH of the soil, the availability of the phosphate carriers added to these soils, and the monetary cost that often determines the type of phosphate fertilizer used on a particular crop or soil.

14. The various deposits that have been discussed, and those which are yet to be exploited, play an important role in the mining, manufacture and cost of producing phosphates for fertilizer.

15. Soil and plant tissue tests for determining critical levels have been developed and are widely used in diagnosing plant needs for phosphorus.

QUESTIONS

The availability of phosphorus is affected by the acidity of the soil. At a soil pH above 6.5, fertilizer phosphorus is precipitated by:

Answer		
1. ( )	1.	Aluminum
2. ( )	2.	Calcium

At a soil pH of 5.0, fertilizer phosphorus is precipitated by:

3.	Hydrogen
4.	Iron
5.	Magnesium
6.	Manganese
7.	Potassium

Answer
3. ( )
4. ( )

Directions for Nos. 5–14. Column A lists several phosphorus materials that are used for fertilizers. Column B lists some P_2O_5 and P percentages in a scrambled fashion. In the answer columns, rearrange the percentages of P_2O_5 and P in an order that most nearly describes the materials listed in column A.

Column A	Answer P_2O_5			Column B P_2O_5	P
5 and 6.				54	32
Bone meal	(	) (	)	25	27
7 and 8.				72	24
Super-phosphoric acid	(	) (	)	20	21
9 and 10.				12	17
Triple superphosphate	(	) (	)	62	11
11 and 12.				48	9
Super-phosphate (concentrated)	(	) (	)	38	5
13 and 14.					
Basic slag	(	) (	)		

True or False:

15. _____Good correlations have been obtained relating phosphorus taken up by plants and some chemical soil tests.

16. _____The highest concentration of phosphorus in the soil solution can be maintained at a pH value of about 6.0 to 6.5.

17. _____As a source of phosphorus for plants, concentrated superphosphate is not as effective as ordinary superphosphate.

18. _____Not more than 10 to 30 percent of the phosphorus added to soil in the form of fertilizer is removed by the crop grown immediately following the application.

19. _____Soil-test methods to assess the plant availability of soil phosphorus have proven to be very useful as a basis for making fertilizer phosphorus recommendations.

chapter 5

Multinutrient Fertilizers

Hill fertilizer experiments on corn have shown that growth and yield responses from phosphorus fertilizer are closely related to the ease with which the fertilizer dissolves in water. Yield increases from sources with different water solubilities of the phosphorus were L-R, 3.8 (255 kg/ha), 7.3 (499 kg/ha), 10.8 (725 kg/ha), 12.7 (853 kg/ha), and 13.5 (907 kg/ha) bushels per acre. Corn yield without phosphate was 76 (5107 kg/ha) bushels per acre. (Reprinted from Iowa Farm Science, Iowa State College, p. 18–82. Oct. 1955).

According to plant nutrient classification systems, the major plant nutrients supplied by fertilizers are nitrogen, phosphorus and potassium. Secondary nutrients are calcium, magnesium and sulfur. These six together may be called macronutrients in contrast to the seven micronutrients: iron, manganese, zinc, boron, copper, molybdenum and chlorine. These seven will be discussed later in a section on micronutrients.

Multinutrient fertilizers are guaranteed to contain more than one of the major plant nutrients, nitrogen, phosphorus or potassium. The presence or absence of secondary nutrients particularly calcium and sulfur, is frequently referred to in discussions of such fertilizers.

Sulfuric acid and rock phosphate are the sources of sulfur and calcium, respectively, in multinutrient fertilizers. These materials are two of the four principal raw materials used to manufacture multinutrient fertilizers:

1. Rock phosphate.
2. Sulfuric acid.
3. Ammonia.
4. Potassium chloride.

The first three and their derivatives are combined in various proportions to produce nitrogen-phosphorus and nitrogen-phosphorus-sulfur fertilizers. The fourth one, KCl, is added to the nitrogen-phosphorus or nitrogen-phosphorus-sulfur compounds to produce multinutrient goods with potassium. Potassium nitrate is a multinutrient nitrogen-potassium fertilizer, and it is discussed in the section on potassium. Raw materials required in the production of multinutrient fertilizers without nitrogen are rock phosphate, sulfuric acid and muriate of potash.

Nutrients Needed in Fertilizer Vary. The fertilizer needs of nonlegume crops growing on arid and semiarid soils of the Aridisol and Mollisol order, with high base saturation influences, are quite different from the fertilizer needs of crops growing on humid leached soils of the Spodosol and Ultisol orders. In contrast to Aridisols and Mollisols with high base saturation, Spodosols and Ultisols are those that represent much of the humid, leached-soil regions of the temperate zones and are often deficient in the bases calcium, magnesium and potassium. In this concept, it is assumed that soils of the Mollisol and Aridisol orders will be able to supply crops with ample calcium, magnesium and potassium from natural soil supplies;

accordingly, such crops will need to be treated with a fertilizer containing only nitrogen, phosphorus and sulfur in order for the plant to be well nourished with the major and secondary essential elements. On the other hand, nonlegume crops growing on leached soils developed as Spodosols and Ultisols must be limed to provide calcium and magnesium and must be improved with a fertilizer containing nitrogen, phosphorus, potassium and sulfur if plants growing on these soils are to be well nourished with the major and secondary nutrients.

Admittedly, this concept is an oversimplification, and there are exceptions since some soils of the Spodosol and Ultisol orders contain an ample supply of one or more of the nutrients, calcium, magnesium and potassium. It is also true that some soils of the Aridisol and Mollisol orders do not contain ample natural supplies of one or more of these bases. However, these exceptional conditions can be located by field experiments and calibrated by soil and plant-tissue testing. For the purpose of establishing bench marks for basic geographical considerations of fertilizer technology and use, the concept of nitrogen-phosphorus-sulfur for soils of the Aridisol and Mollisol orders and nitrogen-phosphorus-potassium-sulfur plus dolomitic limestone for soils of the Spodosol and Ultisol orders will be helpful. (See Figure 7.2 on page 219.)

Beginning of Multinutrient Fertilizer Industry. The discovery of deposits of rock phosphate in South Carolina in 1867 and in Florida in 1887 developed into a superphosphate fertilizer industry in the heart of the southeastern United States. Prior to this development, the fertilizer industry was largely based on the imports of a considerable amount of guano from the Peruvian Islands, and its use was greatest near seaports. An examination of experiment station records of the early twentieth century shows that superphosphate (acid-phosphate) was being adopted as the main ingredient of cotton fertilizers throughout the southeastern United States. In some parts of the region, ordinary superphosphate, 8 to 9 percent phosphorus (18 to 20 percent P_2O_5), is still used as a basic ingredient of mixed (multinutrient) fertilizers. Although technological changes in the fertilizer industry, since World War II, have provided a much wider variety of phosphatic fertilizers, phosphorus is still considered the basic ingredient of mixed (multinutrient) fertilizer manufacture.

Recovery Efficiency of Nitrogen-Phosphorus-Potassium. The efficiency of recovery of fertilizer phosphorus by plants is much lower than the recovery of fertilizer nitrogen and potassium. The recovery of fertilizer nitrogen and potassium by plants amounts to 50 to 90 percent of that added, but only 10 to 30 percent of fertilizer phosphorus added to soil is recovered by growing plants. It is therefore important to study the factors that influence the availability of sources of phosphorus in multinutrient fertilizers, one of the most important of which is the water solubility of the phosphorus compounds in the fertilizer.

Available Phosphorus. Early efforts to evaluate sources of phosphorus fertilizers in the United States took into account only the amount of phosphorus soluble in neutral ammonium citrate, a method prescribed by the Association of Official Analytical Chemists.[1] It is the phosphorus content that is guaranteed on fertilizer labels. This phosphorus is referred to as available phosphoric acid (APA) by fertilizer control officials in the United States. It is a measure of the water solubility of the phosphorus compounds in fertilizer to an extent approximately that of dicalcium phosphate, 0.02 parts in 100 parts of water at about 60°F (15°C).

Solubility of Phosphorus Compounds. There are five compounds that contain most of the phosphorus found in multinutrient fertilizers: diammonium phosphate, monoammonium phosphate, monocalcium phosphate, dicalcium phosphate and tricalcium phosphate. Their solubility in water at 60°F (15°C) is shown in Figure 5.1, which consists of Chart A and Chart B, which has a scale 30 times as large as Chart A. The most soluble compound is diammonium phosphate, 65 parts of which will dissolve in 100 parts of water. This is only half as water soluble as ammonium nitrate and about equal in solubility to potassium chloride, the major fertilizer sources of nitrogen and potassium respectively.

Solubility of Nitrogen and Potassium Compounds. Sulfate of ammonia, a major source of nitrogen and sulfur in fertilizers, has a solubility in water about equal to that of diammonium phosphate. Major fertilizer salts of nitrogen and potassium have a solubility similar to that of diammonium phosphate. Also of interest is the fact that it is illegal in the United States to guarantee potassium in a fertilizer unless it is water soluble. In multinutrient fertilizers, all potassium and nitrogen compounds have a high degree of water solubility. Only fertilizer phosphorus compounds have a low degree of water solubility, as indicated by the bars representing dicalcium and tricalcium phosphate in Chart B in Figure 5.1.

Three Kinds of Nitrogen-Phosphorus Materials. Practically all crops that are fertilized with both nitrogen and phosphorus are fertilized with multinutrient fertilizers. Such fertilizers contain one of three different kinds of nitrogen-phosphorus compounds, ammonium phosphates, nitric phosphates or ammoniated superphosphate. Where nitrogen-phosphorus-potassium mixtures are required, potassium chloride is simply added to the nitrogen-phosphorus compounds.

Table 5.1 lists all of the chemical compounds found in a typical nitric phosphate, in several ammoniated superphosphates and in two ammonium phosphate sulfate grades. Table 5.2 lists the forms and water solubilities of the phosphorus in the three types of nitrogen-phosphorus fertilizers. Both

[1]Association of Official Analytical Chemists, *Methods of Analysis* 12th ed. (Washington, D.C.: AOAC, 1975).

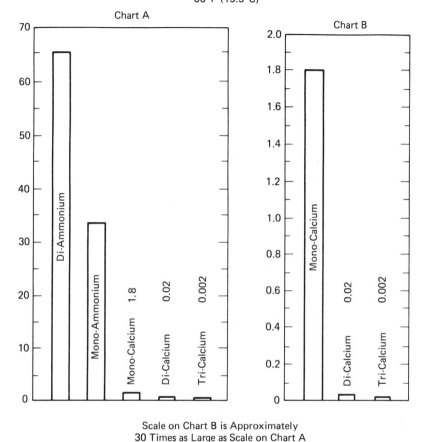

Solubility of Phosphates

Parts of Phosphate Which
Dissolve in 100 Parts of
Water at Approximately
60°F (15.5°C)

Figure 5.1 Comparative solubilities of phosphate compounds found in fertilizers.

tables were prepared from data supplied by the laboratories of the National Fertilizer Development Center of the Tennessee Valley Authority at Muscle Shoals, Alabama.

Data from many places have shown that there is little difference in crop response to sources of potassium or to sources of nitrogen in commercial fertilizers. The three types of commercial multinutrient fertilizers discussed here are the sources of almost all field-crop fertilizers, and the nitrogen and potassium in all three are very similar and very soluble (Table 5.1). However, there is a great deal of difference in the solubility of the phosphorus compounds in the three fertilizers (Table 5.2). The three types of commercial multinutrient nitrogen-phosphorus materials in order

Table 5.1

CHEMICAL COMPOUNDS IN COMMERCIAL FERTILIZERS

	Percent by Weight						
	Nitric Phosphate	Ammoniated Superphosphate				Ammonium Phosphate Sulfate	
Compound	14-14-14	5-20-20	6-24-24	6-12-12	10-10-10	16-20-0	14-14-14
KCl		17	28	4	4		7
KNO_3	1		[1]	[1]	[1]		
K_2SO_4		10	4	2	5		16
KH_2PO_4		3	12	2	1		2
K syngenite[2]		3	2	22	18		
NH_4Cl	15	13	7	11	12		11
NH_4NO_3	4		[1]	[1]	[1]		
$(NH_4)_2SO_4$		4	2	1	3	52	37
$NH_4H_2PO_4$		11	8	8	8	11	18
$(NH_4)_2HPO_4$						20	5
NH_4 syngenite[3]		2	1	10	12		
$CaSO_4$		3	7	7	9		
$CaHPO_4$	26	6	6	2	3		
Apatite	1	14	3	12	9		
Quartz	3	4	1	7	4	9	
Others	42[4]		5[5]				
Total	92	90	86	88	88	92	96

[1]Amount present was not determined.

[2]Potassium syngenite, $K_2SO_4 \cdot CaSO_4 \cdot H_2O$.

[3]Ammonium syngenite, $(NH_4)_2SO_4 \cdot CaSO_4 \cdot H_2O$.

[4]$NH_4NO_3 \cdot 2KNO_3$.

[5]$Ca(H_2PO_4)_2 \cdot H_2O$.

of increasing phosphorus solubility are nitric phosphate, ammoniated superphosphate and ammonium phosphate sulfate.

The description, method of manufacture, handling characteristics, product use and chemistry of these three types of multinutrient fertilizers will be discussed because the raw materials used and the techniques of manufacture employed influence the efficiency of fertilizers as suppliers of plant nutrients for crop growth and development.

Manufacturing Techniques. The three principal techniques used in the manufacture of most multinutrient fertilizers are acidulation, ammoniation and granulation. Acidulation is the process or technique of treating a material with an acid. The most common acidulation process is the treatment of phosphate rock with acid.

Ammoniation is the process or technique of introducing ammonia

150

Table 5.2
FORMS OF P_2O_5 IN COMMERCIAL FERTILIZERS

			Percent of Total P_2O_5		
			Water Insoluble		
			Apatite		
Grade	*Water soluble*	*Dicalcium phosphate*	*Citrate soluble*	*Citrate insoluble*	*Other[1]*
Nitric phosphate					
14-14-14	5	91	1	2	1
Ammoniated superphosphate					
6-12-12	41	8	38	2	11
10-10-10	48	15	31	6	0
5-20-20	49	14	21	5	11
15-15-15	76	10	12	2	0
6-24-24	76	16	4	2	2
Ammonium phosphate sulfate					
16-20-0	88	0	0	0	12[2]
14-14-14	93	0	0	0	7

[1]Assumed to be iron and aluminum phosphates.

[2]$(Al,Fe)PO_4 \cdot xH_2O$ (gel) detected by microscopic analysis.

into superphosphate that forms ammoniated superphosphate. For ammoniation of ordinary superphosphate, the commercial operator works in the range of 6 to 7 pounds (2.7 to 3.2 kg) of ammonia per unit (20 pounds or 9.1 kg of P_2O_5) and 4 to 5 pounds (1.8 to 2.3 kg) per unit in concentrated superphosphate.

Granulation is the process or technique of collecting small size particles into larger granules or pellets of about 4- to 8-mesh (4.3- to 2.3-mm) size. It is a technique that has been widely adopted as a means of improving the handling and storage properties of fertilizers.

AMMONIUM POLYPHOSPHATE—LIQUID FERTILIZER

Description.   11-16-0 (11-37-0)[2] is a clear, nearly neutral solution containing nitrogen in the ammoniacal form and P_2O_5 in the polyphosphate and orthophosphate forms. The source of information on liquid fertilizer of the ammonium polyphosphate form is Report No. T63-1 FII from the Tennessee Valley Authority, Muscle Shoals, Alabama.

[2]The numbers within the parentheses refer to percentages of $N–P_2O_5–K_2O$ and those before the parentheses refer to nitrogen-phosphorus-potassium.

Method of Manufacture. It is made by ammoniating superphosphoric acid containing about 79 percent P_2O_5. This acid is kept hot prior to ammoniation since, at this concentration, it will solidify at ambient temperatures.

Handling Characteristics. Ammonium polyphosphate can be shipped, stored and handled in mild steel equipment. Any general-purpose pump designed for liquid fertilizers may be used to pump 11-16-0 (11-37-0). It can be stored for substantial periods at near-zero temperatures without salting out. However, prolonged storage during the summer months will result in partial hydrolysis of polyphosphate. This may result in a higher salting-out temperature.

Chemical Characteristics. Its physical and chemical properties are:

pH	6.0
Viscosity, centipoises at 75°F (24°C)	80.0
Density, lb/gal (1.72 kg/1) at 75°F (24°C)	11.7
Safe storage temperature	about 0°F (−17°C)
Polyphosphate content, % of total P_2O_5	65.0 minimum
Sulfur content, %	about 0

Product Use. Ammonium polyphosphate material is used in the production of complete liquid fertilizers and for direct application. Since 11-16-0 (11-37-0) contains much of its P_2O_5 in the polyphosphate form, it is more soluble than comparable products made with ordinary phosphoric acid and thus may be used to produce complete fertilizers of higher grades. For example, 11-16-0 (11-37-0) may be blended with urea–ammonium nitrate solution to produce 19-9-0 (19-19-0) having an estimated salting-out temperature under 32°F (0°C), or it may be blended with urea–ammonium nitrate solution, potassium chloride and water to produce 7-10-17 (7-21-21) having an estimated salting-out temperature of 80°F (26°C).

In addition to its use in the production of liquid fertilizers by cold blending, 11-16-0 (11-37-0) is used in formulations with wet-process phosphoric acid to produce sequestered (clear) liquid fertilizers. The property of 11-16-0 (11-37-0) to sequester (essentially keep in solution) the solids that precipitate on ammoniation of wet-process acid makes its use advantageous when the cost of the wet acid is lower than furnace acid. Normally, good sequestration is achieved in the production of nitrogen-phosphorus grades when 20 to 30 percent of the P_2O_5 is supplied as 11-16-0, (11-37-0), and 30 to 40 percent in the production of nitrogen-phosphorus-potassium grades, depending on the amount and type of impurities in the wet acid. One nitrogen-phosphorus grade that is produced using a mixture of 11-16-0 (11-37-0) and a formulation of wet-process phosphoric acid is 10-15-0 (10-34-0). A nitrogen-phosphorus-potassium grade using the mixture plus muriate of potash and nitrogen solution is 5-5-8 (5-10-10).

Crop Response. Crop response to liquid fertilizer applied in bands, or broadcast and worked in, is similar to that obtained with water-soluble dry fertilizers applied in the same way. Band application of liquids is superior to broadcast application because the phosphorus is obviously completely water soluble and banding reduces contact between the soil and fertilizer. Under conditions where water solubility of the phosphorus is important, liquid fertilizers are superior to solid fertilizers containing an appreciable amount of water-insoluble phosphorus. Anthony[3] conducted an experiment comparing concentrated superphosphate with liquid superphosphoric acid. The data indicated that cotton yields were about equal with liquid and solid sources of phosphorus whether applied in the spring or fall.

AMMONIUM PHOSPHATE—SOLID FERTILIZER

Description. Ammonium phosphate is a pelletized white to dark gray material produced as round granules about the size of small vetch seed or number-4 birdshot.

Method of Manufacture. Phosphoric acid is first made by treating finely ground rock phosphate with an excess of sulfuric acid to produce gypsum and phosphoric acid. This is the so-called wet-process method of phosphoric acid manufacture (See Figure 4.3 on page 123). Spent sulfuric acid from the refining of petroleum products by the alkylation process has been found to be satisfactory for the production of wet-process phosphoric acid used in the manufacture of ammonium phosphate fertilizer.

Anhydrous ammonia is added to the liquid phosphoric acid to form technical grade monoammonium phosphate, containing 11 percent nitrogen and 21 percent phosphorus (48 percent P_2O_5) (Figure 5.2).

$$NH_3 + H_3PO_4 \rightarrow NH_4H_2PO_4$$
monoammonium phosphate

More ammonia may be added to wet-process acid to form technical-grade diammonium phosphate, containing 16 to 18 percent nitrogen and 20 to 21 percent phosphorus (46 to 48 percent P_2O_5).

$$2NH_3 + H_3PO_4 \rightarrow (NH_4)_2 HPO_4$$
diammonium phosphate

Ammonia may also be reacted with a mixture of phosphoric and sulfuric acids to form ammonium phosphate sulfate containing 13 to 16 percent

[3]J. L. Anthony, ''Phosphate Applied in Fall Effective in Black Belt,'' *Miss. Farm Res.* 25 (9) (1962).

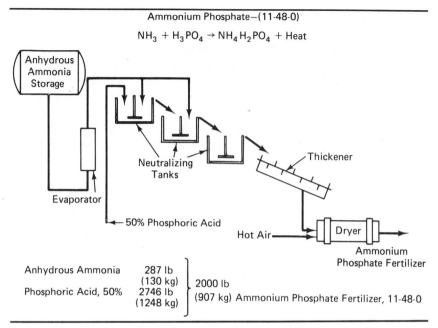

Figure 5.2 Flow diagram of manufacture of fertilizer-grade monoammonium phosphate.

nitrogen, 9 to 15 percent phosphorus (20 to 39 percent P_2O_5) and 7 to 14 percent sulfur.

$$4H_3PO_4 + xH_2SO_4 + (4 + 2x)NH_3 \rightarrow 4NH_4H_2PO_4 \cdot x(NH_4)_2SO_4$$

Phosphoric acid and anhydrous ammonia are introduced into a large reactor (Figure 5.2) where they combine to produce a thick viscous slurry of monoammonium phosphate. A mixture of phosphoric and sulfuric acid can be introduced in varying proportions to make 16-9-0 (16-20-0), 15-13-0 (15-30-0), 13-15-0 (13-39-0) or other grades as desired. These grades are triple salts containing more or less of ammonium sulfate, monoammonium phosphate and diammonium phosphate (Figure 4.3).

From the reactor, the slurry flows to a paddle mixer, called a blunger, where small granules of fertilizer of the same analysis (previously manufactured) are coated with the slurry. Potassium salts may also be introduced here if a nitrogen-phosphorus-potassium grade is desired. It is possible to start the granulating process by "seeding" the slurry in the blunger with small pellets or granules. The blunger contains heavy intermeshing paddles that roll the pellets and coat each uniformly with the product slurry. Without the previously manufactured granules, the slurry will not develop into granules. The freshly coated pellets are discharged from the blunger into an open-flame drier and emerge dry.

From the drier, the pellets go to double-decked screens. The top screen removes the over-sized pellets, which then pass through a crusher and back to the blunger to be built up to product size by additional coats of slurry. The second screen removes the marketable-sized pellets that then proceed to the storage building. Here they are picked up by an overhead crane bucket, deposited in the bagging hopper, transported up the elevator and rescreened. From the last screen, the product moves to the bagger and then to boxcars or trucks at the factory railroad or dock.

In one factory, production capacity is 1,500 tons (1,365 m tons) per day for a certain monoammonium phosphate fertilizer grade. The production capacity for the diammonium phosphate grades in the same operation is around 400 tons (362 m tons) per day.

Diammonium phosphate becomes unstable at temperatures above 300° to 400°F (149° to 204°C); monoammonium phosphate remains stable at much higher temperatures. Thus, the drying temperature for monoammonium phosphate can be much higher than that used to dry diammonium phosphate. However, a TVA process for the production of granular diammonium phosphate at 167°F (75°C) has eliminated this problem.

Ammonium phosphate production has increased more rapidly than any other type of fertilizer since about 1960 when practical and versatile processes for production of economical, high-quality granular products were introduced. The most popular type has been granular diammonium phosphate of 18-20-0 (18-46-0) grade made with wet-process phosphoric acid; more recently 11-24-0 (11-55-0) or 13-22-0 (13-52-0) grades that are primarily monoammonium phosphate have been increasing in popularity. Production is dependable and efficient in large granulation plants with capacities of 700 to 1,000 tons (635 to 907 m tons) per day in single-train units.

Handling Characteristics.   The fertilizer is pelletized, highly concentrated, and free flowing. Tests have shown that after prolonged storage and exposure to humid atmosphere, the relative drillability of the product is equal or superior to any other granulated fertilizer. Bag storage tests show no caking or deterioration after storage for several years.

When 100 pounds (45.4 kg) of a 10-10-16 (10-20-20) grade is shipped, handled, stored and spread on the ground instead of 200 pounds (90.8 kg) of a 5-5-8 (5-10-10) grade, the costs for freight, bags, handling and labor for spreading on the field are reduced by one-half.

Chemical Characteristics.   Fertilizer grade monoammonium phosphate contains 11 percent nitrogen and 48 percent P_2O_5 (21 percent phosphorus). Diammonium phosphate is produced for fertilizer use and is manufactured into several analyses. It usually contains 16 to 21 percent nitrogen and 46 to 53 percent P_2O_5 (20 to 26 percent phosphorus).

A popular grade that has been widely used where soils are of the Aridisol or Mollisol order is 16-9-0 (16-20-0), which is a mixture of 1,150 pounds (522 kg) of ammonium sulfate and 850 pounds (386 kg) of monoammonium phosphate per ton (0.91 m ton). Another grade that is widely used on cotton and other crops on calcareous soils is 11-21-0 (11-48-0).

By varying the proportions of phosphoric acid, sulfuric acid, ammonia and potassium salts, high analyses fertilizers on the $N-P_2O_5-K_2O$ basis such as 12-12-12, 13-13-13, 15-15-15, 10-20-10, 12-24-12, 14-28-14, 6-24-24 and 8-24-24 are produced. Many of these grades have been used on cotton in the Mississippi Valley and on corn and small grains in the Midwest.

Crop Response.   All the fertilizer elements in these materials are water soluble because they are made from liquids and water-soluble salts. This makes them especially useful for quick-growing crops, such as greens, and for early growth and grazing of small grains and pasture grasses. Wheat takes up 50 percent more phosphorus from the fertilizer from ammonium phosphate than it does from ordinary superphosphate by the time the crop is five weeks old.[4] No less important is the use of these soluble materials as starters or side dressers for crops such as corn, cotton, tobacco, fruits and vegetable crops.[5] The apparent influence of phosphorus on earliness of growth is indicated by the responses shown in final crop yields (Figure 5.3). These effects seem to be correlated with the solubility of the phosphorus fertilizer.

Fertilizer placement, especially with row crops, should receive particular attention in order to prevent seed and seedling injury. Diammonium phosphate should not be placed in direct contact with germinating seeds.

AMMONIUM PHOSPHATE NITRATE

Description.   Ammonium phosphate nitrate is a granular material containing 30 percent nitrogen and 4.4 percent phosphorus (10 percent P_2O_5). It is homogeneous; each granule contains uniform amounts of each plant nutrient. Other grades such as 25-12-0 (25-25-0) are manufactured by the Tennessee Valley Authority but present commercial production is largely 30-4.4-0 (30-10-0).

Method of Manufacture.   Ammonium phosphate nitrate is produced from solid ammonium nitrate, anhydrous ammonia and phosphoric acid. About

---

[4]H. G. Dion, J. W. P. Spinks, and J. Mitchell, "Experiments With Radiophosphorus on the Uptake of Phosphorus by Wheat," *Sci. Agric.* 29 (4) (1949).

[5]W. P. Mortensen, A. S. Baker, and G. L. Terman, "Source and Rate of Application of Phosphorus Fertilizers for Vegetable Crops in Western Washington," Washington Agricultural Experiment Station Bulletin 652, 1964.

Figure 5.3   Early response of snap beans in central Maine to three levels of water-soluble phosphorus-0, 14, and 70 percent. (University of Maine)

two-thirds of the phosphoric acid required is reacted with ammonia in a preneutralizer in order to obtain a pH of about 6. The slurry from the preneutralizer is fed to a rotary drum or a pan-type ammoniator-granulator along with solid ammonium nitrate and the remainder of the phosphoric acid and ammonia. The granular product is dried, sized and coated with a diatomaceous-earth conditioner. Figure 5.4 is a flow diagram of processes for production of several high nitrogen fertilizers including ammonium phosphate nitrate.

Handling Characteristics.   The product is screened to minus 6-mesh (3.3 mm) plus 16-mesh (1.2 mm) and conditioned with 3 percent by weight of calcined fuller's earth. It is shipped in bulk or in 50-pound (22.7-kg) bags. Its bulk density is approximately 50 pounds (22.7 kg) per cubic foot (0.028 m³), and it is free flowing. Tests have shown that the drillability of 30-4.4-0 (30-10-0) after exposure to humid atmosphere is equal to that of ammonium nitrate prills. Bag storage tests showed no caking after storage for one month. After a three-month storage period, there was a light bag set that increased slightly after an additional 3 months of storage. Bulk storage tests of several months duration in an unheated building under reasonably dry conditions showed no deterioration of product. However, under moist conditions, the product on the surface of a pile will progressively absorb moisture and deteriorate. Therefore, the bulk material should be protected with a suitable moisture-proof covering, such as plastic sheeting.

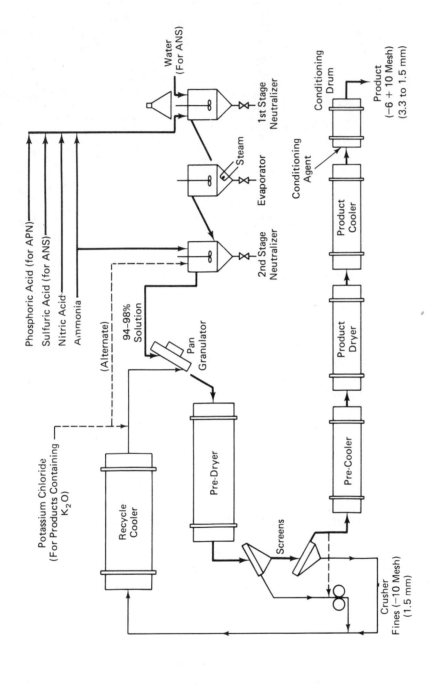

Figure 5.4 Flow diagram of pan-granulation process for production of granular high-nitrogen fertilizers.

Chemical Properties. The nitrogen in 30-4.4-0 (30-10-0) is present as ammonium nitrate and ammonium phosphate; slightly over half of the nitrogen is in the ammoniacal form. The phosphorus is supplied from phosphoric acid and is completely water soluble. The equivalent acidity is about 1,080 pounds (491 kg) of calcium carbonate per ton (0.91 m ton) of the material (1.8 lb $CaCO_3$ × 600 lb nitrogen).

Uses in Manufacturing. Ammonium phosphate nitrate is useful in ammoniation-granulation plants for the production of high-analysis grades such as 21-6-6 (21-14-7), 20-4-8 (20-10-10), 16-4-6 (16-8-8), 18-7-14 (18-18-18), 15-7-12 (15-15-15) and 13-6-10 (13-13-13). From 60 to about 85 percent of the nitrogen in these products may be supplied as 30-4.4-0 (30-10-0). Granulation efficiency is high, and the products exhibit good storage properties when packed in moisture-proof paper bags if they are dried to about 0.5 percent moisture content and conditioned with 2.5 percent by weight of kaolin clay. In many areas, ammonium phosphate nitrate may also be used to economic advantage as a substitute for ammonium sulfate in the production of granular fertilizers of low nitrogen-to-phosphorus ratios.

In addition to its use in granulation processes, 30-4.4-0 (30-10-0) is used in combination with other raw materials of similar particle size in the production of high-analysis grades in bulk blending operations.

Crop Response. Ammonium phosphate nitrate is used for direct application. However, it should be supplemented with potash unless soils are high in available potassium. The high nitrogen-to-phosphorus ratio makes this fertilizer useful on soils that test medium to high in available phosphorus. When rates of application do not exceed 100 lb/acre (112 kg/ha), this material can be applied in the row for crops such as wheat and corn. For higher rates of application, the fertilizer should be applied broadcast or banded to the side and below the seed. It may be used as a top-dressing on pasture and forage crops, but potash also is needed on many soils.

UREA–AMMONIUM PHOSPHATE

Description. Urea–ammonium phosphate is usually in the form of light gray granules about 10-mesh (1.5 mm) in size. It has attractive possibilities as a fertilizer, particularly for rice. As of this writing, it has been manufactured only on a pilot plant basis. Present TVA plans include the completion at Muscle Shoals, Alabama of a urea synthesis reactor and other facilities by 1978 to produce urea–ammonium phosphates (UAP).

Method of Manufacture. The process is reported to involve the reaction of phosphoric acid in a rotary ammoniator with the effluent of a urea reactor. The process consists of two steps. First, urea is made from ammonia and carbon dioxide, but the reaction is incomplete, resulting in an effluent that is a mixture of urea, ammonia, water and carbon dioxide. Second, wet-process phosphoric acid that reacts with the ammonia to form diammonium phosphate is added to the mixture. Granulation in the TVA ammoniator-granulator yields a mixture of urea and diammonium phosphate.

Handling Characteristics. Urea–ammonium phosphates may be used for bulk blending with other granular materials. Distribution with tractor-drawn fertilizer distributors or by truck-mounted distributors are acceptable methods of applying any fertilizer. However, UAP should be disked in, plowed down or soil incorporated by some other method prior to planting. For rice production, UAP can be spread on the muddy seedbed by airplane distributors immediately after the first water is removed. This is usually about 30 to 40 days after planting. If rice is fertilized prior to planting by any nitrogen-phosphorus fertilizer, the material should be plowed down deep to minimize the early stimulation of weeds and grass that compete with the young rice seedlings.

Chemical Characteristics. Urea–ammonium phosphate fertilizers can vary in nitrogen–P_2O_5 ratios from about 1-2 to 2-1. Typical grades are 25-15-0 (25-35-0), 29-13-0 (29-29-0) and 34-8-0 (34-17-0). In the 29-13-0 grade, about 40 percent of the total nitrogen is present as ammonium phosphate, and 60 percent is present as urea. A nitrogen-phosphorus-potassium grade made for high phosphate soils is 27-4-15 (27-9-18).

Crop Response. Since UAP fertilizers consist largely of urea and diammonium phosphate, crop response tends to reflect that of each component. Since both of its components are quite damaging to germinating seeds, placement of UAP with seed should generally be avoided. A second problem is that nitrogen losses as ammonia may be severe. Thus, it is suggested that UAP be incorporated with the soil.

Because rice is such a major food grain in most of the developing nations, research to determine the best fertilizer grades and methods of application should be expanded. Since UAP is high in nutrient content and can be produced from low-cost ingredients in a fairly simple plant, this material offers much promise as a tool to produce food for peace and for the expanding populations.

AMMONIATED SUPERPHOSPHATE

Description. Ammoniated superphosphate is used in fertilizer factories as a powdered gray material with an acid odor. It is sold for addition to the soil as a grayish granular or pulverized material containing nitrogen and phosphorus in varying proportions. The granular material is usually 4- to 8-mesh (4.3 to 2.3 mm) in size.

Method of Manufacture. Aqua or anhydrous ammonia is mixed with ordinary superphosphate, resulting in a chemical reaction. The principal products formed are ammonium sulfate and dicalcium phosphate (Figure 5.5).

$$Ca(H_2PO_4)_2 + CaSO_4 + 2NH_3 \rightarrow 2CaHPO_4 + (NH_4)_2SO_4$$

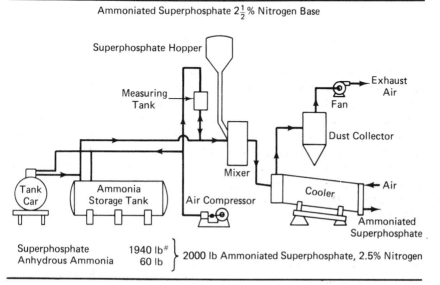

Ammoniated Superphosphate $2\frac{1}{2}$% Nitrogen Base

| Superphosphate | 1940 lb[#] | |
| Anhydrous Ammonia | 60 lb | 2000 lb Ammoniated Superphosphate, 2.5% Nitrogen |

Figure 5.5 Flow diagram of manufacture of ammoniated superphosphate. (pounds ÷ 2.2 = kilograms)

A continuous ammoniator of simple design was developed by the Tennessee Valley Authority and royalty-free licenses were issued beginning in 1954. A considerable portion of ammoniated superphosphate granular fertilizers made in the United States is produced by this process. It is being widely adopted by the foreign fertilizer industry. With it, large proportions of ammonia can be incorporated in superphosphate without excessive losses of ammonia or phosphorus availability in neutral ammonium citrate.[6] The ammoniator is a rotating cylinder of equal diameter and length (3 × 3 ft or 90 × 90 cm) with retaining rings at each end and without flights. The ammoniating medium is introduced through a multiple-outlet slit-type distributor under a rolling bed of solids. Air is blown through the equipment to remove water vapor. The ammoniator has been used suc-

[6]L. D. Yates, F. T. Nielsson, and G. C. Hicks, "TVA Continuous Ammoniator," *Farm Chem.* 117 (7) (1954), 41–46.

cessfully with standard nitrogen solutions and with liquid and gaseous anhydrous ammonia.

In the pilot plant ammoniator, from 6 to 7 pounds (2.7 to 3.2 kg) of neutralizing ammonia per 20 pounds (9.1 kg) of available P_2O_5 (8.7 pounds phosphorus) in the feed is incorporated into a mixture of ordinary superphosphate producing an approximate nitrogen-to-phosphorus ratio of 4:7. When muriate of potash and ammonium nitrate are incorporated, a 6-5-10 (6-12-12) is produced. When concentrated superphosphate is used, 4.5 pounds (2.0 kg) of neutralizing ammonia per 20 pounds (9.1 kg) of P_2O_5 (8.7 pounds phosphorus) is incorporated, producing a product of approximate nitrogen-phosphorus ratio of 8:14. The following reaction occurs:

$$Ca(H_2PO_4)_2 + NH_3 \rightarrow NH_4H_2PO_4 + CaHPO_4$$

The rolling bed in the ammoniator-granulator promotes the formation of granules under favorable conditions of temperature and moisture. When sulfuric acid is introduced under the bed, granular 10-4-8 (10-10-10) fertilizer based on ordinary superphosphate, and granular 10-9-17 (10-20-20) fertilizer based on concentrated superphosphate are made with nitrogen solutions as the only source of nitrogen.

Although the 3×3 feet (90×90 cm) pilot plant ammoniator-granulator can be operated at rates up to only 3 tons (2.7 m tons) per hour, some of the plants that use larger ammoniators, such as 8×16 feet (240×480 cm), produce excellent granular products and effect good ammonia recovery at production rates of 20 tons (18.1 m tons) per hour.

Handling Characteristics. Fertilizer made from ammoniated superphosphate and ammoniated concentrated superphosphate can be granular or pulverant, depending on the process used. The pulverant material can be free flowing and noncaking if a conditioning material, such as finely ground limestone, sand, rice hulls, granite dust, kaolin clay or tobacco stems, is used. Granular goods may be mixed with other granular material.

Granular and semigranular fertilizers made from ordinary pulverant superphosphate, nitrogen solutions and coarse potash can be produced with limited manufacturing facilities. Table 5.3 is a factory work sheet giving raw materials for the manufacture of granular 10-4-8 (10-10-10) with medium-sized plant facilities. Semigranular 4-5-10 (4-12-12) and 10-4-8 (10-10-10) have been produced in plants having limited facilities (Tables 5.4 and 5.5). If kept dry, these materials will not bridge or cake in an ordinary fertilizer drill. They can be broadcast uniformly on the land from truck bodies having sloping sides with auger-type delivery to whirling end-gate spreaders.

Chemical Characteristics. Fertilizer-grade ammoniated ordinary superphosphate that is sold for use in areas having high potassium reserves in the soil contains around 4 percent nitrogen and 16 percent P_2O_5 (7 percent

Table 5.3

MANUFACTURE OF GRANULAR 10-10-10 WITH MEDIUM-SIZED PLANT FACILITIES [1]

Weight (lb)[2]		Raw Material	N	P_2O_3	K_2O
H_2O					
5.8	85	H_2SO_4(93%) 66° B(7% H_2O)	0	0	0
1.5	290	Ammonium sulfate 21% N(0.5% H_2O)	61	0	0
16.0	320	Nitrogen solution[3] 45% N(5% H_2O)	144	0	0
71.8	1025	Superphosphate 19.7% P_2O_5(7% H_2O)	0	202	0
1.5	323	Muriate of potash(coarse) 62% K_2O(0.5% H_2O)	0	0	200
96.6	2043		205	202	200
43.0	43	H_2O Loss—heat of ammoniation			
53.6	2000	Net after steam loss			

[1]Without rotary cooler, but with equipment for handling H_2SO_4 and a conveyor belt long enough that the material will be exposed to the ambient environment for cooling and drying.

[2]Pounds ÷ 2.2 = kilograms.

[3]Containing 5% H_2O, 25% NH_3 and 70% NH_4NO_3. The 224 (70% of 320) pounds of ammonium nitrate makes a liquid muddy phase at 200°F. This is important because it assists in the production of granules. See *Fertilizer Technology and Use*, 1st ed. (Madison, Wisconsin: Soil Science Society of America), p. 356.

phosphorus). Fertilizer-grade ammoniated concentrated superphosphate that is sold in similar areas contains around 8 percent nitrogen and 32 percent P_2O_5 (14 percent phosphorus). About one-half of the phosphorus in both of these materials is water soluble, but all of the phosphorus is citrate soluble and considered available to plants. Both of these fertilizers

Table 5.4

MANUFACTURE OF SEMIGRANULAR 4-12-12 WITH LIMITED MANUFACTURING FACILITIES USING A BASE MATERIAL FORMULA 4.1-12.5-12.5

Weight (lb)[1]		Raw Material	N	P_2O_5	K_2O
H_2O					
9 (5%)		183 Nitrogen Solution (45%)[2]	82	0	0
89 (7%)		1270 Superphosphate (19.7%)	0	250	0
2 (0.5%)		403 Muriate of Potash (coarse) (62%)	0	0	250
3 (0.5%)		164 Dolomitic Limestone (coarse)	0	0	0
103		2020	82	250	250
20		20 H_2O Loss—heat of ammoniation			
83		2000 = One Net Ton With 4.15% H_2O			

[1]Pounds ÷ 2.2 = kilograms.

[2]Containing 5% H_2O, 25% NH_3 and 70% NH_4NO_3 and analyzing 45% nitrogen.

Table 5.5

MANUFACTURE OF SEMI-GRANULAR 10-10-10

		Weight (lb)[1]			
	Raw Material		N	P_2O_5	K_2O
1600	4.1-12.5-12.5		67	200	200
400	Ammonium nitrate (33.5%)		134	0	0
2000			201	200	200

[1]Pounds ÷ 2.2 = kilograms.

contain sulfur and calcium, although the 4-7-0 (4-16-0) contains about twice as much calcium and three to four times as much sulfur as the 8-14-0 (8-32-0).

Fertilizers having concentrations higher than 10-4-8 (10-10-10) or 8-5-10 (8-12-12) cannot be made with an ammoniator-granulator using ordinary superphosphate. Concentrations beyond 10-9-17 (10-20-20) are probably not feasible with this equipment using concentrated superphosphate.

Crop Response. Like ammonium phosphate sulfate, the ammoniated superphosphates are acid forming in the soil. A physiological acidity equivalent to 5 pounds (2.3 kg) of calcium carbonate per pound (0.45 kg) of nitrogen is developed because the nitrogen in both is present largely in the form of $(NH_4)_2SO_4$. In humid regions, acid soils are developed as a result of leaching, erosion and crop removal. Thus, lime must be added to replace the calcium requirement of crops and to neutralize the small amount of acidity developed by the acid-forming nitrogen. Until the end of World War II, ammoniated superphosphate was the principal inorganic source of all multinutrient fertilizers used in the United States. It has been successfully used on every agricultural soil low in fertility and on every crop in need of fertilizer.

NITRIC PHOSPHATE

Description. Nitric phosphate is a granular gray to white material that is usually screened to about 8-mesh (2.3-mm) size.

Method of Manufacture. Interest is stimulated in the acidulation of phosphate rock with nitric acid when the possibility of a sulfur shortage or an expanded capacity for nitric acid exists. This method has been employed in Europe since about 1928. There are probably four or five factories in the United States presently capable of producing nitric phosphates. The basic process consists of treating phosphate rock with nitric acid followed by ammoniation of the resultant mixture. The reactions may be expressed as follows:

$$Ca_{10}(PO_4)_6F_2 + 14HNO_3 + 3H_2O \rightarrow 3Ca(H_2PO_4)_2 \cdot H_2O +$$
$$7Ca(NO_3)_2 + 2HF \uparrow$$

$$Ca(H_2PO_4)_2 + Ca(NO_3)_2 + 2NH_3 \rightarrow 2CaHPO_4 + 2NH_4NO_3$$

A mixture of sulfuric and nitric acids has been employed to treat phosphate rock and this has overcome the disadvantage of the nitric phosphates on soils deficient in sulfur.

$$Ca_{10}(PO_4)_6F_2 + 12HNO_3 + 4H_2SO_4 \rightarrow 6H_3PO_4 + 4CaSO_4 +$$
$$6Ca(NO_3)_2 + 2HF \uparrow$$

The products of the above reaction are ammoniated:

$$6H_3PO_4 + 4CaSO_4 + 6Ca(NO_3)_2 + 2HF + 13NH_3 \rightarrow 5CaHPO_4 +$$
$$NH_4H_2PO_4 + 4CaSO_4 + 12NH_4NO_3 + CaF_2$$

Several modifications of the process have been described by the Tennessee Valley Authority.[7] One such process, a pilot plant development, utilizes an ammoniator-granulator to convert a mixture of phosphate rock, nitric and phosphoric acids, ammonia and potassium chloride to granular, high-nitrogen multinutrient fertilizers (Figure 5.6). The products of this process, when dried, are fertilizers containing dicalcium phosphate, ammonium nitrate, potassium chloride and a small amount of ammonium phosphate. Compared with the other processes this modified nitric phosphate process has several advantages, although the products contain no sulfur. It is more versatile because it permits the production of several high-analysis granular products. The production rate is comparable with the ammoniated superphosphate process but is not nearly as high as that of the monoammonium phosphate process. The proportion of water-soluble phosphorus (ammonium phosphate) in the ammoniated product may be increased by the use of a greater proportion of phosphoric acid. A product made in Germany by a process similar to this is a 12-5-17 (12-12-21) fertilizer.

Handling Characteristics. The fertilizer is granulated, highly concentrated and free flowing. Tests have shown that the product remains uncaked during prolonged storage under normal warehousing conditions when packaged in multiwall paper bags having one asphalt-laminated ply. The savings in labor, freight, bags, handling and spreading resulting from the use of high-analysis goods are inherent in the nitric phosphate fertilizers.

[7]"Introduction and Use of TVA Nitric Phosphates," Tennessee Valley Authority, National Fertilizer Development Center Circular Z–45, 1974.

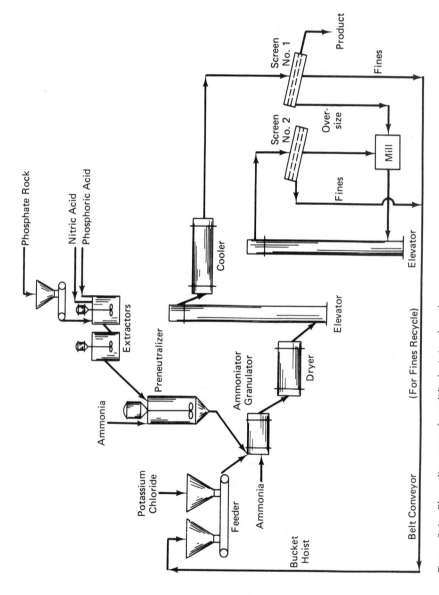

Figure 5.6   Flow diagram of modified nitric phosphate process.

166

Chemical Characteristics. Fertilizer-grade nitric phosphate usually contains 12 to 17 percent nitrogen and 22 to 35 percent P_2O_5 (10 to 15 percent phosphorus). If the proportions of nitric, sulfuric or phosphoric acids as well as ammonia and potassium chloride are varied, high-analysis fertilizers such as 11-5-9 (11-11-11), 12-5-10 (12-12-12) and 14-6-12 (14-14-14) can be made. About 98 percent of the phosphorus in nitric phosphate is soluble in neutral ammonium citrate, and 5 to 40 percent is water soluble.

Crop Response. Experimental tests by scientists in several southeastern states have shown that the response of cotton to nitric phosphates, made with nitric and sulfuric acids, is as good as that to ammoniated superphosphates and ammonium phosphates of similar nitrogen-to-phosphorus ratios. Tests comparing 18 to 78 percent water solubilities of nitric phosphates with corn in India demonstrated different degrees of effectiveness with different soils, but the nitric phosphate containing 18 percent water-soluble phosphorus was inferior to all others at the same level of application in all soils.[8]

COMPARATIVE CROP RESPONSE TO SOURCES OF PHOSPHORUS IN THREE KINDS OF MULTINUTRIENT FERTILIZER

Table 5.6 gives the results of data on cotton response to nitric phosphate, ammoniated superphosphate and ammonium phosphate in high-analysis fertilizers.[9] All of the soils had a pH of less than 5.5, except the Hunt Clay which had a pH of 6.2. The fertilizers were banded about 3 inches (7.5 cm) below the seed.

Based on the increased yield of cotton from a mixture containing concentrated superphosphate as 100, a nitric phosphate with sulfur 11-5-9 (11-11-11) produced the highest yields and was rated 120. The better cotton yield increases obtained with nitric phosphates containing calcium and sulfur indicate that these materials are at least equal to the crop-producing efficiency of ammonium phosphate and ammoniated superphosphate on these acid soils.

Fuller et al.[10] studied the absorption of phosphorus by cotton in the field from two radioactive nitric phosphate fertilizers of about 17 and 44 percent water-soluble phosphorus and an aqueous solution of H_3PO_4

[8]A. K. Rishi and N. N. Goswami, *J. India Soc. Soil Sci.* 22(4) (1974), 365–70.

[9]U. S. Jones, "Phosphate Fertilizers," Mississippi Agricultural Experiment Station Bulletin 503, 1953.

[10]W. H. Fuller, T. C. Tucker, E. W. Carpenter, and J. L. Abbott, "Absorption of Phosphorus by Cotton from Fertilizers at Different Water Solubilities as Related to Stage of Growth at the Time of Application," *Soil Sci. Soc. Amer. Proc.* 27 (1963), 397–401.

Table 5.6
COTTON RESPONSE TO NITRIC PHOSPHATE, AMMONIATED SUPERPHOSPHATE AND AMMONIUM PHOSPHATE IN HIGH-ANALYSIS FERTILIZERS

Analysis and Description of Fertilizers Added at a Rate of 40 lb Nitrogen (18.2 kg Nitrogen), 40 lb P_2O_5 (18.2 kg P_2O_5) and 40 lb K_2O (18.2 kg K_2O).	Yield Increases Over Check of Seed Cotton at:								Relative Efficiency (super = 100)
	Newton 1950–52		Brooksville 1950–52		Holly Springs 1950–52		Raymond 1951		
	lb/acre	kg/ha	lb/acre	kg/ha	lb/acre	kg/ha	lb/acre	kg/ha	
10-0-10 check, no phosphate	407[1]	456	87	97	794	889	824	923	–
9-9-9, same as 11-11-11 40 M[2] with lime	901	1009	544	609	254	284	792	887	121
11-11-11 40 M[2] nitric phosphate with sulfur	935	1047	586	656	222	249	600	672	120
10-10-10 ammoniated superphosphate with sulfur	788	883	527	590	215	241	392	439	103
14-14-14 nitric phosphate without sulfur	774	867	518	580	181	203	488	547	101
11-11-11 10 M[3] nitric phosphate with sulfur	763	855	453	507	181	203	616	690	99
13-13-13 ammonium phosphate with sulfur	720	806	551	617	175	196	440	493	99
15-15-15 nitric phosphate, low H_2O solubility phosphorus, no sulfur	780	874	475	532	127	142	600	672	98
15-15-15 nitric phosphate, high H_2O solubility phosphorus, no sulfur	673	754	568	636	82	92	672	753	96
7-7-7, same as 10-10-10 with lime	737	825	392	439	240	268	576	645	96
Concentrated superphosphate plus 40 lb nitrogen and 40 lb K_2O	854	956	578	647	126	141	184	206	100
L.S.D. 0.05	81	91	174	195	81	91	241	270	

[1]Check yields are total yields per acre. All others are increases over the check.
[2]40 M means ground to pass a 40-mesh (0.5-mm) screen.
[3]10 M means ground to pass a 10-mesh (2-mm) screen.

168

supplemented with NH_4NO_3 at rates of 0, 20 and 60 lb/acre (0, 22.4 and 67.2 kg/ha) of phosphorus. The experiment was on a calcareous soil. The data on the uptake of phosphorus indicated that the phosphorus of the nitric phosphates is absorbed preferably from the early application (4/19), but more phosphorus was absorbed from the wholly soluble H_3PO_4 at the late application (7/19). It appears that the reaction products of the H_3PO_4 with the $CaCO_3$ of the calcareous Arizona soil are more available to the plant than the phosphorus of the nitric phosphates since the more mature plant took proportionately more phosphorus from this source.

Ensminger[11] investigated the influence of water solubility of fertilizer phosphorus on cotton response. He found that ammoniated superphosphate, on the average, increased yields only 85 percent as much as did ordinary superphosphate with an equal amount of nitrogen mixed with it. Wright et al.[12] demonstrated (Figure 5.7) that 100 pounds (45 kg) of phosphorus from ordinary superphosphate could be degraded to an equivalent of 20 pounds (9 kg) of phosphorus by ammoniating the superphosphate with 7 percent ammonia. Both Ensminger and Wright et al. indicated that 100 percent water-soluble phosphorus was excellent quality phosphorus but that 40 to 60 percent water-soluble phosphorus was satisfactory for most crops and soils.

Wright et al.[13] conducted 49 field experiments with corn and wheat

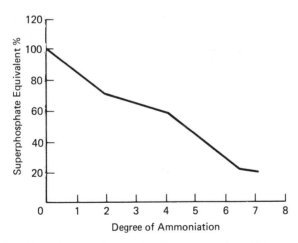

Figure 5.7   The effect of ammoniation of ordinary superphosphate on phosphorus availability to cotton in terms of superphosphate equivalents. (From Wright, Lancaster, and Anthony, 1963.)

[11]L. E. Ensminger, "Need for Water-Soluble Phosphorus in Fertilizer," *Highlights of Agr. Res.* 6 (2) (1959).

[12]B. Wright, J. D. Lancaster, and J. L. Anthony. "Availability of Phosphorus in Ammoniated Ordinary Superphosphate," Mississippi Agricultural Experiment Station Technical Bulletin 52, 1963.

[13]Wright, Lancaster, and Anthony, "Availability of Phosphorus."

for grain and forage as well as tests with cotton to study the influence of the ammoniation of superphosphate on the availability of phosphorus. The data showed that there was a good response to added phosphorus for each crop (this is necessary to detect differences) and that the ammoniation of ordinary superphosphate reduces the effectiveness of the phosphate with every crop tested and with each increment of ammoniation.

Ensminger's work,[14] in addition to that reported for cotton, included studies with wheat forage as the test crop to measure the influence of ammoniation of superphosphate, which reduces phosphorus solubility. Results showed that yields of wheat forage increased with increasing phosphorus solubility at the first clipping. The effect was less pronounced for the second clipping. Grain yields were not affected.

Very precise studies of the importance of water solubility of phosphates were reported by Dr. John Pesek of Iowa State University Agronomy Department.[15] In 20 experiments over a period of four years, mixed (multinutrient) fertilizers with higher percentages of water-soluble phosphorus were superior as hill- or row-applied starters for corn (Table 5.7) and for short-season crops such as oats. Oats were also used as a test crop in Mississippi to measure differences in multinutrient fertilizers with results similar to those in Iowa (Table 5.8).

Vegetables have been the test crop in a number of experiments with different multinutrient fertilizers in Washington, Maine and Michigan. On

Table 5.7

FERTILIZER PHOSPHATES APPLIED IN HILL FOR CORN IN HOWARD COUNTY, IOWA ON FLOYD SILT LOAM (pH 6.2)[1]

N		P_2O_5		K_2O		P_2O_5	$CORN\ YIELD$	
						Water soluble %		
lb/acre	kg/ha	lb/acre	kg/ha	lb/acre	kg/ha		bu/acre	kg/ha
30	33.6	0	0	30	33.6		75.8	5,094
30	33.6	30	33.6	30	33.6	2	79.6	5,349
30	33.6	30	33.6	30	33.6	14	83.1	5,584
30	33.6	30	33.6	30	33.6	40	86.6	5,819
30	33.6	30	33.6	30	33.6	90	88.5	5,947
30	33.6	30	33.6	30	33.6	100	89.3	6,001
L.S.D. .05							2.0	134

[1]Adapted from J. R. Webb and J. T. Pesek, "An Evaluation of Phosphorus Fertilizers Varying in Water Solubility: I. Hill Application for Corn," *Soil Sci. Soc. Am. Proc.* 22 (1958), 533–538.

[14]Ensminger, "Need for Water-Soluble Phosphorus."

[15]John Pesek, "Water-Soluble Phosphate For Corn," *Iowa Farm Sci.*, October 1955, 18–82; and "Significance of Water-Soluble Phosphatic Fertilizers," *Crop Life*, Jan. 31, 1955, 64–68.

Table 5.8

OAT FORAGE YIELDS AT BROOKSVILLE, MISSISSIPPI (HUNT CLAY, pH = 6.8) WITH THREE
MULTINUTRIENT FERTILIZERS MADE BY THREE DIFFERENT METHODS OF MANUFACTURE
RESULTING IN DIFFERENT LEVELS OF WATER SOLUBILITY OF PHOSPHORUS.[1]

Fertilizer Added to Give 60 lb Nitrogen, 60 lb P_2O_5 and 60 lb K_2O (67.2 kg/ha of Nitrogen, P_2O_5 and K_2O).	Oat–Hay Yield	Water-Soluble P_2O_5	
	lb/acre	kg/ha	%
460 lb/acre (515.2 kg/ha) 13-13-13 ammonium phosphate	5851	6653	95
545 lb/acre (610.4 kg/ha) 11-11-11 nitric phosphate	4973	5570	18
600 lb/acre (672.0 kg/ha) 10-10-10 ammoniated superphosphate	4381	4906	20
600 lb/acre (672.0 kg/ha) 10-0-10 check—no phosphate	1401	1569	
L.S.D. .05	588	659	

[1]Adapted from U. S. Jones, Mississippi Agricultural Experiment Station Bulletin 503.

western Washington acid soils, vegetable crops, in order of decreasing response to band-applied fertilizers having a high proportion of their phosphorus in water-soluble form were cucumbers, pole beans, potatoes and sweet corn. At a given rate of band-applied phosphorus fertilizer, crop yields decreased with decrease in content of water-soluble phosphorus. Response decreased in the following order: ammonium phosphates, superphosphates, ammoniated superphosphate and nitric phosphate medium to high in water solubility, ammoniated superphosphate and nitric phosphate low in water solubility, dicalcium phosphate, calcium metaphosphate, fused tricalcium phosphate and phosphate rock.[16] Yields of crops tended to be higher with increasing amounts of sources high in water-soluble phosphorus.

Contents of phosphorus in the leaves during the early growth period were closely correlated with water-soluble phosphorus contents of the fertilizers and with final vegetable crop yields. These findings for vegetable crops by Mortensen et al. coincide with similar unpublished results on phosphorus content of cotton leaves obtained by Dr. Glen Hardy at the University of Arkansas. On central Maine acid soils, results indicated a general increase in early growth of snap beans and sweet corn with an increase from 0 to 87 percent in water solubility of applied phosphorus. This was reflected in better root growth and larger leaf area (Figure 5.3). Percentage of phosphorus in the plants and yields were not consistently affected by water solubility of the fertilizers that were applied in row sidebands to supply 22 pounds (11 kg) of phosphorus (50 pounds or 22.7 kg of P_2O_5) per acre (0.4 ha). In general, however, results indicated a need for water-soluble phosphorus, especially on soils testing low in this nutrient.[17]

[16]Mortensen, Baker, and Terman, "Source and Rate of Application."

[17]D. T. Lewis, "Effectiveness for Snap Beans of Phosphate Fertilizers Varying in Percent of Water-Soluble P_2O_5," M.S. Thesis (Orono, Maine, University of Maine, June 1962).

On an acid, Fox sandy loam in Michigan, results showed a general increase in phosphorus content of field peas and corn as the percent phosphorus soluble in water increased. The fertilizer was pulverant and banded, and the plant phosphorus content (percent) was determined after 2 month's growth.[18]

The results of many phosphorus fertilizer experiments in 15 western United States showed that ordinary and concentrated superphosphate, ammonium phosphate and liquid phosphoric acid were more effective than calcium metaphosphate, which was more effective than dicalcium phosphate. Tricalcium phosphate and rock phosphate were the least effective of all the sources tested.[19] In the southeastern states similar results were obtained by Jones[20] in Mississippi, Ensminger[21] in Alabama and Long[22] in Tennessee.

Data reported by Ensminger[23] on the average water solubility of 1,272 fertilizer samples in 1964 from 10 states in the southeastern United States showed that 20 percent of the samples collected did not contain as much as 40 percent of their phosphorus in water-soluble form. The data also showed that in Texas and Oklahoma, where most of the fertilizer sold is ammonium phosphate based, most all of the fertilizer had a high water solubility. In contrast, it was evident that the fertilizers sold in Kentucky, Tennessee and Virginia were somewhat low in solubility.

BULK-BLENDED FERTILIZER

Description. Bulk blending is physically mixing solid fertilizer materials into multinutrient mixtures in contrast to the homogeneous mixing of two or more nutrient carriers in a slurry and then drying them so that an individual particle contains the two or more nutrients. For example, a 10-9-25 (10-20-30) fertilizer may be mixed in a slurry of nitrogen solutions,

[18]R. L. Cook, K. Lawton, L. S. Robertson, and C. M. Hansen, "Phosphorus Solubility, Particle Size and Placement, as Related to the Uptake of Fertilizer Phosphorus and Crop Yield," *Proc. 32nd National Joint Committee on Fertilizer Application* (1956), pp. 49–61.

[19]H. B. Peterson, L. B. Nelson, and J. L. Pashal, "A Review of Phosphate Fertilizer Investigations in 15 Western States Through 1949," United States Department of Agriculture Circular 927, 1953.

[20]U. S. Jones, "Phosphate Fertilizers."

[21]L. E. Ensminger, "Response of Crops to Various Phosphate Fertilizers," Alabama Agricultural Experiment Station Bulletin 270, 1950.

[22]O. H. Long and B. P. Hazlewood, "Fertilizer Experiments with Cotton," Tennessee Agricultural Experiment Station Bulletin 220, 1951.

[23]L. E. Ensminger, "Water-soluble Phosphorus Content of Fertilizer Collected in Ten Southeastern States in 1964," *Commer. Fert. Plant Food Ind.*, October 1966.

triple superphosphate and muriate of potash; the slurry is then granulated and dried to produce granules, each one of which contains some nitrogen, superphosphate and muriate of potash. The bulk blend is a physical mixing of, for example, ammonium nitrate pellets, granular superphosphate and coarse muriate of potash. Each individual particle of this blend contains only one of the ingredients of the mixture.

Bulk blending was started in 1947, the same year that anhydrous ammonia use was suggested for direct application to soil. Woodford County, Illinois was the locale of the first bulk-blend operation. It rapidly developed into a new system of handling, marketing and applying fertilizers in the United States. It appears to have considerable potential in developing countries that are unable to justify their own manufacturing facilities yet need some flexibility in providing the plant nutrients required for crop production.

Considerations Regarding Location. Fertilizer bulk-blending plants provide a convenient and economical means of mixing dry fertilizer materials to produce specified ratios and grades of varying nutrient percentages. See Table 5.9 for the number of pounds of raw material carriers needed for the desired percent of nutrient in the multinutrient fertilizer. A principal virtue of blending is the ability of the blender to produce an unlimited number of ratios and grades to suit individual needs—frequently determined by soil testing. Blending plants usually serve a marketing area of 12 to 15 miles (19.3 to 24.1 km) or less. Often the bulk-blending enterprise is incorporated within a firm having other farm supply and/or farm product buying enterprises to provide a well-rounded business serving farmers' needs. The location of the blending plant is an important determinant of its success in the area that it is supposed to serve.

Manufacture and Handling. Solid fertilizer materials such as triple super-phosphate, diammonium phosphate, ammonium nitrate, potassium chloride and urea are used in bulk blending.[24] The processing steps involved are (Figures 5.8 and 5.9):

1. Unloading raw materials into storage.
2. Transporting raw materials to weighing device.
3. Mixing or blending raw materials.
   a. In selecting raw materials for a bulk blend, you should take care to select materials that do not react with one another. For example, it would not be advisable to blend urea with triple superphosphate since these materials react to release water that causes wetness and bridging in the fertilizer applicator, creating difficulties in applying the product. Urea and ammonium nitrate

[24]E. A. Harre, "Bulk Blending Materials," *Fert. Prog.* 4 (1973), 18, 22.

Table 5.9
DESIRED PERCENT OF CONSTITUENT IN COMPLETE GOODS

NUMBER OF POUNDS OF RAW MATERIAL FOR DESIRED PERCENT

Percent in raw Materials	½	1	1½	2	2½	3	3½	4	4½	5	5½	6
1	1000	2000										
2	500	1000	1500	2000								
3	333	667	1000	1333	1667	2000						
4	250	500	750	1000	1250	1500	1750	2000				
5	200	400	600	800	1000	1200	1400	1600	1800	2000		
6	167	333	500	667	833	1000	1167	1333	1500	1667	1833	2000
7	143	286	429	571	714	857	1000	1143	1286	1429	1571	1714
8	125	250	375	500	625	750	875	1000	1125	1250	1375	1500
9	111	222	333	444	556	667	778	889	1000	1111	1222	1333
10	100	200	300	400	500	600	700	800	900	1000	1100	1200
11	91	182	273	364	455	545	636	727	818	909	1000	1091
12	83	167	250	333	417	500	583	667	750	833	917	1000
13	77	154	231	308	385	462	538	615	692	769	846	923
14	71	143	214	286	357	429	500	571	643	714	786	857
15	67	133	200	267	333	400	467	533	600	667	733	800
16	63	125	188	250	313	375	438	500	563	625	688	750
17	59	118	176	235	294	353	412	471	529	588	647	706
18	56	111	167	222	278	333	389	444	500	556	611	667
19	53	105	158	211	263	316	368	421	474	526	579	632
20	50	100	150	200	250	300	350	400	450	500	550	600
21	48	95	143	190	238	286	333	381	429	476	524	571
22	45	91	136	182	227	273	318	364	409	455	500	545
23	43	87	130	174	217	261	304	348	391	435	478	522
24	41	83	125	167	208	250	292	333	375	417	458	500
25	40	80	120	160	200	240	280	320	360	400	440	480
26	38	77	115	154	192	231	269	308	346	385	423	462
27	37	74	111	148	185	222	259	296	333	370	407	444
28	36	71	107	143	179	214	250	286	321	357	393	429
29	34	69	103	138	172	207	241	276	310	345	379	414
30	33	67	100	133	167	200	233	267	300	333	367	400
31	32	65	97	129	161	194	226	258	290	323	355	387
32	31	63	94	125	156	188	219	250	281	313	344	375

NUMBER OF POUNDS OF RAW MATERIAL FOR DESIRED PERCENT

Percent in raw materials	7	8	9	10	11	12	13	14	15	16	17	18
33	30	61	91	121	152	182	212	242	273	303	333	364
34	29	59	88	118	147	176	206	235	265	294	324	353
35	29	57	86	114	143	171	200	229	257	286	314	343
36	28	56	83	111	139	167	194	222	250	278	306	333
37	27	54	81	108	135	162	189	216	243	270	297	324
38	26	53	79	105	132	158	184	211	237	263	289	316
39	26	51	77	103	128	154	179	205	231	256	282	308
40	25	50	75	100	125	150	175	200	225	250	275	300
41	24	49	73	98	122	146	171	195	220	244	268	293
42	24	48	71	95	119	143	167	190	214	238	262	286
43	23	47	70	93	116	140	163	186	209	233	256	279
44	23	45	68	91	114	136	159	182	205	227	250	273
45	22	44	66	89	111	133	156	178	200	222	244	267
46	22	43	65	87	109	130	152	174	196	217	239	261
47	21	43	64	85	106	128	149	170	191	213	234	255
48	21	42	62	83	104	125	146	167	187	208	229	250
49	20	41	61	82	102	122	143	163	184	204	224	245
50	20	40	60	80	100	120	140	160	180	200	220	240
1												
2												
3												
4												
5												
6												
7	2000											
8	1750	2000										
9	1556	1778	2000									
10	1400	1600	1800	2000								
11	1273	1455	1636	1818	2000							
12	1167	1333	1500	1667	1833	2000						
13	1077	1231	1385	1538	1692	1846	2000					
14	1000	1143	1286	1429	1571	1714	1857	2000				
15	933	1067	1200	1333	1467	1600	1733	1867	2000			
16	875	1000	1125	1250	1375	1500	1625	1750	1875	2000		
17	824	941	1059	1176	1294	1412	1529	1647	1765	1882	2000	
18	778	889	1000	1111	1222	1333	1444	1555	1667	1778	1889	2000

Table 5.9 (continued)

Percent in raw materials	7	8	9	10	11	12	13	14	15	16	17	18
19	737	842	947	1053	1158	1263	1368	1474	1579	1684	1789	1895
20	700	800	900	1000	1100	1200	1300	1400	1500	1600	1700	1800
21	667	762	857	952	1048	1143	1238	1333	1429	1524	1619	1714
22	636	727	818	909	1000	1091	1182	1273	1364	1455	1545	1636
23	609	696	783	870	957	1043	1130	1217	1304	1391	1478	1565
24	583	667	750	833	917	1000	1083	1167	1250	1333	1417	1500
25	560	640	720	800	880	960	1040	1120	1200	1280	1360	1420
26	538	615	692	769	846	923	1000	1077	1154	1231	1308	1385
27	519	593	667	741	815	889	963	1037	1111	1185	1259	1333
28	500	571	643	714	786	857	929	1000	1071	1143	1214	1286
29	483	552	621	690	759	828	897	966	1034	1103	1172	1241
30	467	533	600	667	733	800	867	933	1000	1067	1133	1200
31	452	516	581	645	710	774	839	903	968	1032	1097	1161
32	438	500	563	625	688	750	813	875	938	1000	1063	1125
33	424	485	545	606	667	727	788	848	909	970	1030	1090
34	412	471	529	588	647	706	765	824	882	941	1000	1059
35	400	457	514	571	629	686	743	800	857	914	971	1029
36	389	444	500	556	611	667	722	778	833	889	944	1000
37	378	432	486	541	595	649	703	757	811	865	919	973
38	368	421	474	526	579	632	684	737	789	842	895	947
39	359	410	462	513	564	615	667	718	769	821	872	923
40	350	400	450	500	550	600	650	700	750	800	850	900
41	341	390	439	488	537	585	634	683	732	780	829	878
42	333	381	429	476	524	571	619	667	714	762	810	857
43	326	372	419	465	512	558	605	651	698	744	791	837
44	318	364	409	455	500	545	591	636	682	727	773	818
45	311	356	400	444	489	533	578	622	667	711	756	800
46	304	348	391	435	478	522	565	609	652	696	739	783
47	298	340	383	426	468	511	553	596	638	681	723	766
48	292	333	375	417	458	500	542	583	625	667	708	750
49	286	327	367	408	449	490	531	571	612	653	694	735
50	280	320	360	400	440	480	520	560	600	640	680	720

¹Pounds ÷ 2.2 = kilograms.

materials should not be mixed. Urea, in combination with nitric phosphate, will form a very hygroscopic (water-attracting) mixture that will become too damp in the application equipment to be applied satisfactorily. Care should be taken in the use of diammonium phosphate together with superphosphate since diammonium phosphate will react with triple superphosphate. The diammonium phosphate ammoniates the monocalcium phosphate of the triple superphosphate to form dicalcium phosphate and monoammonium phosphate with the release of some water. Since this reaction is slow, no difficulty should arise with the use of diammonium phosphate in combination with triple superphosphate if materials are applied soon after they are mixed.

   b. Uniform mixing in a well-designed mixer.
   c. A good blend can be handled in conveyors, hoppers and bagging machines without becoming segregated.
4. Either bagging the blended materials or feeding them to a bulk-spreading truck.
5. Transporting the blended product to the field.

Advantages of Bulk Blending.   The reasons why blending is attractive are:

1. Economy. The finished raw materials of nitrogen, phosphate and potassium fertilizers can be produced in large economic plants in different areas and combined in the market area.
2. Versatility. Practically any grade or ratio can be produced.
3. Convenience. Bulk-spreader truck can carry freshly blended material directly to field. No bagging or storage of fertilizers is necessary. No farmer labor is required.

Disadvantages of Bulk Blending.   The pitfalls in blending are:

1. Materials must have approximately the same particle size. Segregation can be prevented by closely matching the particle size of each raw material used in the blend. Differences in shape or density have only a very small effect on the tendency of a blend to segregate.
2. Materials must be chemically compatible. Do not blend urea and ammonium nitrate. Do not blend urea and triple or ordinary superphosphate.

Product Use and Placement.   Bulk-blended fertilizers are used for the same crops and with the same results as homogeneous mixtures of the same nutrient content. They can be, and are, applied with equipment that places them precisely in the ground, to the side and below the seed. On the other

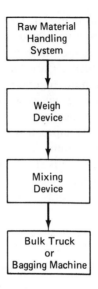

Figure 5.8   Materials flow diagram for bulk blending plant.

hand, by far the largest proportion of bulk blends in the United States are applied broadcast on the surface of the soil and then worked in by plowing, disking or other methods. For the broadcast method of application, consideration must be given to supplementing the fertilizers with starter or pop-up fertilizers at planting and side-dressing or top-dressing fertilizers later in the growing season.

Starter and Pop-up Fertilizers.   These are added with a fertilizer attachment on the planter and are used to supplement a regular fertilizing program or a very fertile soil by providing a quick start or pop-up of the seedling. They usually contain water-soluble ingredients high in phosphorus. They stimulate early shoot and root growth that permits rapid development of the root system so that it can contact more soil and quickly utilize the fertile soil or broadcasted fertilizer that develops the crop to an early, high-yielding maturity.

Starter solutions may be added to transplanting water in order to reduce loss of plants, to stimulate early growth and to supplement the regular fertilizing program.

Side-dressing and Top-dressing Fertilizers.   Side-dressing is defined as placing the fertilizer to the side, about 6 inches (15 cm) from the stalk, vine or trunk of the growing plant. This is done with a fertilizer hopper or dispenser. Liquid or solid fertilizer may be used. For example, nitrogen may be side-dressed to corn when it is about knee high (Figure 2.10 on page 57).

Top-dressing is defined as spreading the fertilizer over the top of plants usually at the period coinciding with the greatest need of the

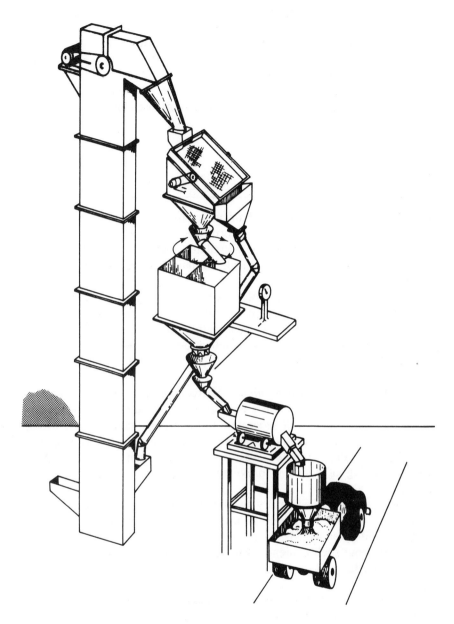

Figure 5.9 Plant with rotary mixer at an elevated position.

growing plant. For example, wheat and rice are top-dressed with nitrogen just prior to the boot stage when the developing seed in the stalk can be felt with the fingers; this is the time of greatest need for nitrogen. When wheat land is too wet for ground equipment and when rice is flooded, aerial application of fertilizer is commonly employed to supply nitrogen when it is needed.

Side-dressing and top-dressing of crops other than rice are most widely used in areas of high rainfall and coarse sandy soils. On loamy or heavier soils, annual crops may be efficiently fertilized by adding all the nutrients prior to planting.

Fertilizing rice is often postponed until after planting and emergence in order to minimize the weed problem, which imposes more severe strains on fertilized than on unfertilized crops.

BULK DISTRIBUTION OF FLUID AND DRY FERTILIZER

Bulk distribution refers to mechanical handling and application of dry fertilizer materials in nonpackaged form, i.e., in large lots. Dry fertilizer, in this context, usually refers to granulated materials, both bulk blends and homogeneous mixtures. Practically all fluid fertilizer is applied by bulk distribution.

The importance of this subject is that handling and spreading fertilizers in bulk introduces the possibility of nonuniform application of nutrients. It is the purpose of this section to examine the causes and agronomic effects of nonuniform application that might arise from this method of fertilizer distribution.

Causes of Nonuniform Distribution. Segregation of materials and poor spreading techniques are the two causes of nonuniform distribution. Frequently bulk blends will separate into the original components through handling. Research has shown that segregation can be minimized by carefully matching granule sizes. For fluid suspensions, segregation through settling of solids is prevented by agitation before and during application.

The spreading of either dry or fluid fertilizers calls for skill in handling equipment. If poor technique is used, uneven distribution of fertilizer nutrients will often be visible in the crop. The farmer is usually quite concerned about such evidence. The effects of poor spreading techniques apply equally to solid, liquid or suspension forms.

Agronomic Effect of Nonuniform Distribution. Segregation of particles results in nonuniform nutrient mix in fertilizer, with a variation in the nutrient ratio. While this creates problems for quality control inspection and analysis, it would have to be quite serious before causing noticeable agronomic effects or losses in the farmer's profits. This is primarily a problem with blended solids, although it can also occur with suspensions if proper agitation is not obtained.

The agronomic importance of nonuniform spreading depends largely on the magnitude of the deviation from the average rate, the shape of the response curve or surface, the rate of nutrients added and the level of soil fertility. With all these variables to consider, it is difficult to predict the yield or profit effects of nonuniform application in advance. Studies indicate that wide deviations can be tolerated without serious yield or profit losses. This would not be true for sites of low nutrient levels or when average rates of application are low.

FLUID MULTINUTRIENT FERTILIZERS

Fluid mixed fertilizers, as a matter of definition, are those fertilizers that can be pumped and handled as liquids.[25] Single nutrient fluids, such as nitrogen solutions, have been considered in an earlier outline. There are three main categories of fluid fertilizers.

1. Clear liquids. These are true solutions and are limited to rather low analyses since salting out will occur at low temperatures if fertilizer grades are higher than 8-8-8, 5-10-10 or similar grades. See the section entitled ''Ammonium Polyphosphate—Liquid Fertilizer.''
2. Suspension fertilizers. These are mixtures of liquids and finely divided solids in which the solids do not settle rapidly and can be redispersed readily by agitating to give a uniform mixture. Certain types of clay are usually added as suspending agents.
3. Slurry fertilizers. These are mixtures of liquids and finely divided solids in which the solids settle rapidly in the absence of agitation and form a firm layer in the bottom of the tank. This layer is difficult to resuspend.

FLUID AND DRY FERTILIZERS COMPARED

From a crop production point of view, there should be no difference between fluid and dry fertilizers when the same source and amounts of nutrients are used and the fertilizer is applied in the same manner to the soil.

Major Plant Nutrients—Nitrogen, Phosphorus and Potassium. Muriate of potash (KCl) is the most common source of potassium in all fertilizers. However, some dry fertilizer, suspension and slurry fertilizers may contain potassium in the form of sulfate of potash, sulfate of potash-magnesia or potassium nitrate. Muriate of potash is almost always the potassium source of the clear liquid fertilizers now on the market.

[25]H. V. Rogers, ''Fluid Fertilizers,'' Clemson University Extension Circular 498, 1968.

The nitrogen in fluid fertilizers is usually in the form of ammonium nitrate or urea, along with diammonium phosphate and monoammonium phosphate. The nitrogen in dry fertilizer can come from many sources. Most of the granular fertilizers have as much of the nitrogen in the ammoniacal form as possible because it is the least expensive source. Other sources are ammonium nitrate, sodium nitrate, ammonium sulfate, monoammonium phosphate, diammonium phosphate and urea.

Except for dry fertilizers containing nitrogen in a slowly soluble form, there is not much difference in crop response between liquid and solid nitrogen sources in fertilizers. Contrary to public opinion, organic nitrogen sources seldom rank more than 75 percent as effective as inorganic sources, such as ammonium nitrate, in long-term tests with field or vegetable crops.

The main difference between fluid and dry fertilizers is related to the solubility of the phosphorus. Practically all potassium and inorganic nitrogen in both fluid and dry fertilizer is 100 percent water soluble. Obviously, the phosphorus in clear liquid fertilizer is 100 percent water soluble. In dry fertilizer, the phosphorus sources may vary in water solubility from 5 percent in nitric phosphates and about 30 percent in some highly ammoniated superphosphates to almost 100 percent in diammonium or monoammonium phosphates (Table 5.1). However, both the water-soluble and the citrate-soluble phosphates are considered available to the plant. In suspension and slurry fertilizers, at least part of the phosphorus is more likely to be in the water-soluble form, but some water-insoluble phosphorus sources can be used.

Much research concerning phosphorus has been reported. Scientific results substantiate the fact that water solubility is a desirable feature.

Secondary Plant Nutrients—Calcium, Magnesium and Sulfur. Until 1967, the clear liquid fertilizer sold contained very little calcium, magnesium or sulfur. It is possible to put a small amount of magnesium and sulfur in the solution, but the sources that will remain in the solution are relatively expensive. On acid soils, a well-planned liming program can supply adequate calcium and magnesium to crops, but it may be necessary to add sulfur to clear liquid fertilizers to avoid sulfur deficiencies of field and vegetable crops grown on sandy soils.

Soils vary in sulfur content. Some have adequate amounts to last for several years, and some have lesser amounts, especially in the topsoil. Some sulfur is brought down in rainfall. However, where high rates of fertilizer are added on sandy soil and where there is heavy crop removal, some additional sulfur may be necessary.

Suspension and slurry fertilizers may contain sulfur, magnesium and calcium since these elements from rather inexpensive sources can be added to the mix. Dry fertilizers can vary considerably in the amount of calcium, magnesium and sulfur they contain, depending on the analysis and the source of nutrients. Some dry fertilizers, especially some of the granular blends made from diammonium phosphate, may not contain any more secondary nutrients than liquid fertilizers.

Micronutrients. When the essential metal micronutrients zinc, iron, manganese and copper are added to clear liquid fertilizers, precipitation often occurs as a reaction with phosphates.[26] (See Table 8.20 for comparison in solubility of some micronutrient compounds in clear liquid fertilizers.) Chelates of the metal micronutrients that can be mixed with liquids without causing precipitation are available; however, they cost more than the metal micronutrient salts that are not chelated. Boron in a soluble source can be added to liquid fertilizer very inexpensively.

Dry fertilizers may or may not contain micronutrients. It is, however, possible to add them to all dry fertilizers. As in the clear liquid fertilizers, some of the micronutrients may react with the phosphates and end up in less available forms.

Suspension or slurry fertilizers may contain the same source of micronutrients as those found in dry fertilizers. Even if they are precipitated (caused to separate from the solution or suspension), they still remain in suspension and can be distributed throughout the mix.

Advantages of Clear Liquid Fertilizers

1. Labor saving since one man can operate a farmer-owned or rented tank or the fertilizer can be custom applied.
2. No segregation of particles.
3. Water-soluble phosphate.
4. Herbicides and insecticides can be blended.

Disadvantages of Clear Liquid Fertilizers

1. A farmer cannot count the bags; he must trust the operator and the meter.
2. Some solutions may salt out at lower temperatures.
3. Because of the danger of salting out, high-analysis grades cannot be mixed. Thus, high rates must be used to supply needed plant food.
4. Inexpensive sources of secondary nutrients must come from a source other than mixed fertilizer.

[26]J. J. Montvedt, "Agronomic Response to Micronutrient in Fluid Fertilizers," *Proc. Nat. Fert. Solns. Assoc.*, (1973) 23–25.

Advantages of Suspension or Slurry Fertilizers

1. Labor saving in the same respect as clear liquid fertilizers.
2. No segregation of particles if well agitated.
3. Secondary nutrients and micronutrients from relatively inexpensive sources can be mixed with the fertilizer.
4. Herbicides and insecticides can be blended satisfactorily.
5. Since salting out is no problem, high-analysis grades can be formulated.

Disadvantages of Suspension or Slurry Fertilizers

1. Suspension and especially slurry fertilizers are not as easy to handle as clear liquids. Agitation is essential and pumps and nozzles must be adapted to these more viscous (thick) fluids. There are still some problems that have not been satisfactorily solved.
2. Storability is not as good as for most liquids. Slurries can settle and cause serious problems in storage.

No mention has been made of comparative prices, distance to the fertilizer plant, availability of nurse tanks and other factors that the individual farmer will have to consider in choosing between a fluid and a dry fertilizer.

SUMMARY

1. The four principal raw materials used in fertilizer manufacture are rock phosphate, sulfuric acid, ammonia and potassium chloride.
2. The three principal types of multinutrient fertilizers are nitric phosphate, ammoniated superphosphate and ammonium phosphate.
3. Three techniques used in fertilizer manufacture are acidulation, treating a material with acid; ammoniation, introducing NH_3 into superphosphate; and granulation, collecting small-size particles into large granules or pellets.
4. Three ways to produce multinutrient fertilizers are granulation of homogeneous mixtures, bulk blending of granular ingredients and mixing fluid ingredients.
5. The way a fertilizer is made affects its quality and its availability to plants.
6. Fertilizer is granulated only to improve on its handling characteristics and not to stop up farmers' distributors.
7. Pulverulent and powdered mixtures have been replaced by granular and fluid fertilizers.

8. In the United States, 15 percent of multinutrient fertilizers are fluid; 40 percent of all fertilizers (including single nutrient ones) are fluid, and 60 percent are solid.

9. Nearly all nitrogen and potassium compounds found in multinutrient fertilizers have high water solubility; all phosphorus compounds do not.

10. The phosphorus in granulated, homogeneous mixtures is usually less soluble in water than phosphorus in fluid or bulk-blended mixtures.

11. Comparative crop response to phosphorus fertilizers of varying solubilities show better yields for water-soluble forms.

12. Bulk-blended fertilizer has many advantages but creates the possibility of particle segregation resulting in nonuniform application of nutrients.

13. The agronomic effects of nonuniform application as a result of either segregation or poor spreading techniques would be important primarily with a combination of low soil nutrient levels and low rates of application.

14. Fluid fertilizers are applied either broadcast or banded, as desired.

15. Research shows that the physical form of the nutrient, dry or fluid, has no measurable effect on its agronomic properties.

16. Fluids are used where relatively low-analysis fertilizers such as 9-4-7 (9-9-9) and lower are profitable and where farmers are seeking to reduce labor requirements to a minimum and are concerned with uniformity of fertilizer application.

Problem: Given a soil recommendation of 80 pounds (36.4 kg) of nitrogen, 40 pounds (18.2 kg) of P_2O_5 and 40 pounds (18.2 kg) of K_2O per acre (0.4 ha), which of the following fertilizers and how many pounds per acre (kg/ha) would you apply? 10-20-0; 20-10-10; 10-10-10. Express these grades on a nitrogen-phosphorus-potassium basis.

Problem: Use Table 5.9 to compute the amounts of materials necessary for a ton (0.91 m ton) of 4-12-12 using the following carriers:

	lb of material	kg of material
Ammonium sulfate, 21% nitrogen	381	173.2
Superphosphate, 20% P_2O_5	1200	545.5
Muriate of potash, 60% K_2O	400	181.8
Filler	19	8.6
	2000	909.1

Explanation: Go into Table 5.9 under "Percent in Raw Material" in left column at 21, read right until you reach 4 under "Desired Percent of Constituent"; this shows 381 pounds (173.2 kg) of 21 percent nitrogen are required to give to a ton

(0.91 m ton) of fertilizer 4 percent nitrogen. Go into Table 5.9 under "Percent in Raw Material" in left column at 20; read right until you reach 12 under "Desired Percent of Constituent"; this shows 1200 pounds (545.5 kg) of 20 percent P_2O_5 is required to give to a ton (0.91 m ton) of fertilizer, 12 percent P_2O_5. Repeat for 60 percent (30 percent × 2) muriate of potash. Go into left column under "Percent in Raw Material" until you reach 30, read right until you reach 12 under "Desired Percent of Constituent"; this shows 800 pounds (363.6 kg) of 30 percent K_2O are required to give to a ton (0.91 m ton) of fertilizer, 12 percent K_2O. Since the material contains 60 percent K_2O, it will require one-half as much (800 ÷ 2) or 400 pounds (181.8 kg) of material to give the 12 percent in 4-12-12.

Problem: Compute the amount of materials necessary for a ton (0.91 m ton) of 4-12-12 using the following carriers:

Ammonium sulfate, 21% nitrogen
Superphosphate, 46% P_2O_5
Muriate of potash, 60% K_2O
Filler

QUESTIONS

1. The four principal raw materials used to manufacture multinutrient fertilizers are:
   a.

   b.

   c.

   d.

2. The three principal techniques used in manufacture of multinutrient fertilizers are:
   a.

   b.

   c.

3. The five principal phosphate compounds found in fertilizers are:
   a.

   b.

   c.

   d.

   e.

4. Practically all multinutrient fertilizers are made with one of three different kinds of nitrogen-phosphorus materials. These are:
   a.

   b.

   c.

5. In multinutrient fertilizers, the potassium and nitrogen compounds are usually completely water _____,but several of the phosphorus compounds are _____in water.

6. Phosphorus compounds in fertilizers can be classified conveniently into three groups on the basis of solubility. These classifications are:

   a.

   b.

   c.

True or False:

7. _____Ammoniated concentrated superphosphate and diammonium phosphate fertilizer made with rock phosphate, ammonia and wet-process phosphoric acid contain about 1 to 2 percent sulfur.

8. _____The U.S. official chemical test for available potassium in fertilizers is based on its solubility in water.

9. _____The U.S. official chemical test for available phosphorus in fertilizers is not based on its solubility in water.

chapter 6

Potassium—
The Catalyst

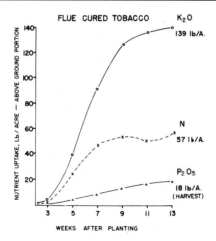

Tobacco (*Nicotiana spp*) is not a food but is a member of the family of nightshade plants that include Nicotiana and various species of Solanum such as egg plant and potato. Note that the uptake of potassium is greater than that of nitrogen or phosphorus and that the potassium demand is greater during the seventh to eighth week after planting. Lb/A × 1.12 = kg/ha. (Raper, C.D., Jr., and C. B. McCants. Tob Sci. 10:109, 1966.)

Potassium is mined from deep deposits on the earth, from old salt lakes and from the depths of the sea. Indeed, the Dead Sea, described in the Book of Genesis, is the site of a potassium mining operation. Potassium is found in the by-product brines of seawater that has been distilled into fresh water by atomic-reactor–driven steam coils that also produce electricity. It is found in all living matter and has been used as a soil amendment in the form of animal manures, plant residues and wood ashes since the dawn of history. However it has been recognized as an element essential for plant growth only since the beginning of the nineteenth century. More recently, it has been found to be essential to the health and well being of all animals. Its functions in all living things, plant and animal, can be described in five summary statements:

1. Potassium is the most abundant metal cation in cells, perhaps helping to regulate salt-water balance.
2. Potassium activates or stimulates enzyme activity.
3. Potassium is essential to nerve and muscle activity.
4. Potassium is used in the storage and release of energy.
5. Potassium deficiency can be associated with malnutrition and disease.

As an essential element for plant growth, potassium has presented a challenge to those who have attempted to establish its exact function in plants. Unlike nitrogen, phosphorus, sulfur, calcium and magnesium, potassium does not enter into permanent organic combinations in plants but exists for the most part as soluble inorganic and organic salts. About 99 percent of the potassium in dried plant tissue can be removed by washing with rainwater. Therefore, it is difficult to assign to it roles closely related to inorganic metabolism. Thus, the term *catalyst*, which means accelerator of reactions, is frequently applied. It may be thought of as a "chemical policeman," keeping the flow of life's chemistry moving. Scientists over the years have shown that potassium is essential for various growth processes in living cells.

For example, evidence is now available indicating that potassium influences a great many physiological processes, including the following:

1. Cell division.
2. Photosynthesis—formation of carbohydrates.
3. Translocation of sugars.
4. Reduction of nitrates and subsequent synthesis into proteins.
5. Enzyme activity.

190

It is generally recognized that, directly and indirectly, potassium is a factor in the assimilation of carbon dioxide by plants. Early studies demonstrated a close relationship between carbohydrates and potassium level: plants well supplied with potassium contain more carbohydrates than plants deficient in this element. For example, a decrease in carbohydrates resulted when potassium was withheld from tomato plants in solution culture.

One plant characteristic influenced by potassium that has been widely studied is the structure of stems and cell walls. It is generally accepted that stiffness of straw or stalk is related to an adequate supply of potassium within the plant. The development of lignin and cellulose is related to the level of carbohydrate accumulation. However, if carbohydrates are used in protein synthesis as rapidly as they are produced, as when nitrate supplies are high, then cell walls may be thin and stems weak even though the plant tissues contain adequate levels of potassium.

It is possible to demonstrate an increase in nitrate reduction when potassium-deficient tomato plants are placed in a solution of potassium salts.

High potassium concentrations are associated with actively growing plant tissues. When soil potassium is insufficient for optimum growth, potassium is commonly transported from more mature to meristematic tissues; consequently, older leaves exhibit deficiency symptoms earliest. Chlorosis followed by necrosis appears first around the edges and tips of leaves and eventually works its way all the way to the midrib.

The younger foliage may later show starvation symptoms. Potassium deficiency symptoms are characterized by a sharp contrast between chlorotic necrosis and healthy green areas of the leaves of many crops. Classical examples are most often observed in the grasses, most notably in corn. In later stages of potassium starvation, leaf edges become necrotic, the tissue disintegrates, and the leaf presents a ragged appearance. This condition is often called leaf scorch.

Forms of Potassium in Soils. Based upon degree of availability, soil potassium can be grouped into three categories: (1) difficultly available, (2) slowly available, and (3) readily available (Figure 6.1).

The difficultly available portion is only slightly soluble and often makes up 90 to 98 percent of the total soil potassium. This potassium is found in the crystalline structures of primary minerals, such as orthoclase feldspars and muscovite mica. These minerals, when exposed to the various weathering processes, gradually undergo decomposition over a period of geological time. With this decomposition, there is a release of potassium ions that may be:

1. Lost in drainage waters.
2. Taken up by living organisms.
3. Held as an exchangeable ion on surrounding clay particles.
4. Converted to one of the slowly available forms of soil potassium.

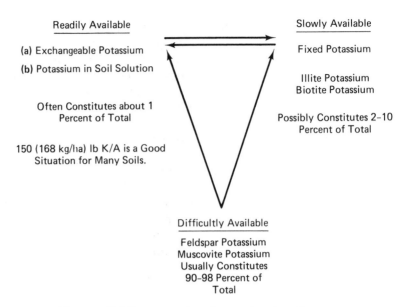

Readily Available

(a) Exchangeable Potassium

(b) Potassium in Soil Solution

Often Constitutes about 1
Percent of Total

150 (168 kg/ha) lb K/A is a Good
Situation for Many Soils.

Slowly Available

Fixed Potassium

Illite Potassium
Biotite Potassium

Possibly Constitutes 2–10
Percent of Total

Difficultly Available

Feldspar Potassium
Muscovite Potassium
Usually Constitutes
90–98 Percent of
Total

Figure 6.1 Three availablity categories of potassium in soils.

This form of potassium makes little contribution toward meeting the seasonal requirements of plants for potassium.

Slowly available potassium constitutes about 2 to 10 percent of the total in mineral soils; biotite mica and illite are examples of this form.

The readily available portion makes up about 1 percent of the total. Exchangeable potassium and potassium in soil solution constitute this form. An adequate level of exchangeable potassium in many soils is about 150 lb/acre (168 kg/ha) furrow slice.

An equilibrium exists between readily available and slowly available potassium (Figure 6.1). Readily available potassium is easily absorbed by plants and is also easily extracted from the soil by weak acids and other exchangeable cations. Slowly available potassium is not readily absorbed by plants, but it is much more available to plants than the potassium present in the difficultly available primary minerals. Slowly available potassium can be extracted from soils by the use of strong acids.

The availability of these three forms of soil potassium may be likened to three kinds of money. The readily available is like money in your pocket, the slowly available is like money in the bank, and the difficultly available is like frozen assets. The transformation of difficultly available forms to more available forms is very slow, amounting to no more than a few pounds per acre (kg/ha) per year.

Fertile loam soils commonly contain 1 to 2 percent total potassium. The amount is greater than that of any other of the major nutrient elements. In virgin soils, frequently there is 5 to 10 times as much total potassium as phosphorus or nitrogen. The amount of total potassium in mineral soils varies from about 4,464 to 44,642 pounds per acre (5,000 to

50,000 kg/ha) furrow slice. Since readily available potassium constitutes about 1 percent of the total, mineral soils may contain from 44.6 to 446.4 pounds per acre (50 to 500 kg/ha) of readily available potassium. Most of this is in the exchangeable form, with the remainder being in soil solution.

Potassium Requirements of Plants. The potassium requirement varies among the stages of plant growth. Most crop seeds contain from 0.1 to 1.0 percent potassium. This quantity is sufficient only for germination and very early development. Corn seedlings were analyzed 18 and 27 days after planting in soils to which no fertilizer was added. At the time of the first sampling, it was found that the seedlings had absorbed 8 to 10 times as much potassium from the soil as was originally present in the seeds. A week later, the seedlings had removed 25 times as much potassium as there had been in the seed kernels. These data point to a very rapid uptake of soil potassium by corn in the seedling stage.

Vegetative growth is characterized by a progressive increase in the amounts of inorganic elements, such as potassium, absorbed. In tobacco, on the twenty-first day from transplanting, potassium was being absorbed at a rate of 0.1 pound per acre (0.1 kg/ha) per day. A maximum rate of uptake of 1.75 pounds per acre (1.96 kg/ha) per day occurred 49 days after transplanting, but the rate declined to 0.8 pounds per acre (0.9 kg/ha) per day after 63 days.

Potassium is easy to apply, and it increases yields and quality of crops on deficient soils. It is available to most farmers, has an essential role in the plant and increases the value of the crop to which it is applied.

A soil test by a qualified technician can determine the level of potassium in the soil and suggest the amount needed. A soil containing 0.2 milliequivalent of readily available potassium per 100 grams is considered to have an adequate level. This amounts to 156 lb/acre (175 kg/ha) of potassium.

To supply a potassium-deficient soil, several kinds of potassium materials are available (Table 6.1). A description of these materials, how they are made, their chemical and handling characteristics and how they are used will be discussed in the following sections.

MINING, EXTRACTION AND LOCATION OF POTASH RAW MATERIALS

The leading world producers of potash are Canada, the United States, the Soviet Union, West Germany, East Germany and France, probably in that order. Israel, Spain and the Congo also produce some potash. There appears to be ample supplies of raw materials in the world.

The North American potash industry is relatively young compared with other segments of the fertilizer industry. The first large-scale production of muriate of potash was initiated in 1916 at Searles Lake, California.

Table 6.1
ANALYSIS OF REPRESENTATIVE SAMPLES OF POTASH MATERIALS

Constituent	Muriate of Potash (KCl) Percent	Potassium Nitrate (KNO₃) Percent	Sulfate of Potash (K₂SO₄) Percent	Sulfate of Potash-Magnesia (K₂SO₄·2MgSO₄) Percent	Potassium Carbonate (K₂CO₃) Percent
Potash (K₂O)	60–62.5	44–45	50–52	21–22	63–64
Potassium	50.34	36.94	41.34	18.14	52.91
Sodium	1.13		0.76	1.08	
Sulfur	0.11	0.29	17.66	22.73	variables
Magnesium	0.11	0.23	0.70	11.19	variables
Chlorine	47.39	1.14	2.07	1.54	0.42
Nitrogen		12.96			
Moisture	0.21	2.01	0.52	0.30	variables
Other	0.71	46.43	36.95	45.02	46.61
Total	100.00	100.00	100.00	100.00	100.00

Sylvinite deposits in New Mexico were discovered 10 to 15 years later, and Carlsbad, New Mexico has since become the center of potash fertilizer production in the United States. Since 1950, ore reserves near Moab, Utah and in Saskatchewan, Canada have been developed. Canadian deposits are enormous, and annual output of potash fertilizer is climbing rapidly. In fact, Canadian production now supplies more potash to the United States than does domestic production.

Potash fertilizers are also recovered from salt marshes near Wendover, Utah. The quantity produced, however, is so small that it is insignificant relative to the total potash industry.

Characteristics of Major Producing Areas. The New Mexico deposits occur as a part of a vast salt basin formed during the Permian period over 200 million years ago. Potash minerals are found 650 to 2,500 feet (198 to 762 m) beneath the surface of the earth. Sylvinite and langbeinite are the sources of most of the potash fertilizers produced in this area. Sylvinite is a mixture of KC1 and NaC1, and langbeinite is $K_2SO_4 \cdot 2MgSO_4$. Sylvinite is mined from a depth of 800 to 1,800 feet (244 to 549 m) and langbeinite at a level of about 850 feet (259 m).

Potash mineral beds near Moab, Utah are composed principally of sylvinite and carnallite, a potassium-magnesium chloride. The ore being exploited is at a depth of about 2,700 feet (823 m).

In Saskatchewan, Canada, potash minerals occur at depths up to 8,000 feet (2,438 m). Principal minerals are sylvinite and carnallite. These deposits are mined at depths from 3,600 to more than 5,000 feet (1,097 to 1,524 m). The shaft method is used at depths of 3,000 to 3,600 feet (914 to

194

to 1,097 m), and the solution method is employed at depths of 5,300 feet (1,615 m). Nearly all North American potash reserves are here.

Brine sources of potash are being exploited at Searles Lake, California and Wendover, Utah. The Searles Lake reserve covers about 34 square miles (88 km²), and potash is currently being recovered from brine that contains about 3 percent K_2O. The Utah brines are very small in extent and contain about 2 percent K_2O.

Methods of Mining and Extraction. Mining techniques for bringing raw potash materials to the surface involve both solid ore recovery by conventional shaft procedures and solution mining by sinking wells to the ore body with subsequent dissolution.

Shaft Mining (Solid Ore Recovery). Two principal methods are employed: (1) the conventional coal-mining technique of face undercutting, drilling and blasting and (2) a continuous-mining method that utilizes specially developed mining machines. In the first method, the loosened ore is loaded in shuttle cars and hauled to the foot of an elevator shaft where it is crushed to a size below six inches (15.2 cm) before being carried to the surface.

The development of continuous-mining machines ranks as one of the major advances in potash mining. This equipment cuts the ore directly from the mine face, thus doing away with undercutting, drilling, blasting and crushing. These machines develop 200 horsepower and operate with a 4,160-volt motor. They mine 5 tons (4.55 m tons) of ore per minute. Conveyor belts are used to transport the crushed ore to the bottom of the shaft.

Solution Mining. In this method, wells are sunk to the ore body. Warm brines containing controlled amounts of KCl–NaCl at a temperature of 135°F (57°C) are pumped down to the bed, dissolving the potassium salt and returning the potassium-laden brine to the surface for refining. The well is 7 inches (17.8 cm) in diameter at the surface.

Solution-mining techniques are being employed by at least one company in Canada. Wells extend to depths over 5,000 feet (1,524 m). Herein lies one advantage over shaft mining: it is difficult to mine profitably with shafts below 3,600 to 4,000 feet (1,097 to 1,219 m). Also, a greater total extraction of ore is possible with solution-mining techniques.

Extraction from Brines. At Searles Lake, brine is pumped several miles (kilometers) to the refining plant from wells sunk 70 to 130 feet (21 to 39 m). Wells may have a life of several years before the composition of the brine becomes unsatisfactory. Most brines from upper layers contain 5 percent KCl whereas the lower layers are 30 percent KCl.

In Utah, brine is pumped from a network of more than 50 miles (80 km) of canals, 3 feet (0.9 m) wide and 14 feet (4.2 m) deep.

Dead Sea brines contain 0.0067 lb/gal (11.5 g/l) of KCl. When the brines crystallize, they become carnallite, which is later refined.

POTASSIUM CHLORIDE (MURIATE OF POTASH)

Description.   Potassium chloride (KCl) is refined red or grayish red (it may be red, pink, gray or white) salt. When the trace amount of iron in the colored salt is removed, it looks and tastes much like table salt. It is manufactured into fine, standard, coarse and granular grades. The coarser materials mix well with granular nitrogen-phosphorus compounds to form nitrogen-phosphorus-potassium–blended multinutrient fertilizers.

Liquid multinutrient fertilizers require the recrystallized, white KCl, which is of a high degree of purity and it is also used in some granulated nitrogen-phosphorus-potassium homogeneous solids.

Method of Manufacture.   A large percentage of soluble potassium salts used in the United States is of Canadian origin. In 1943, potassium salt deposits were discovered in Saskatchewan, Canada, and today a substantial part of world production is mined there. This salt deposit extends over an area 350 miles (564 km) long and 120 miles (193 km) wide. Valuable underground reserves also occur in the area of central Utah.[1] In addition, surface deposits occur at Searles Lake, California; Salt Lake, Utah; and Wendover, Utah. Prior to World War I, most of the potassium fertilizer salts were mined in Germany and France.

Potassium chloride (muriate of potash) is prepared through refining techniques for removing KCl from underground ores as brines. Two processes are used for the purification of KCl from potash ores; flotation and crystallization. Both processes involve a separation of NaCl from KCl. Crystals of these two chemicals are commonly found interlocked in a mineral mixture called sylvinite. The flotation process is by far the most prevalent, and the greatest proportion of the agricultural muriate of potash is produced by flotation beneficiation.

In the *flotation* process, small quantities of special reagents called flotation agents are added to the KCl–NaCl mixture to coat or film the KCl particles selectively. The treated mixture is then agitated in a mechanical cell in such a manner that air bubbles are introduced and a frothing condition develops. The finely divided air bubbles attach themselves to the filmed KCl particles and float them to the surface of the cell where the froth, rich in selected KCl, is skimmed off. Different flotation reagents are discussed by Alexsandrovich.[2]

[1]E. R. Ruhlman, "Potash," in *Industrial Minerals and Rocks* 3rd ed., (New York: American Institute of Mining, Metallurgical and Petroleum Engineers, 1960), pp. 669-680.

[2]K. M. Alexsandrovich, *Sov. Chem. Ind.* (Engl. Transl.), 7(2) (1975), 900–2.

Figure 6.2 illustrates the flotation process and the major production steps used in underground and surface operations at a potash mine and refinery.

Solid ore recovery and processing is described below and will be better understood if the sequence of numbers in Figure 6.2 is matched with the numbers in parentheses in the description that follows. The loosened ore is loaded in shuttle cars (1). The cars shuttle 15-ton (13.7-m ton) loads to a conveyor belt that carries the ore to the base of the shaft (2). There, the ore is crushed to a maximum size of six inches (15.2 cm). Ore storage bins here and on the surface assure uninterrupted operations. Continuous-mining machines are employed in all recently constructed facilities. These machines grind the ore from the lode body, thus eliminating blasting, crushing and shuttle cars; conveyor belts move the ore to shafts for lifting above ground. At the shaft, the ore is hoisted to the surface in buckets in 20-ton (18.2-m ton) loads (3). The shaft, divided by a wall, has a second hoist for men and equipment. Fresh air is piped down one side and exhaust air up the other.

On the surface, the ore moves from building to building by conveyor. It first goes through the crushing circuit (4) at the mill where it is reduced to about ⅛-inch (0.32-cm) particles. In this size, the ore pieces are either KCl or NaCl. In a scrubber (5), brine is added to make a slurry. In a

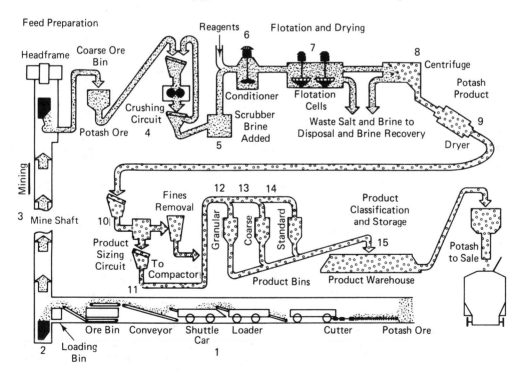

Figure 6.2 How potash is mined and refined. (Courtesy of Texas Gulf Sulfur, New York, N.Y.)

conditioner (6), the ore is mixed with the flotation reagent as described previously. Potash is skimmed from the surface of flotation cells (7), centrifuged to remove the brine (8), and dried in a rotary kiln dryer (9). The final processing is the product-sizing circuit (10): the product is screened, and fine particles are compacted (11). Granular (12), coarse (13) and standard (14) sizes are stored separately, first in product bins, then in huge warehouses ready for shipment.

The *crystallization* process employed in separating potassium chloride from sodium chloride is largely dependent on their different solubilities in hot and cold water. The solubility of potassium chloride increases rapidly with a rise in temperature whereas the solubility of sodium chloride varies only slightly. Cool brine saturated with both salts is heated to 212°F (100°C) and passes over the finely ground ore. The ore solution is subsequently cooled slowly; the KCl precipitates and leaves the NaCl in the cool supernatent liquid.

Secondary Processing. Increasing demand for potash products in coarse and granular forms has resulted in the development and application of techniques for transforming finely divided materials into products of large particle size. Two methods of converting fine muriate of potash into coarser particles are fusion and compaction. Compaction is the newer process and is more widely used.

Fusion Process. Finely divided KCl from flotation recovery is heated to 1391°F (755°C) to form a melt, which is solidified by cooling. The flakes thus formed are crushed and screened to give the desired particle size.

Compaction. This newer technique is essentially one of feeding variously sized muriate particles at temperatures ranging from 200° to 250°F (93° to 121°C) between pressure rolls [300 tons (273 m tons)] in order to compress it into sheets [⅛ to $^{1}/_{16}$ inch (0.32 to 0.16 cm) thick] or briquets, which are crushed and screened to desired sizes. Brine or water may be added prior to compacting. Heat-compacting rollers will assist the process.

Handling Characteristics. Fine crystalline potassium chloride is not very hygroscopic. It flows freely and does not cake excessively as long as it is kept in a dry place. When it is mixed with ammonium salts, ammonium chloride, which is hygroscopic, may be formed. Unless the mixture is properly cured or granulated, it absorbs moisture from the atmosphere and cannot be applied successfully with ordinary fertilizer distributors. Ammonium chloride is a very active chemical ingredient, causing poor physical condition of fertilizers. Standard and granular potassium salts can be distributed evenly with ordinary fertilizer distributors. They mix well with semigranular nitrogen-phosphorus products to produce free-flowing mixed fertilizers. Thus, there is an increasing demand for granular, dust-free potassium products that store well and are free flowing.

Chemical Characteristics.   Typical analyses of potassium chloride fertilizer salts are given in Table 6.2. Muriate of potash may contain 48 to 62.5 percent K_2O (39 to 51 percent potassium) and about 47 percent chlorine. The most common product on the North American market is 60 percent potash (49.9 percent potassium); this grade makes up 80 percent of the production in North America. The name "muriate" is derived from the term, muriatic acid, which is the common name for hydrochloric acid. All potassium fertilizer salts are soluble in water and may contain traces of bromine, sulfur, magnesium, calcium, sodium, iron and boron. Sea water contains 0.04 percent potassium in the form of KCl.

Crop Response.   It has been estimated that at least 78 percent of the potassium salts consumed in the world is used in the form of potassium chloride. This, in itself, indicates the wide acceptance of this material as a fertilizer. Over 90 percent of all processed potassium is consumed as fertilizer.

Muck, peat and sands are usually the most potassium-deficient soils. Soils in arid regions are usually the most potassium-sufficient soils,

Table 6.2
TYPICAL ANALYSES OF POTASSIUM CHLORIDE FERTILIZER SALTS
PRODUCED BY THE POTASH COMPANY OF AMERICA

Constituent	Potassium Chloride		Manure Salts
	Standard[1] percent	Granular[2] percent	Percent
Potash, (K_2O)	(61.00)	(61.00)	(25.00)
Potassium chloride, (KCl)	(96.56)	(96.56)	(39.58)
Potassium	50.64	50.64	20.75
Sodium	1.10	0.92	22.87
Calcium	0.02	0.02	0.12
Magnesium	0.06	0.12	0.25
Chlorine	47.64	47.33	54.08
Sulfate (SO_4)	0.08	0.10	0.65
Bromine, etc.	0.07	0.09	0.03
Water-insoluble material	0.35	0.85	1.10
Moisture	0.03	0.03	0.15
Total	99.99	100.00	100.00

[1]20 to 65 mesh (0.8-mm to 0.2-mm size opening)
[2]6 to 20 mesh (3.3-mm to 0.8-mm size opening)

containing 400 lb/acre (448 kg/ha) or more of readily available potassium. The most potassium-deficient soils in the United States are those in the Southeast. The Corn Belt and the southeastern part of the United States consume most of the potassium chloride produced in this country. It is unfortunate that soluble potassium ore deposits in the United States are so remote from the agricultural areas of maximum use.

Total world demand for potassium has been growing at a rate of about 5 percent annually, about as fast as any plant food nutrient. Potassium chloride is usually the most inexpensive carrier of potassium on the fertilizer market.

POTASSIUM SULFATE

Description.   Potassium sulfate (K_2SO_4) is refined beige- to pink-colored salt. It is manufactured into fine, granular or semigranular grades. It may be mixed with a slurry of other materials to form compound fertilizers or used in granular blends.

Method of Manufacture.   Potassium sulfate is manufactured by a number of processes from a number of different raw materials. Four of them, known as the Langbeinite, Trona, Mannheim and Hargreaves processes, are discussed. The Langbeinite process consists of dissolving langbeinite, a double sulfate of potassium and magnesium, in water and subsequently adding a concentrated solution of potassium chloride in accordance with the following equation:

$$K_2SO_4 \cdot 2MgSO_4 + 4KCl \rightarrow 3K_2SO_4 + 2MgCl_2$$

Potassium sulfate precipitates and is separated from solution by centrifuging.[3] The wet cake is dried, screened and sent to storage.

In the Trona process for the manufacture of potassium sulfate, KCl is first reacted with burkeite, $Na_2CO_3 \cdot 2Na_2SO_4$, to produce a high potassium glaserite cake, $Na_2SO_4 \cdot 3K_2SO_4$. The glaserite is separated and then treated with high-purity KCl brine. This converts into solid potassium sulfate, which is filtered from the NaCl brine.

$$Na_2CO_3 \cdot 2Na_2SO_4 + 4KCl \rightarrow 2K_2SO_4 + 4NaCl + Na_2CO_3$$

The potassium sulfate is washed, dried and sent to fertilizer-grade K_2SO_4 bulk storage.Typical analyses of sulfate of potash produced by this method are given in Table 6.1.

[3]E. C. Kapusta, and N. C. Wendt, "Advances in Fertilizer Potash Production," in *Fertilizer Technology and Usage*, ed., M. H. McVickar *et al.* (Madison, Wis: Soil Science Society of America, 1963), pp. 219–223.

In addition to production from burkeite of Searles Lake and the langbeinite ore of New Mexico, potassium sulfate is also manufactured in the United States from potassium chloride and sulfuric acid by the Mannheim furnace process. Manufacture of potassium sulfate in this way is confined to areas in the Southwest where the co-product, hydrochloric acid, finds local use in the treating of oil wells.

The reaction between potassium chloride and sulfuric acid to yield potassium sulfate proceeds in two stages. Potassium acid sulfate is formed in the first stage, which is exothermic and is initiated at normal temperatures.

$$KCl + H_2SO_4 \rightarrow KHSO_4 + HCl$$

In the second stage, which is endothermic, the acid sulfate is converted into the normal sulfate by further reaction with potassium chloride under the influence of external heat.

$$KCl + KHSO_4 \rightarrow K_2SO_4 + HCl$$

In commercial practice, both stages of the reaction are conducted in a Mannheim-type salt cake furnace. Potassium chloride and sulfuric acid in the required proportions are fed in separately and continuously from the top and near the center. The reactions proceed toward completion as the mixture gradually is moved outward through the hotter zones [1112° to 1292°F (600° to 700°C)] to the circumference and outlet by slowly rotating plows. The hot potassium sulfate product requires cooling before being bagged or sent to storage. The hydrogen chloride evolved passes out of the furnace at the top and is absorbed in water to form hydrochloric acid. The K_2SO_4 contains little free acid. A typical analysis of potassium sulfate manufactured by this method is given in Table 6.3.

Potassium sulfate is also manufactured from kainite, $KCl \cdot MgSO_4 \cdot 3H_2O$, in Italy, Germany and the United States by the Hargreaves process, involving the reaction of KCl with sulfur as follows:

$$2KCl + SO_3 + H_2O \rightarrow K_2SO_4 + 2HCl$$

Handling Characteristics.   Sulfate of potash handles and stores well. Its physical characteristics and handling qualities are better than KCl because it is less hygroscopic. Sulfate of potash, 50 percent K_2O, is quoted as granular or standard grades. In the standard grade, it mixes well with other fine ingredients of powdered mixed fertilizer. In granular grade, it is adapted to bulk, granular mixes.

Chemical Characteristics.   The product as sold in the United States contains 90 to 95 percent potassium sulfate. It analyzes 52 percent K_2O (43 percent potassium), 16 percent sulfur, and not more than 2.5 percent chlorine. A typical total analysis shows traces of sodium, calcium, magnesium, chlorine and bromine (Table 6.3).

Crop Response. The yield response of most crops on potassium-deficient soils will be the same with potassium chloride and potassium sulfate. The quality of tobacco, beets, Irish potatoes and corn may be improved by substituting sulfate of potash for at least some of the potassium needs of deficient soils on which these crops are grown. Although small quantities of chlorine, for example, 18 to 27 lb/acre (20 to 30 kg/ha), may be beneficial to most crops, beets, Irish potatoes, tobacco and perhaps corn are sensitive to large quantities of chlorides, such as the amounts present in a ton per acre (2,240 kg/ha) of ordinary mixed fertilizers like 4-5-10 (4-12-12) and 5-4-8 (5-10-10) that are added to the soil at or near planting time.

The major market for sulfate of potassium is in the fertilizer industry of the United States eastern seaboard. The farmers in this region demand tobacco fertilizers that contain at least 50 percent of the total nitrogen in nitrate form; 2 percent magnesium, at least one-half of which should be water soluble; not more than 2 to 4 percent chlorine; 6 to 8 percent calcium in available form; and 2 to 3 percent, but not more than 4 percent, sulfur. For side-dressing tobacco with nitrogen and/or potassium, all of the nitrogen should be in the nitrate form, and the potassium should be either in the nitrate or sulfate form. From this description of fertilizers for tobacco, it may be deduced that four kinds of potassium salts are used in tobacco fertilizers, viz., KCl, K_2SO_4, KNO_3 and water-soluble magnesium from $K_2SO_4 \cdot 2MgSO_4$.

Table 6.3
TYPICAL ANALYSES OF POTASSIUM
SULFATE MANUFACTURED FROM
POTASSIUM CHLORIDE AND SULFURIC
ACID

Constituent	Percent
(Potash, K_2O)	(52.00)
(Potassium sulfate, K_2SO_4)	(96.20)
Potassium	43.17
Sodium	0.94
Calcium	0.02
Magnesium	0.06
Chlorine	1.80
Sulfate (SO_4)	53.65
Bromine, etc.	0.02
Water-insoluble material	0.32
Moisture	0.02
Total	100.00

Description. Sulfate of potassium-magnesia ($K_2SO_4 \cdot 2MgSO_4$) is a white-to peach-colored material. It is offered in powdered or granulated form. It may be mixed with other plant food chemicals to form powdered or granulated mixed fertilizer containing water-soluble magnesium. It is seldom applied to the soil as an individual material.

Method of Manufacture. Sulfate of potassium-magnesia is manufactured by washing any double sulfate of potassium and magnesium in fresh water. When the chloride salts are washed away, residual langbeinite is left. Although langbeinite is soluble, the rate at which it dissolves is quite slow. This slow rate, coupled with the fast solubility rate of the chloride salts, is the basis for the process. The product is dried, screened, ground and bagged.

Briefly, the process[4] consists of the following steps; the numbers in parentheses refer to Figure 6.3. Mine run ore is dry crushed by hammer mills operating in a closed circuit with vibrating screens (1). The crushed ore is then screened to make a separation at 10-mesh (2.0 mm) (2). The 10-mesh (2.0-mm) coarse material is sent to a washing tumbler where it is mixed with a portion of the wash water (3). The tumbler discharge is mixed with the 10-mesh (2.0-mm) fine fraction, and the balance of the wash is pumped to the first of two classifiers operating in series (4). The clear feed water is added to the second classifier with the solid langbeinite from the first unit (5). The overflow from the second classifier flows by gravity to the washing tumbler (3) and then is pumped to the first classifier with the new feed (4). The overflow from the first classifier is treated in a cyclone to remove suspended fines, and the liquor is rejected as waste (4). The rejected liquor contains more than 20 percent of sodium chloride whereas the residual solid phase analyzes 96 to 98 percent langbeinite. The solid product is centrifuged (6), dried (7), and marketed under the trade name of Sul-Po-Mag® by International Minerals and Chemical Corporation. Sulfate of potash-magnesia is also manufactured by the Duval Corporation, which sells it under the trade name K-mag®.

Handling Characteristics. Sulfate of potash-magnesia handles and stores well. Its physical characteristics and handling qualitites are better than most soluble potassium salts. Since it is offered as a granular or powdered material, it can be used in blending, mixing or other solid fertilizer manufacturing processes.

Chemical Characteristics. Sulfate of potassium-magnesium mineral is a double salt, $K_2SO_4 \cdot 2MgSO_4$, and the commercial grade is sold under

---

[4]K. D. Jacob, ed., *Fertilizer Technology and Resources in the United States.* Agronomy: A Series of Monographs, Vol. II (New York: Academic Press, 1953).

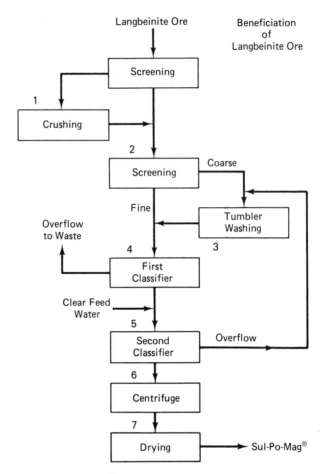

Figure 6.3 Flow diagram for the manufacture of sulfate of potassium magnesium from langbeinite.

several registered trade names. It contains not less than 21 percent K_2O (18 percent potassium), 18 percent MgO (11 percent magnesium), and 23 percent sulfur. According to the Association of Official Analytical Chemists' definition, it should not contain more than 2.5 percent chlorine.

Product Use. It is used in agriculture as a primary source of water-soluble magnesium and also as a supplemental source of sulfate of potash.

POTASSIUM NITRATE

Description. The agricultural grade of potassium nitrate is of uniform particle size, free flowing, noncaking, and with a screen size approximately in the range of $-10+35$ mesh ($-2+0.5$ mm). Although nitrate of potash in the pure form is a colorless crystal, the fertilizer product has impurities that impart a white appearance to the product. A powdered grade is available for use in manufacturing mixed fertilizer.

Method of Manufacture.   In the early twentieth century, potassium nitrate was crystallized out of bat guano. In a plant that has been constructed at Vicksburg, Mississippi, nitrate of potash is now produced by a chemical reaction between nitric acid and agricultural muriate of potash (Figure 6.4). It employs a process combining potassium chloride and nitric acid in a packed tower that produces nitrate of potash as well as a co-product, liquid chlorine, for industrial uses:

$$4KCl + 4HNO_3 + O_2 \rightarrow 4KNO_3 + 2Cl_2 + 2H_2O$$

Molten KNO_3 is removed from the bottom of the tower, and the gaseous by-products are withdrawn from the top (Figure 6.5). A similar process is employed by Israel Mining Industries near the Dead Sea.

Handling Characteristics: Conditions to be observed in storing and handling nitrate of potash are the same as for nitrate of soda and are much less stringent than for ammonium nitrate. Depending on humidity and

Figure 6.4   Muriate of potash, stored in the large conical bin on the extreme right, and nitric acid, stored in the large upright tank on the extreme left, are combined to form chlorine and nitrate of potash in the reactor tower just to the left of the conical storage bin. Liquified chlorine is stored for shipment in the horizontal pressure vessel. (Photo, courtesy of Southwest Potash Corporation, New York, N.Y.)

Figure 6.5 From the bottom of the reactor tower in Figure 6.4, the molten solution of nitrate of potash goes to the crystallizer shown here. The product is crystallized from solution and dried for commercial handling in the low building to the left of the crystallizer. (Photo, courtesy of Southwest Potash Corporation, New York, N.Y.)

length of storage, a crust may form on top of agricultural nitrate of potash storage piles just as on muriate of potash, but the crust easily breaks up in the course of normal handling, as it does in muriate of potash. In mixed fertilizers, a powdered grade of nitrate of potash does not contribute to caking. Nitrate of potash, being only slightly hygroscopic and granulated, can be spread conveniently by truck, fertilizer distributor or airplane.

Chemical Characteristics. Nitrate of potash is soluble in water to the extent of from 13 parts per hundred at 32°F (0°C) to 246 parts per hundred at 212°F (100°C). In cold water, nitrate of potash is less soluble than muriate of potash but more soluble than sulfate of potash. In water of room temperature, nitrate of potash and muriate of potash exhibit about the same solubility, which is much higher than that of sulfate of potash. In water above room temperature, the solubility of nitrate of potash is not only greater than that of muriate of potash and sulfate of potash but it also increases rapidly as the temperature rises while those of muriate of potash and sulfate of potash remain relatively unchanged.

Fertilizer-grade KNO_3 contains about 13 percent nitrogen and 44 percent K_2O (36 percent potassium). Pure nitrate of potash is crystalline, colorless, and contains about 14 percent nitrogen and 38 percent potassium. Its molecular weight is 101.10, and its specific gravity is 2.11, which makes it about twice as heavy as water. As a chemical compound, it is neutral, and its nitrogen-to-potassium oxide ratio is roughly 1:3.

Crop Response. Up to 1959, nitrate of potash had been used as a fertilizer principally as the Chilean nitrate of soda-potash containing 15 percent nitrogen and 14 percent K_2O (11 percent potassium). High-analysis nitrate of potash of the kind now being produced in this country had previously found very little use as a fertilizer because of its cost. Only a few hundred tons of such nitrate of potash were imported annually for the formulation of specialty fertilizers. In recent years, tonnage has rapidly increased in the high-value crop region. Its agronomic value has been well demonstrated. On cotton, tests have shown that KNO_3 is equal in value to other sources of potassium and nitrogen in equivalent quantities.

Potassium is needed in large quantities for the growth and nutrition of the tobacco plant, and a relatively high content of potassium in cured tobacco is desirable for good smoking quality. Its relative lack of chlorine and sulfur is an important advantage of nitrate of potash as a tobacco fertilizer. In fact, because of the pronounced effect of chlorine on the burning characteristics of tobacco, some states regulate the percentage of chlorine permissible in tobacco fertilizers. Although the adverse effect of excessive sulfur on tobacco quality does not presently appear to be as severe as that of chlorine, there is a growing interest in means of reducing the quantity of sulfur applied in fertilizers.

Another advantage of nitrate of potash as a tobacco fertilizer is that its nitrogen is in the nitrate form. The growth and development of tobacco is known to be adversely affected when the major portion of the available nitrogen is in the ammonium form. It has been found desirable in tobacco preplant fertilizers to use about one-half of the nitrogen in the nitrate form.

The practice of applying part of the tobacco fertilizer before transplanting and the remainder as a side-dressing, usually 2 to 3 weeks after transplanting, has the advantage of reducing the injury to the transplants that may occur when high rates of preplant fertilizer are used. Side-dressing also provides a means for replenishing the potassium and nitrogen that may have been removed from the root zone by leaching from the sandy soils on which flue-cured tobacco is generally grown. Potassium and nitrogen are the elements usually recommended in side-dressing fertilizers, and it is desirable that, for side-dressing, all of the nitrogen be in the nitrate form.

Nitrate of potash has also been used successfully as a source of nitrogen and potassium for potatoes, tomatoes, corn, citrus and carnations. The latter are grown in nutrient solutions added to gravel and vermiculite. The suggested solution for this purpose includes nitrate of potash as the sole source of potassium. Nitrate of potash is also used in water-soluble lawn fertilizers.

Potassium Metaphosphate. Potassium metaphosphate, $K_6(PO_3)_6 \cdot H_2O$, is made by reacting P_2O_5 with KCl at high temperatures in a process similar to that used to manufacture calcium metaphosphate. It is only slightly soluble in water but is readily soluble in an ammonium citrate solution. It is considered an available source of phosphorus and potassium to plants and is of interest because, when in its pure form, it contains about 100 percent plant food having a composition of 60 percent P_2O_5 (26 percent phosphorus) and 40 percent K_2O (33 percent potassium). It is manufactured on a limited commercial scale, as of this writing.

Potassium Carbonate. Although it is still obtained from the ashes of plants, potassium carbonate, K_2CO_3, which was first designated potash, is manufactured from KCl. The first step is the electrolysis of KCl brine to produce potassium hydroxide (KOH), which is crystallized out of solution at the cathode. Potassium carbonate, K_2CO_3, is made by neutralizing caustic potash (KOH) with CO_2 gas. Liquid, 48 to 52 percent, K_2CO_3 or solid $2K_2CO_3 \cdot 3H_2O$ containing 83 to 85 percent K_2CO_3 is offered. The liquid product may find use in liquid fertilizers, and limited quantities of the solid have been used as a nonchloride form of potassium for tobacco. It is comparatively expensive.

A different product, reported to be under consideration in France, is produced by a nonelectrolytic process consisting of the direct carbonation of KCl in the presence of a water solution of an organic amine.

PREDICTION OF POTASSIUM NEEDS AND RESPONSES

Soil Tests. Soil tests for potassium have received considerable attention in most areas of the United States. Correlation of chemical methods with greenhouse-grown plants has become standard procedure. A number of studies have been made using a bioassay with either plants or microorganisms to predict yield of crops in the field. Other studies have attempted to correlate chemical methods with field response; this comparison meets the goal for prediction purposes but is the most difficult to define and understand. The present state of the chemical assessment of soil potassium is generally limited to predicting response to applied potassium in terms as broad as low, medium, high and possibly very high "available" potassium. All laboratory and greenhouse-type methods are based on having soil samples that adequately represent a field. As a consequence, the proper sampling procedure deserves priority for validating the soil test. In the final analysis, the most accurate assessment of crop response to potassium, or any other nutrient, can only be made by growing the specific crop in a specific environment.

In contrast to biological soil tests, chemical procedures are quite rapid and do not require special facilities for plant or organism growth. Chemical soil tests for potassium involve the two important steps of extraction of some fraction of the potassium in the soil and quantitative determination of the potassium in the soil extract. There is a great diversity in extraction techniques for available potassium among soil-testing laboratories, although many of them use either a weak acid or ammonium acetate. In 13 state laboratories in the southern region of the United States, six different extracting solutions are employed.[5] All laboratories determine the concentration of potassium in the soil extract using atomic absorption or flame-emission photometry. In arid regions of Oklahoma, Puerto Rico and Texas, particularly where saline soils are found, distilled water is added to make a paste, and the water is removed by vacuum. The filtrate is analyzed for potassium. Thus, the choice of procedure for determination of potassium will depend on the nature of the soils under study.

Response as Related to Soil-Test Levels. The Soil-Test Work Group of the National Soil Research Committee[6] published one of the most comprehensive studies on potassium soil-test methods and growth-chamber values using rye as a test plant. Soil samples from 39 laboratories were tested for potassium, and these data were compared with rye uptake values. These results showed that correlation of soil test with rye uptake values was 0.483 when the soil-test potassium was determined by extracting with the ammonium ion. This was the highest correlation coefficient of any of the soil-test methods. None of them were highly correlated with rye uptake values.

In another study,[7] the exchangeable potassium level of a New York soil was related to the yield response of alfalfa to added potassium. Little yield response was obtained when the amount of soil potassium was greater than 80 lb/acre (90 kg/ha) furrow slice; the correlation coefficient was 0.755. Some states indicate a likelihood of response to added potassium if the soil-test level is below a certain value—e.g., 0.2 milliequivalent per 100 grams of soil. Figure 6.6 shows the interaction of potassium fertilization with the response of wheat to nitrogen and phosphorus where nitrogen, phosphorus and potassium soil levels were all low.[8] The important

[5]W. E. Sabbe, et al., "Procedures Used in State Soil-Testing Laboratories in the Southern Region of the United States," Southern Cooperative Series Bulletin 190, 1974.

[6]Soil Test Work Group of the National Soil Research Committee, "Soil Tests Compared with Field, Greenhouse and Laboratory Results," North Carolina Agricultural Experiment Station Technical Bulletin 121, 1956.

[7]D. J. Lathwell and M. Peech, "Interpretation of Chemical Soil Tests," Cornell University Agricultural Experiment Station Bulletin 995, 1964.

[8]O. H. Brensing and J. Q. Lynd, "Soil Fertility Studies for Improved Wheat Production in Eastern Oklahoma, 1957–1960," Oklahoma State University Experiment Station Bulletin B–594, 1962.

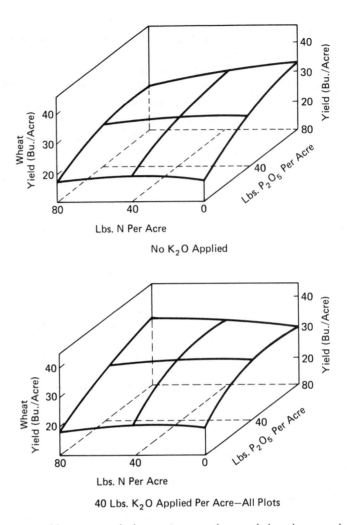

Figure 6–6. Yield response of wheat to increased rates of phosphorus and nitrogen without potassium and with 40 lb. of K_2O added per acre on Dennis silt loam, Wagoner County, Okla. Three-year average 1957–59. Soil test P=2–12 lb/A; K=80–140 lb/A. Lb per acre × 1.12 = KG/ha. BU per acre × 67.2 = KG/ha.

interaction influence of potassium on this soil is noted when three-dimensional diagrams are compared. Note that no increases were obtained with nitrogen additions at any of the P_2O_5 combinations when no K_2O was added. But with the addition of 40 lb/acre (45 kg/ha) of K_2O, yields were increased with all rates of nitrogen when combined with 40 to 80 pounds per acre (45 to 90 kg/ha) of P_2O_5. On this Dennis silt loam, the soil-test phosphorus was 2 to 12 pounds per acre (2.24 to 13.4 kg/ha), and the soil-test potassium was 80 to 140 pounds per acre (90 to 157 kg/ha).

Plant Analyses. Plant potassium concentrations in the various types and parts of plants usually range on the dry basis from about 0.1 percent in seeds up to 10 percent in young leaves. Two types of plant analyses, entire plant or plant part, are used to evaluate the nutrient status of a crop. The most valuable procedure involves analyzing a sample that is representative of either the entire plant (tops and roots) or, more frequently, the entire above-ground plant. Uptake values or percent potassium is used to assess the soil and fertilizer contributions to the plants. In a study on alfalfa,[9] the yield response was greater than 20 percent from the application of potassium fertilizer when the potassium content of the plant from the check plots was less than 1.25 percent. The correlation coefficient (r) was 0.788.

Diagnosis of nutrient adequacy or inadequacy is often performed by sampling some specific plant part and relating the concentration of nutrient found there to levels established by previous experimentation. Most of the levels of adequacy have been established in nutrient culture experiments similar to those outlined by Ulrich.[10]

Critical Levels. It is commonly believed that the best plant tissue to use for measuring the potassium status of the cotton plant is the petiole of the recently matured leaf. There is a slight deviation from this for different growth stages of cotton and for different species of deciduous plants. Soil and plant scientists warn against using immature leaves at the top of the plant. Some specify using the petiole of the basal leaf until early bloom and thereafter the petiole of the first mature leaf. A search for a critical level of potassium in cotton leaves or petioles indicates that there is a scarcity of data that will allow a tissue test to be evaluated in quantitative terms.

Blaser's plant analysis research on alfalfa at Virginia Polytechnic Institute and State University indicates that a potassium content of 2.0 to 2.5 percent is required for maximum yields. Hanway's research at Iowa State University showed little or no yield increase with corn when the content of the leaf opposite and below the ear at silking time was above 2.0 percent potassium. In California, it was found that the potassium deficiency level was 0.5 percent for pears when the spur leaves were assayed from a sample taken between June 15 and July 15; it was also found that the potassium deficiency level was 1.0 percent for peaches when the matured leaves near the base of current growth were assayed from a sample taken between June 15 and July 30.[11]

[9]Lathwell and Peech, "Interpretation of Tests."

[10]A. Ulrich, "Plant Analysis—Methods and Interpretation of Results," in *Diagnostic Techniques for Soils and Crops*, ed. H. B. Kitchen (Washington, D.C.: The American Potash Institute, 1948), pp. 199–230.

[11]F. S. Fulmer and M. E. McCollam, "Ask the Leaf," in *Fight Hidden Hunger* (Washington, D.C.: The American Potash Institute, 1964), pp. 14–17.

Petiole potassium values have been found to drop during periods of moisture stress and to rise following moisture applications.[12] This points out the fact that any growth factor that is out of balance tends to influence the other growth factors. The best use of tissue testing is repeated sampling and testing of the field throughout the season.[13]

Plant Analysis as a Guide to Fertilization. Plant analysis for potassium is often most useful as a guide for next year's fertilization program. However, if samples are obtained early enough in the growing season when foliar application or side-dressing is still possible, tissue analysis may be used effectively to assess the present situation.

SUMMARY

1. Potassium is an essential element for plant growth, but its exact function in plants is difficult to establish because it does not enter into permanent organic combinations there.

2. It exists for the most part in plants as soluble inorganic and organic salts. It acts more as a catalyst and has been described as the "chemical policeman keeping the flow of life's chemistry moving."

3. It functions in the production, storage and release of energy as carbohydrates; activates enzymes; regulates salt-water balance in cells; and is the most abundant metal cation in cells.

4. An excellent illustrated presentation of "Roles of Potassium in Plants" was published in the Southern Potash Newsletter by The American Potash Institute in 1965.

5. The potassium content of soils varies, the amounts in mineral soils ranging from about 4,464 to 44,642 lb/acre (5,000 to 50,000 kg/ha) furrow slice.

6. Based on degree of availability to plants, soil K can be grouped into three categories: difficultly available potassium, making up 90 to 98 percent of the total; slowly available potassium constituting 2 to 10 percent of the total; and readily available potassium making up 1 percent of the total potassium.

7. The principal fertilizer compounds containing potassium are potassium chloride, potassium sulfate, potassium-magnesium sulfate and potassium nitrate. They are of equal agronomic effectiveness as sources of potassium for crops. An analysis of representative samples is given in Table 6.1.

8. Potassium sulfate is an excellent source of sulfur as well as potassium.

9. Potassium-magnesium sulfate is often used to correct magnesium and sulfur as well as potassium deficiencies.

[12]W. W. Masters, C. W. Wendt, C. Harvey and A. W. Young, "Nutrient Assimilation by Cotton as Measured by Petiole Analysis," *Agron. Abstr.* (1962), p. 38.

[13]J. B. Jones, "Plant Analysis Handbook for Georgia," University of Georgia Agricultural Extension Service Bulletin 735, 1974.

10. Potassium nitrate has the lowest salt index of all potassium products and is an excellent source of nitrate nitrogen.

11. The potassium chloride component of total United States potash production decreased and then stabilized at about 78 percent in 1970.

12. Over 90 percent of all processed potash in the United States is consumed as fertilizer.

13. About three-fourths of potash consumed in the United States is imported from Canada.

Illustrative Problem: A soil scientist tells you that your soil contains 0.1 milliequivalent of exchangeable potassium per 100 grams. How many pounds of muriate of potash (60 percent K_2O) must be added per acre to bring the level up to 0.2 milliequivalent potassium per 100 grams? (See Figure 1.8). Atomic weight of potassium = 39 grams; K × 1.20 = K_2O; 1 acre furrow slice weighs 2,000,000 lb.

> 1 gram equivalent = 39 g
> 1 milligram equivalent = 39 milligrams
> Milligram equivalent = milliequivalent = meq
> 0.1 meq = 0.1 × 39 = 3.9 milligrams
> 3.9 milligrams/100 grams
> 0.0039 grams/100 grams
> 0.0039 pounds × 20,000/100 pounds × 20,000
> 78.00 pounds/2,000,000 pounds

Answer: For 0.2 meq, you need to add 78 more pounds of potassium.

$$78 \times 1.20 = 93.6 \text{ pounds } K_2O$$

$$0.60x = 93.6 \text{ pounds}$$

$$X = \frac{93.6}{0.60} = 156 \text{ pounds per acre muriate of potash (KCl)}$$

156 × 1.12 (factor to convert lb/acre to kg/ha) = 175 kg/ha

QUESTIONS

1. The three availability categories of potassium in soil are:

   a. _____
   b. _____
   c. _____

2. Exchangeable potassium is _____ available.

3. Illite potassium is _____ available.

4. Over 90 percent of all processed potassium is consumed as _____.

True or False:

5. _____Tobacco needs about 17.8 to 26.7 lb/acre (20 to 30 kg/ha) of chlorine for maximum yield.

6. _____Too much chlorine in tobacco creates an undesirable burning quality in the cured leaf.

7. _____Only about 65 percent of the total annual production of potash is used in agriculture.

8. _____The official chemical test for potassium in fertilizers is based on its solubility in water.

9. _____Potassium carbonate, first obtained from plant ashes, has about 64 percent K_2O and is the highest analysis potash fertilizer that has been produced.

10. _____As a source of potassium for crops, nitrate of potassium is much more effective than muriate of potassium.

Directions for Nos. 11–18. Column A lists several potassium compounds that are used for fertilizer. Column B lists some K_2O and potassium percentages in scrambled fashion. In the answer columns, rearrange the percentages of K_2O and potassium in an order that most nearly describes the fertilizers listed in Column A.

Column A	Answer		Column B	
	K_2O	Potassium	K_2O	Potassium
11 and 12. KNO_3	( )	( )	22	53
13 and 14. KCl	( )	( )	64	37
15 and 16. K_2SO_4	( )	( )	44	41
17 and 18. $K_2SO_4 \cdot 2MgSO_4$	( )	( )	60	18

Problems:

19. A soil containing 0.2 milliequivalent of potassium per 100 grams contains: (atomic weight of potassium = 39)

_____pounds per acre

_____kilograms per hectare

20. A soil containing 0.15 milliequivalent of potassium per 100 grams would require _____pounds per acre of muriate of potash to increase the potassium content to 0.25 milliequivalents of potassium per 100 grams. Atomic weight of potassium = 39; muriate of potash = 50 percent potassium.

chapter 7

Calcium, Magnesium and Finely Ground Limestone

Pimiento Pepper

Pimiento Peppers with Varying Degrees
of Blossom-End Rot

Calcium Content %

0.22   0.20   0.18

Pimiento peppers showing varying degrees of blossom-end rot, a disorder related to the calcium content of the fruit. Fruits which contained 0.18% Ca developed about 25% rot, fruits containing 0.21–0.22% Ca had 2–10% rot and those containing 0.24% Ca were free of the disorder. A similar relationship between blossom-end rot and calcium content exists with tomato fruits. (Hamilton, L.C. and W.L. Ogle. Proc. Amer. Soc. Hort. Sci. 80: 457-461, 1962.)

About 1825, Edmund Ruffin determined that marl would improve the growth of clover on the soils of Tidewater, Virginia. He suggested an acidity test and noted that calcium and magnesium were important plant nutrients. We know today that they are essential and may be supplied by marl or dolomitic limestone. In addition to supplying essential plant nutrients, finely ground limestone will also neutralize the acidity of mineral soils in humid regions, almost all of which are acid and indeed were acid when they were first cleared by people like Edmund Ruffin. Cultivation, crop residues and fertilization with nitrogen and sulfur all have taken their toll and have contributed to the natural soil acidity brought about through leaching of bases by excess water in a humid region.

From 1943 to 1949, tests were conducted in Mississippi[1] to determine the changes in pH and the lime requirement of three light-textured soils over a six-year period under corn and cotton culture. The six-year average yields per acre (hectare) were 400 lb/acre (448 kg/ha) of lint and 35 bu/acre (2,352 kg/ha) of corn. Although these yields are not high, they were close enough to the average annual yields in the hill section of Mississippi to give an idea of the magnitude of lime loss. In 1943, these three fields were limed to neutralize the soil acidity and to bring the pH up to 7.0. It required 2,400 lb/acre (2,688 kg/ha) for the Prentiss fine sandy loam, 2,800 lb/acre (3,136 kg/ha) for the Ruston sandy loam and 3,400 lb/acre (3,808 kg/ha) for the Grenada silt loam. After six years of continuous cropping from 1943 through 1949, the lime required to bring the pH back to 7.0 was calculated. On the Prentiss fine sandy loam, 1,280 lb/acre (1,433 kg/ha) was required. On the Ruston sandy loam, 1,560 lb/acre (1,747 kg/ha) was required. On the Grenada silt loam, 1,920 lb/acre (2,150 kg/ha) was required. Annual losses of lime from 1943 to 1949 were obtained by dividing the total loss by six. The annual lime losses in these years amounted to 213 lb/acre (238 kg/ha), 260 lb/acre (291 kg/ha) and 320 lb/acre (358 kg/ha) per year respectively on the fine sandy loam, sandy loam and silt loam soils (Table 7.1).

Chemically pure lime is CaO but, as commonly used, refers also to $CaCO_3$ and $Ca(OH)_2$. Lime as used here refers to any of these compounds, with or without magnesium and other limestone-derived materials, that are added as an amendment in order to neutralize acid soils.

In addition to neutralizing soil acidity and supplying calcium and magnesium, lime has a considerable influence on the availability of other plant nutrients (Figure 7.1). There is a general trend of relation of a soil's reaction on pH and associated factors to the availability of its plant nutrient elements. Each element can be thought of as a band whose width

---

[1]U. S. Jones and C. Dale Hoover, "Lime Requirement of Several Red and Yellow Soils as Influenced by Organic Matter and Mineral Composition of Clays," *Soil Sci. Soc. Am. Proc.* 14 (1949), 96–100.

216

Table 7.1
CHANGES IN pH AND LIME REQUIREMENT OF VARIOUS SOILS OVER THE PERIOD 1943–1949 UNDER CORN AND COTTON CULTURE

Soil Type	Initial pH	Lime Added To Give pH 7 1943 Lb/Acre[1]	pH 1943	pH 1945	pH 1947	pH 1949	Lime Required To Give pH 7 1949 Lb/Acre[1]
Prentiss fine sandy loam	5.0	2400	7.0	6.9	6.8	6.1	1280[2]
Ruston sandy loam	5.3	2800	7.0	6.6	6.4	5.9	1560
Grenada silt loam	5.0	3400	6.9	6.7	6.6	6.0	1920

[1]Lb/acre × 1.12 = kg/ha

[2]Recent data indicate that this much lime may be required each year by a 10 to 12 ton/acre corn silage or Coastal Bermuda hay crop.

at any particular pH value indicates the relative influence of this pH value and associated factors on the presence of the element in readily available forms, but not to the actual amount present, this being influenced by other factors such as cropping and fertilization.

In Figure 7.1, it is noted that high soil pH or alkalinity tends to make nitrogen and phosphorus less available to plants. Also, acidity or low soil pH tends to make these same nutrients, as well as potassium, sulfur, calcium and magnesium, less available to growing roots. All of these essential nutrients have a satisfactory degree of availability between a pH of 6.0 and 6.5.

Acid-forming fertilizer is not the most important, but it is one of the several factors contributing to the development of soil acidity in humid regions. According to a paper by W. H. Pierre in the *1938 Yearbook of Agriculture,* "Soils and Men," 1.8 pounds (0.8 kg) of lime is required to neutralize each pound (0.45 kg) of nitrogen added to soil as anhydrous ammonia, ammonium nitrate, urea and nitrogen solutions made from one or more of these materials. Other workers have established different values, but it is certain that the acidifying effect of modern nitrogen fertilizers is considerable, and good soil management in humid regions requires that this fact be considered.

Soils classified as Spodosols, Ultisols and Oxisols in humid regions are often deficient in calcium and other bases. On the other hand, soils in arid and semiarid regions of the Mollisol and Aridisol orders often have a sufficient supply of calcium and other basic plant nutrients. The "Lime Line" (Figure 7.2) divides the lime-requiring regions in the United States from those where liming is not generally needed.

There are soil conditions where so much lime is present that it is very detrimental to plant growth. Soil that is irrigated for many years with hard water can accumulate an excessive amount of bases, and subsequently its capability to support healthy plant growth is limited. Examples are the soils in Arkansas, Louisiana and California where rice is irrigated with

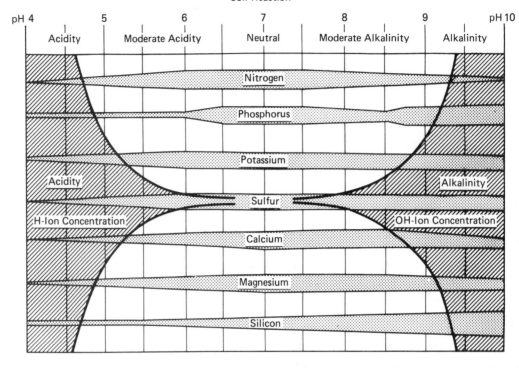

Soil Reaction

Figure 7.1   Diagram indicating the influence of soil reaction on the solubility of silicon, the three major plant nutrients, N, P, and K and the three secondary nutrients, Ca, Mg, and S. The width of the bands indicates more or less solubility of the element as influenced by conditions associated with pH and not necessarily to actual amounts present.

water containing large quantities of calcium bicarbonates. Such conditions have been diagnosed by soil and water assays for pH and carbonates and by plant tests for micronutrients. The solution to the problem is usually the addition of a small quantity of zinc, manganese or iron.[2]

Salting of the land, resulting from the evaporation of irrigated water, is a serious problem in some parts of the western United States and in modern Iraq. In ancient Mesopotamia between 2400 and 1700 B.C., the same salting undermined the agriculture and almost destroyed civilization as it was known at that time. This tragic fact of nature gave powerful impetus to the eclipse of Sumerian culture and language, which was the earliest known civilization in the history of the human community.[3]

[2]B. R. Wells et al., "Effect of Zinc on Chlorosis and Yield of Rice Grown on Alkaline Soil," Arkansas Agricultural Experiment Station Report Series 208, 1973.

[3]T. Jacobsen and R. M. Adams, *Salt and Silt in Ancient Mesopotamian Agriculture*.

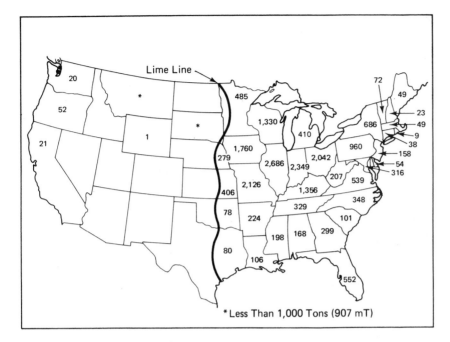

Figure 7.2   The lime line divides the U.S. into the east, made up predominantly of Ultisols and Spodosols which need lime, and the west made up predominantly of Aridisols and Mollisols which do not. The figures on the map indicate the consumption of liming materials in the recent past.

Without rather sophisticated soil science and drainage engineering, correcting saline soil conditions is all but impossible; however, correcting soil acidity is very practical and economical. The purpose of the following discussion is to present some facts about naturally occurring calcium carbonate and related minerals that are widely used to correct acidity in soils of humid regions.

Liming acid soils is a desirable practice because it:

1. Corrects soil acidity.
2. Supplies calcium and magnesium.
3. Speeds the decay of organic matter and the liberation of plant nutrients.
4. Increases the availability of residual and applied phosphorus.
5. Increases fixation of nitrogen by soil and plant organisms.
6. Increases crop yields.
7. Improves the physical properties of soils.
8. Reduces the activity of injurious substances in the soil.

Soil tests for lime requirement, as described in the section on calcium, are about the most accurate tests available, and the suitable pH for most field and vegetable crops will be about 6.0 to 6.5. The amount of lime to add is determined by the pH, the texture and organic matter content of the soil. Soil testing takes the guesswork out of lime requirements. Soil scientists can suggest a sound liming program where needed. Farmers in humid regions should not attempt to grow most field, forage, pasture or vegetable crops or to use phosphorus, potassium or nitrogen until they have found through a soil test whether the land is acid and needs lime. If it is acid, a sufficient amount of dolomitic lime to raise the pH level to 6.0 to 6.5 should be added. This will increase the supply of calcium and magnesium and will "sweeten" the soil. Research has clearly shown that the availability of phosphorus is influenced by soil acidity (Figure 7.3) and that one pound (0.45 kg) of fertilizer phosphorus produced more pasture herbage at pH 6.3 than did three pounds (1.4 kg) at pH 5.2.[4] The pH level that is

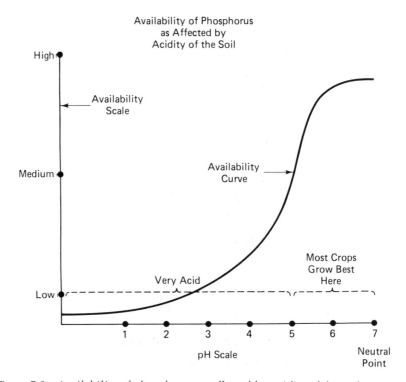

Figure 7.3 Availability of phosphorus as affected by acidity of the soil.

[4]U. S. Jones, "Phosphate Fertilizer," Mississippi Agricultural Experiment Station Bulletin 503, 1953.

most conducive to plant growth is also the level that is most satisfactory for the growth of beneficial soil microorganisms. *Truly, the proper use of lime may be the backbone of permanent agriculture in humid regions.*

Where magnesium is plentiful, there is very little difference among the various kinds of finely ground agricultural lime, but there is a lot of difference between using no lime and using two tons per acre (4.48 m tons/ha), or even 1,000 pounds per acre (1,120 kg/ha) (Table 7.2).[5]

When the soil pH drops below 5.8 (Table 7.5), it is a positive signal that the storehouse or supply of bases, such as calcium and magnesium, is getting low and needs to be replenished. This can best be done by adding agricultural lime as required.

Lime is lost from the soil in a number of ways—crop removal, leaching and soil erosion. A good crop of clover, for instance, absorbs bases equal to 150 pounds (68 kg) of lime, corn 100 pounds (45 kg) and soybeans 80 pounds (36 kg). An acre (0.4 ha) of cotton stalks, leaves, lint, seed and burrs removes 100 to 200 pounds (45 to 90 kg) or as much as alfalfa, one of the most lime-loving crops. Fortunately, only the cotton fiber and seed are taken off the land.

Water passing through light-to-medium-textured row crop soils in a humid region washes out 200 to 300 pounds (90 to 136 kg) of lime per year over and above that returned to the soil by crop residues (Table 7.1). Erosion removes not only lime and other nutrients, but the whole topsoil as well.

High rates of nitrogen are increasing yields per acre (ha) and

Table 7.2
RELATIVE LIMING VALUE OF 1000 POUNDS EACH OF FOUR DIFFERENT LIMING MATERIALS MEASURED BY WHITE CLOVER FORAGE AND SEED YIELDS ON FOUR LIGHT- TO MEDIUM-TEXTURED ACID SOILS IN MISSISSIPPI

	Dry Forage Lb/Acre[1]	Seed lb/Acre[1]
No lime	1834[2]	107
Calcium lime	2162	150
Calcium silicate slag	2132	132
Dolomitic lime	2194	140
Basic slag	2101	152
LSD for all sources	N.S.	N.S.

[1]Lb/acre × 1.12 = kg/ha

[2]Lime significantly increased the yield of forage and seed. No statistically significant difference was found among lime sources.

[5]W. D. Edwards, "The Role of Lime and Minor Elements on White Clover Forage and Seed Yields," M. S. Thesis (State College, Mississippi: Mississippi State University, 1953).

improving crop production efficiency all over the world. However, these high yields result in the removal of much lime from the soil. In sandy soils of humid regions, as much as one-half of this nitrogen washes through the soil profile, carrying with it bases such as calcium and magnesium. Although, this is definitely adding to the lime deficit, the calculated amount of loss attributable to acid-forming fertilizers is greatly exceeded by crop removal, leaching and soil erosion.

A lime program on acid soils is needed in order to maintain production efficiency, increase yields per acre (ha) and produce better quality crops.

Many factors influence the lime requirement of the soil. Loamy acid soils and those with high organic matter content need more lime than do sandy acid soils with little organic matter. A soil test will tell what the soil needs.

Sources of Lime. There are many kinds of agricultural lime made by many processes and with different chemical and physical properties. The common forms of lime include the following materials:

1. Calcitic limestone (calcium carbonate and impurities).
2. Dolomitic limestone (calcium carbonate and magnesium carbonate plus impurities).
3. Marl ($CaCO_3$, clay and organic matter).
4. Quick lime or burned lime (CaO):

$$CaCO_3 + heat \rightarrow CaO + CO_2$$

5. Water-slaked lime or hydrated lime ($Ca(OH)_2$):

$$CaO + H_2O \rightarrow Ca(OH)_2$$

6. Air-slaked lime ($CaCO_3$):

$$CaO + CO_2 \rightarrow CaCO_3$$

7. By-product materials such as sugar beet refuse, paper mill sludge, tannery waste, wood ashes, marble dust, oyster shells, gas work waste and various slags from industry.

Agricultural dolomitic limestone is a gray to white finely ground material. Some of it is a by-product of zinc-refining processes that use ground mother dolomite from the Knox formation in Tennessee. Deposits of dolomite are widely distributed in the United States.

In addition to agricultural uses, limestone is employed as a building material or foundation in the form of marble and rock. It is also used as a chemical because of its alkaline nature and its ability to neutralize acids.

It is for this chemical purpose that it is used on crop land. The acid-neutralizing value based on 100 percent $CaCO_3$ calcitic lime is given for several liming materials in Table 7.3.

Limestone chiefly contains calcium carbonate, which was derived from an accumulation of ancient organic remains such as sea shells and other foraminifers. Marl, an earthy, crumbly deposit found near seashores or former seashores and consisting of clay mixed with calcium carbonate, is sometimes used as lime for the land. Sugar-factory sludge, basic slag from the steel industry, marble dust, oyster shells and wood ashes are also used to a limited extent.

Chalk, a soft limestone, is used as an agricultural liming material or to write on a blackboard. It is the parent material of the white cliffs of Dover. Underlying the black lands of Texas, Mississippi and Alabama it is known as Selma chalk.

Carrara marble, a beautiful white hard limestone, is used for statues and church altars, and is found in the Appenine Mountains near Carrara, Italy.

When finely ground, limestone appears to be gray, white or buff in color, depending on the color of the original rock. Some limestones have a beige tinge, because of iron impurities.

Handling Characteristics. Over 90 percent of the lime used in the United States for treatment of acid soils is finely ground limestone, consisting for the most part of calcium carbonate or a mixture of calcium and magnesium carbonates. It should be fine enough for 90 percent to pass a 10-mesh (1.9-mm) screen, 50 percent through 50-mesh (0.35 mm) and 25 percent to pass

Table 7.3
LIMING MATERIALS AND ACID NEUTRALIZING VALUE

Liming Materials	Neutralizing Value, $CaCO_3$ Equivalent, %	Equivalent to One Ton of Lime, lb.[1]
Calcitic lime	100	2000
Burned lime	178	1120
Builders' hydrated lime	134	1490
Magnesium carbonate	119	1675
Dolomitic limestone	108–95	1850–2100
Ground shells	88–80	2200–2500
Calcium silicate slag	80–71	2500–2800
Basic slag	71–67	2800–3000
Blast furnace slag	74–67	2700–3000
Flue dust	96	2100
Marl	70–40	2800–5000
Rock phosphate	7	
Wood ashes	40	5000

[1]Lb ÷ 2.2 = kg

100-mesh (0.14 mm). Ten mesh means 10 wires per running inch (2.54 cm), or 100 openings per square inch (6.45 cm²); 50-mesh means 50 wires per running inch (2.54 cm) or 2,500 openings per square inch (6.45 cm²). The agricultural liming materials specifications for fineness and neutralizing value for each state may be obtained from the Agricultural Stabilization and Conservation Program, United States Department of Agriculture office in each state capitol.

Regardless of the source, agricultural lime after mining is ground or pulverized in hammer mills or similar equipment in order for it to be useful on the land. Much limestone that falls below the standards of fineness noted above is being used because it is relatively inexpensive as a by-product. When the amounts added to the land are properly adjusted to take into consideration the lesser quantities of fine material, good results from the coarser lime may be expected. Much of this coarse material, 6 to 8-mesh (3.3 to 2.3 mm), is being used today as a filler in granulated fertilizer. As an added ingredient to granulated multinutrient fertilizer, the coarse material should not be given the same value as the finer material used for direct addition to the soil and as a filler in pulverant fertilizer. It should not, however, be discounted entirely because, with intimate contact to roots, it will eventually become a source of plant nutrients for crops.

One solution to the problem of coarse granular lime would be to granulate finely ground limestone. One product being considered for use as a filler in granulated fertilizer is made from ground limestone, 87 percent of which will pass a 200-mesh (0.074-mm) screen and 74 percent of which will pass a 325-mesh (0.03-mm) screen. Potassium chloride is added to the finely ground limestone in amounts up to 20 percent. The product is dried at 300°F (149°C). The material is 6- to 8-mesh (3.3 to 2.3 mm) in size, and abrasion loss is 4.4 percent when this material is shaken on a 16-mesh (1.2-mm) screen. Experiments have been conducted using K_2SO_4 in lieu of KCl with the view to improve the stability of the product and to reduce the abrasion loss.

As is true of fertilizer, granular lime spreads more efficiently than the pulverized material. However, fertilizer nitrogen, potassium and to some extent phosphorus dissolve in the soil in time for the growing roots to absorb them, but this is not the case with coarse limestone. Until a fine particle of limestone comes in contact with acid soil particles or the carbonic acid released by growing roots, it remains in a rather inert state in the soil, as it did in the quarry and as a building stone. The larger the particle of limestone is, the less opportunity it has to be dissolved because much less surface area is exposed to acid soil particles and growing roots (Table 7.4).

Various mechanisms have been devised to spread lime evenly on the land. Among these are spreaders attached to truck bodies (Figure 7.4) and those pulled by tractor equipment or mounted on tractors (Figure 7.5). For best results, agricultural lime should spread evenly on the land and

Table 7.4
EFFECTS OF LIMESTONE FINENESS AND TYPE OF LIMESTONE ON SOIL pH

Limestone Fineness, % passing 60-mesh (0.25-mm) size opening	Type of Limestone	Initial Soil pH	Soil pH 2 years After Application of Limestone	
			2.5 ton/acre[1]	5.0 ton/acre[1]
45	Calcitic	5.7	6.0	6.4
80	Calcitic	5.4	6.5	6.8
45	Dolomitic	5.3	5.7	6.2

[1]Ton/acre × 2.24 = m ton/ha.

then plowed or disked in until it is intimately mixed with the furrow slice of soil.

Chemical Characteristics. The power of limestone to correct soil acidity is determined by its fineness and content of calcium carbonate or magnesium carbonate. Consequently, the neutralizing value of different kinds of lime varies with the percentages of these compounds. It is customary to consider pure calcium carbonate as a basis for comparison, giving it a *neutralizing value* of 100. The purity of the liming material is then

Figure 7.4 Lime and fertilizer spreader mounted on truck body.

225

Figure 7.5 Lime and fertilizer spreader adapted for mounting on tractor.

expressed in terms of the percentage of calcium carbonate equivalent (Table 7.3).

To illustrate the concept of calcium carbonate equivalent, consider a sample of pure $MgCO_3$. Since the molecular weight of $MgCO_3$ (84.3) is *lighter* than the molecular weight of $CaCO_3$ (100) and since a *molecule of MgCO_3* will neutralize exactly the same amount of acid as a *molecule of CaCO_3*, one should understand readily that pound for pound (kg for kg) $MgCO_3$ can neutralize more acid than $CaCO_3$. This is true because there are more $MgCO_3$ molecules in a pound (kg) than there are $CaCO_3$ molecules.

To be exact, a pound (0.45 kg) of $MgCO_3$ will neutralize 1.19 times as much acid as a pound (0.45 kg) of $CaCO_3$. This factor is obtained by dividing the gram equivalent weight of $CaCO_3$ (50) by the gram equivalent weight of $MgCO_3$ (42), $50 \div 42 = 1.19$. Expressed on a percentage basis, this means that pure $MgCO_3$ has a neutralizing value of 119 percent, compared with 100 percent for pure $CaCO_3$. Thus, for a material containing 50 percent $MgCO_3$ (neutralizing value = 119) and 50 percent $CaCO_3$ (neutralizing value = 100), the neutralizing value of the mixture would be

Note the neutralizing value of dolomitic limestone in Table 7.3. These principles are expressed in chemical equations as follows:

$$CaCO_3 + H_2SO_4 \rightarrow CaSO_4 + H_2CO_3; \text{ mol. wt. of } CaCO_3 = 100$$

$$MgCO_3 + H_2SO_4 \rightarrow MgSO_4 + H_2CO_3; \text{ mol. wt. of } MgCO_3 = 84$$

This method of determining the neutralizing value of limestone depends on the reaction of an acid with the carbonates of the limestone, and, therefore, the greater neutralizing value of $MgCO_3$, contained in the dolomitic limestone, is measured chemically.

Marl, soft limestone and chalk contain varying amounts of clay, muck, sand or other impurities (Figure 7.6). It is not uncommon to find samples with 40 to 70 percent calcium carbonate equivalent. Limestone rock usually contains impurities of clay; consequently, samples of ground limestone seldom, if ever, approach 100 percent calcium carbonate. Usually, good quality agricultural calcitic lime contains 85 to 95 percent calcium carbonate. Good quality agricultural dolomitic lime usually contains 95 to 100 percent calcium carbonate equivalent.

Limestone composed mainly of calcium carbonate is called high calcium or calcitic limestone; those having more than 10 percent of their value in the form of magnesium carbonate are called dolomitic limestones. True dolomite contains about 12 percent magnesium. Dolomitic limestone supplies about ten times as much magnesium to soils in the United States as all other magnesium carriers put together.

Ground dolomite, as commonly sold for agricultural lime, contains the equivalent of about 12.8 percent magnesium and 17 percent calcium. This material is a double carbonate of calcium and magnesium. Nearly all calcitic limestones contain small amounts of magnesium. Magnesium carbonate itself is 10 times more soluble than calcium carbonate. Yet the double carbonate of calcium and magnesium (dolomite) is much less soluble than either carbonate singly. See the difference in soil pH change 2 years after application in Table 7.4. Because magnesium has a smaller atomic weight than calcium, it is not uncommon for dolomite to assay as high as 105 percent calcium carbonate equivalent.

Crop Response to Lime. In general, liming acid soils is a beneficial and profitable practice. However, Irish potatoes, tobacco, red top and strawberries appear to be benefitted little, if at all. Watermelons, blackberries, blueberries and azaleas do better on acid soils (Table 7.5) and, indeed, may be markedly injured by liming. The latter crops need the calcium and magnesium contained in liming materials, but the influence of the alkalinity produced appears to be detrimental. This is most likely caused by the fact that manganese, iron, zinc, boron, and perhaps other micronutrients are in limited supply and are made unavailable by the alkalinity produced. In the case of potatoes, acidity helps to control the potato scab organism; in

Figure 7.6 Top photo shows scarifier or ripper breaking up Selma chalk in Cedar Bluff, Mississippi. Lower photo shows truck dumping rock on feeder for movement to grinder and subsequently to open freight car.

flue-cured tobacco, moderate acidity improves the quality of the cured leaf; in peanuts, it helps control the fungus disease, Pythium.

Considerable soil fertility research in the past century has contributed a great amount of information on crop response to lime in humid regions of the United States. A brief statement on the expected response of various crop commodities grown on acid soils to lime is given below.

Soybeans.   It is the consensus of those concerned with soybean production on acid soils that yield increases from proper liming may range from 3 to

Table 7.5

229
Liming Acid Soils

APPROXIMATE SOIL pH RANGES SUITABLE FOR VARIOUS PLANT SPECIES

pH 4.8–5.2 (acidity)	pH 5.8–6.2 (moderate acidity)		pH 6.3–6.7[1] (nearly neutral)
Azaleas	Apples	Peaches	Alfalfa
Blackberries	Beans, lima	Peanuts	Asparagus
Blueberries	Beans, snap	Peppers, Pimiento	Cabbage
Grass, carpet	Beans, velvet	Pine, yellow	Carrots
Grass, centipede	Cantaloupes	Radishes	Ladino—grass
Hydrangia, blue	Corn	Rice	Lettuce
Irish potatoes	Cotton	Small grain	Onions
Juniper, Irish	Cowpeas	Sorghum	Peas
Pine, longleaf	Crimson clover	Soybeans	Red clover
Watermelon (< 5.8)	Cucumber	Squash	Spinach
	Grass, most kinds	Strawberries	Sweet clover
	Greens, mustard, etc.	Sudan grass	Timothy
	Iris, blueflag	Sweet potatoes	White clover
	Kale	Tobacco	
	Lespedeza	Tomatoes	
	Parsnips	Trefoil, birdsfoot	
	Red top	Vetch	

[1]On highly organic soils, it is not advisable to lime to a high pH. Good plant growth on these soils may be obtained where the pH is around 5.5. Approximate pH ranges refer to those suitable for mineral soils in humid regions where lime is often added. Many of these species, for example alfalfa, apples, iris, cotton, rice, sorghum, sweet clover and timothy, grow well on calcareous soils with pH values up to 7.5 to 8.0.

8 bushels per acre (202 to 538 kg/ha). Correcting soil acidity is the most critical need in achieving significant soybean yield increases throughout the southeastern and mid-western United States.

Cotton. Research and field tests would suggest that where lime is necessary and properly applied, an increase of 100 pounds per acre (112 kg/ha) of lint cotton can reasonably be expected. Recent research shows that cotton seedlings are especially sensitive to less than optimum soil pH. See Rios and Pearson[6] for a discussion of calcium needs of cotton roots and Lancaster[7] for a discussion of magnesium nutrition of cotton.

Peanuts. There are many fields in the southeastern United States where it would be reasonable to expect that liming could increase peanut yields by 500 pounds per acre (560 kg/ha). Much progress has been made by

[6]M. A. Rios and R. W. Pearson, "The Effect of Some Chemical Environmental Factors on Cotton Root Behavior," *Soil Sci. Soc. Am. Proc.* 28 (1964), 232–235.

[7]J. D. Lancaster, "Magnesium Status of Blackland Soils of Northeast Mississippi for Cotton Production," Mississippi Agricultural Experiment Station Bulletin 560, 1958.

growers in removing soil acidity as a hindrance for achievement of top yields. See Garren and Hallock[8] for a discussion of calcium nutrition of peanuts.

Tobacco. On-the-farm tests for a four-year period during 1964-67 indicate that about $27 to $30 per acre (0.4 ha) additional return can be expected from the use of 1,000 to 2,000 pounds per acre (1,120 to 2,240 kg/ha) of lime where soil test suggests this rate of lime application. Often, the addition of lime for tobacco will not only improve the tobacco but also the crops that follow it. The optimum pH for tobacco is less than for many other crops, but there is much evidence to show that less than optimum pH will prevent maximum yields. This is true of both flue-cured and burley tobacco. See Clemson University's Fertilizer Recommendations[9] for a discussion of calcium and magnesium needs of flue-cured tobacco.

Corn. Where lime is needed, the response may range from an additional 5 to 10 bushels per acre (336 to 672 kg/ha) of corn from proper liming. Continued evidence shows that soil acidity is a severe deterrent to maximum yields of this all-important crop. High nitrogen rates will continue to create the need for proper liming of corn in humid regions. See Underwood and Miller[10] for a discussion of magnesium needs of field and sweet corn.

Forages (Legume and Grasses). Much research throughout the humid regions of the United States suggests that a 50 to 75 percent increase in air-dry hay or forage could be realized from proper liming. Legumes are very responsive to lime, and they will last only a short time if the soils are not properly limed. Even pure grass that is fertilized heavily with nitrogen will respond to lime. High-yielding crops, such as the Bermuda grasses, or other crops used for silage, namely forage sorghum and corn, will respond to proper liming. Generally, all crops used as forage, hay, silage or grazing pastures, will respond to lime in humid regions. They are most responsive when other nutrients are adequately supplied. In many instances, the response to lime will not be as great as described above unless adequate phosphate and potash are applied. See Brupbacher and Sedberry[11] for a

[8]K. H. Garren and D. L. Hallock, "Calcium Helps Prevent Fungus in Peanuts," *Agric. Res.*, March 1968.

[9]"Fertilizer Recommendations for 1979," Clemson University Extension Circular 476.

[10]N. R. Underwood and J. R. Miller, "Effect of Soil pH on the Availability of Magnesium to Corn from Magnesium Sulfate and High Magnesium Liming Materials," *Soil Sci. Soc. Am Proc.* 31 (1967), 390–93.

[11]R. H. Brupbacher and J. E. Sedberry, Jr., "Effects of Magnesium and Sulfur on Growth and Chemical Composition of Clover on Fourteen Coastal Plain Soils in Louisiana," Louisiana Agricultural Experiment Station Bulletin 599, 1965.

and Grunes and Mayland[12] for critical forage magnesium levels for animals.

Small Grains. It is reasonable to expect that in some cases 3 to 8 bushels (202 to 538 kg/ha) of additional oats, barley or wheat may be realized where soil acidity is limiting yields. Seedling wheat, barley or oats may die as a result of severe acidity. High-yielding wheat varieties with their capability of responding to up to 200 pounds (90.9 kg/ha) of nitrogen per acre will need particularly close attention to proper pH in the topsoil and to excessive aluminum content of the subsoil.

Vegetable. High-value crops, such as sweet potatoes, tomatoes, snap beans, cabbage, pickling cucumbers and lettuce, will certainly respond to proper liming according to soil test. Where so much is at stake with a high-value crop, it is particularly important that soil acidity not be a deterrent to maximum, high-quality, profitable yields.

Fruits. Important fruit crops, such as apples, peaches, blueberries, strawberries and dewberries, should be limed only at the suggestion of a soil test. Acid-loving crops, such as blueberries, should be managed only with a soil test to avoid any possibility of overliming.

Turf, Lawns and Home Gardens. It will pay the homeowner to take a soil sample and to have it tested in order to keep an accurate pulse on the soil-acidity problem that may occur in connection with his lawn or garden.[13]

Forestry Resources. Although lime is not used in any widespread fashion on tree establishment or other forest production, vivid evidence has come to light recently of the need for soil testing to maintain a proper pH in the State Tree Nurseries.[14] Soil acidity in some instances was so severe that it was impossible to produce uniform-sized, vigorous seedlings.

Trellised Tomatoes. At present, it is suggested that the soil used for trellised tomatoes be limed to pH 6.5; also, an additional ton (0.9 m ton) per acre of limestone is suggested to cope with the acidity caused by the high nitrogen use that is so necessary for the achievement of high yields. See Hamilton and Ogle[15] for a discussion of calcium and Jones and Jones[16] for a discussion of magnesium needs of tomatoes.

[12]D. L. Grunes and H. F. Mayland, "Controlling Grass Tetany," Leaflet No. 561 (Washington, D.C.: United States Department of Agriculture, 1975).

[13] "Fertilizer Recommendations."

[14] "Fertilizer Recommendations."

[15]L. C. Hamilton and W. L. Ogle, "The Influence of Nutrition on Blossom-End Rot of Pimiento Peppers," *Proc. Am. Soc. Hort. Sci.* 80 (1962) 457–461.

[16]U. S. Jones and T. L. Jones, "Influence of Polyethylene Mulch and Magnesium Salts on Tomatoes Growing on Loamy Sand," *Soil Sci. Soc. Amer. J.* 42, 1978.

Modern fertilizer technology is placing at least equal emphasis on the manufacture of ammonium phosphate and calcium phosphates. This intensifies the consideration of calcium as a plant nutrient. Because all liquid multinutrient fertilizers and 50 percent or more of modern, solid multinutrient fertilizers are almost devoid of calcium, it is no longer feasible to assume that calcium will be supplied as an incidental ingredient of fertilizers.

Essentiality of Calcium.   Calcium is an essential element for both plant and animal growth. The formation of teeth, tusks, bones and cell walls in animals is dependent on an ample supply of calcium. The living system is an outstanding example of cooperation and interdependence: each nutrient has its part to play but none has complete independence. Calcium and phosphorus, the structural components of man and other higher animals, make up most of the mineral material of the body. They are intimately deposited in the protein framework of bone and tooth cells to create a hard tissue able to bear weight and pressure. Many people in the world eat food that satisfies the nutrient requirements of calcium, but some do not. History has frequently recorded a nutrient deficiency disease called rickets. Rickets is caused by a deficiency of calcium and vitamin D.

While no widespread deficiencies of calcium in food have been reported recently, it is nonetheless an essential element for plant growth. Unless it is present in ample quantities in the soil or added as fertilizer or as finely ground limestone, it will become deficient, particularly for crops that utilize substantial quantities annually, such as legumes.

A relatively large part of the calcium content of plants is located in the leaves. It is relatively immobile in plant tissue once it has been laid down. The above-ground portion of most mature grain crops and grasses contains from 0.25 to 0.5 percent calcium. It occurs in sizeable quantities, up to 4 percent, in tobacco leaves. The above-ground parts of cotton, soybean and alfalfa plants average about 2.0 percent calcium. Peanut seeds contain about 0.5 percent calcium.

Calcium pectate is a structural component of plant cell walls. Calcium has a beneficial effect on the permeability of cytoplasmic membranes. Root membranes, in particular, break down in the absence of calcium. Without a certain minimum concentration of calcium in water surrounding plant roots, plant root cells break down and exude inorganic salts and organic compounds into the ambient water. Above the minimal concentration of calcium, this does not happen.

Subsoils low in calcium and high in aluminum content are not penetrated by growing roots. Rios and Pearson[17] in 1964 showed that cotton roots failed to grow in subsoils low in calcium.

[17]Rios and Pearson, "The Effect of Chemical Factors."

Calcium functions in plants as follows:

1. Promotes early root formation and growth.
2. Improves general plant vigor and stiffness of straw.
3. Influences intake of other plant foods.
4. Neutralizes poisons produced in the plant.
5. Encourages grain and seed production.
6. Increases calcium content of food and feed crops.

Prediction of Calcium Needs.   There is no clearly defined leaf symptom of calcium deficiency. Nutrient solutions low in calcium have been shown to cause collapsed leaf petioles in young cotton and soybeans. Later, soybeans exhibit necrosis and brown dead spots on the edges of leaves. A malady knows as "crinkle leaf" occurs on cotton, turnip greens and leafy vegetables growing on acid soils; it is caused by an excess of manganese, up to 1500 ppm in cotton leaves. About 40 ppm of manganese is normal for cotton leaves. Liming acid soils usually cures crinkle leaf. Calcium deficiencies, which are characteristic in several vegetable crops, are blossom-end rot of tomatoes and pimiento peppers, cavity spots in carrots, brown heart of escarole, black heart of celery, and internal tip burn of cabbage.

A deficiency of calcium and other bases, notably magnesium, in soil results in acid soil conditions. A sufficiently accurate method for determining lime requirement involves the change in pH of a buffer solution when a sample of soil is added to it. The change in pH is proportional to the lime needs.

Crop Response.   The addition of calcium as a nutrient is suggested for several crops. For peanuts, when soil calcium is low, it is suggested that calcium be added by liming before planting or from gypsum or basic slag dusted on at the blooming stage. Rates of about 600 lb/acre (672 kg/ha) of gypsum are usually suggested. It has been demonstrated that calcium is needed most in the pegging zone where the peanut seeds are developing.

For tomatoes and pimiento peppers, when soil calcium is low, it is suggested that calcium be added to the soil as lime or gypsum before planting or sprayed on foliage while the fruit is developing in a water solution containing 0.25 percent calcium as calcium chloride. Calcium nitrate is also suggested as a side-dressing when water-soluble calcium and nitrogen are both needed.

For flue-cured tobacco, when calcium is very low in the soil or when the pH is below 5.5, it is recommended that adjustments be made through the application of dolomitic limestone. The amount to be added should be determined by soil analysis, and it should be an amount that will not raise the soil pH above 6.2. Soils with pH above 6.2 are not generally used nor recommended for tobacco production because it has been observed that high rates of added calcium, about 275 lb/acre (308 kg/ha), produced cured leaves with good color but poor quality. Leaves were starchy, thick, stiff and sometimes papery. High levels of calcium are also associated with excessively rapid burning of the leaf. For these reasons, the calcium content in tobacco fertilizers should be about 6 to 8 percent. Normally, 40 to 60 pounds per acre (45 to 67 kg/ha) of calcium from fertilizers is adequate for tobacco.

Incidence of bitter pit, storage breakdown and scald was higher in low calcium McIntosh apples. Foliar sprays supplying 11.96 lb/acre (13.4 kg/ha) of $CaCl_2$ in each of 7 applications increased flesh calcium and markedly reduced storage breakdown and scald.[18]

Sources of Calcium. Most state recommendations in the United States call for the use of limestone as the least expensive calcium source. A tremendous amount of data indicates that finely ground limestone is an effective material for correcting or preventing calcium deficiencies as well as for reducing soil acidity and its detrimental effects on plant growth.

Under certain conditions of neutral to high soil pH, a source of calcium that is soluble and is not alkaline or does not leave a serious alkaline residue in soils is most desirable for certain crops. Examples include the addition of calcium nitrate to tomatoes, pimiento peppers, tobacco and potatoes growing on slightly acid to alkaline soils. For the control of fungus-rotted pods of peanuts and maintenance of optimal fruit yield, experiments have shown that the best rate of gypsum, a fairly soluble calcium sulfate, is about 1,800 pounds per acre (2,016 kg/ha).[19] Fungus attack was significantly reduced when pods contained 0.20 percent or more calcium.

Where soluble, nonlime-type sources of calcium are needed, one of those sources listed in Table 7.6 may be selected. These include superphosphates, calcium nitrate, calcium chloride and gypsum.

MAGNESIUM—THE KEYSTONE OF CHLOROPHYLL

Modern fertilizer technology is rapidly replacing the use of seed meals and 20 percent superphosphate as sources of nitrogen and phosphorus for multinutrient fertilizers. Finely ground dolomitic limestone as a filler in multinutrient fertilizers is rarely used today. In the past, these materials supplied magnesium as an incidental ingredient of fertilizers. Today,

[18]M. Drake, W. J. Bramlage, and J. H. Baker, "Effects of Foliar Calcium on McIntosh Apple Storage Disorders," *Commun. Soil Sci. Plant Anal.* (1978), in press.

[19]Garren and Hallock, "Calcium."

Table 7.6

CHEMICAL CHARACTERISTICS OF CALCIUM IN SOME COMMONLY USED FERTILIZER AND LIMING MATERIALS

Source	Common Name	Formula[1]	Calcium, %	H_2O Solubility g/100 g @ 25°C
Burned lime		CaO	70	0.12
Hydrated lime	Slaked lime	$Ca(OH)_2$	50	0.16
Calcitic lime	Limestone	$CaCO_3$	36	0.01
Oyster shell lime		$CaCO_3$	34	0.01
Dolomitic lime	Dolomite	$CaCO_3 \cdot MgCO_3$	17	> 0.01
Basic slag	Open hearth	$(CaO)_5 \cdot P_2O_5 \cdot SiO_2$	29	> 0.01
Gypsum	Land plaster	$CaSO_4 \cdot 2H_2O$	22	0.24
Calcium nitrate	Cal-Nitro®	$Ca(NO_3)_2 \cdot 2H_2O$	20	100.00
Superphosphate	Ordinary	$Ca(H_2PO_4)_2 \cdot CaSO_4$	20	1.00
Superphosphate	Triple	$Ca(H_2PO_4)_2$	13	1.8
Rock phosphate	Apatite	$Ca_5(PO_4)_3 \cdot F$	33	0.002
Calcium chloride		$CaCl_2$	36	100.00

[1]Approximate only; source materials contain impurities, more or less waters of hydration and extraneous materials.

however, unless magnesium is specifically added as dolomitic limestone, magnesium sulfate, sulfate of potash-magnesia or magnesium oxide, the chance that any multinutrient fertilizer contains any appreciable amount of magnesium is slim indeed.

Essentiality of Magnesium. Magnesium is an essential element for both plant and animal growth. Although magnesium deficiency in humans has not been known to be widespread, it has been defined in lactating animals grazing winter pastures of small grain as grass tetany, grass staggers, wheat poisoning or hypomagnesemia. It develops in periods of high magnesium requirements, low magnesium absorption across the gut and low magnesium intake. Magnesium deficiency is manifested in the animal by low blood-serum magnesium and in acute cases by tetany, an extreme muscle spasm and subsequent death. The condition may be the result of a chronic deficiency of magnesium in the diet or an acute deficiency of magnesium available to the tissues. The exact cause of hypomagnesemia is not clear, but it appears to be a combination of factors, such as prolonged feeding on feeds or pastures deficient in magnesium, an agent in the animal or diet that ties up magnesium, an imbalance of certain minerals, climatic stress or stress of lactation. One of the difficulties associated with an acute case is the sudden onset of the disease and the death of the animal before help can be obtained. Mortality of visibly affected animals is usually high, i.e., 20 to 30 percent.

Under range conditions, 0.05 gal (200 ml) of a 50-percent magnesium sulfate solution injected under the skin gives a high level of blood magnesium in 15 minutes.[20] Some veterinarians suggest that an intravenous injection of a combination of calcium-magnesium gluconate solution is indicated. One preparation that has been used successfully contains about 1.0 percent calcium, 0.8 percent phosphorus and 3.0 percent magnesium in a 25-percent dextrose solution. The usual dose for an adult animal is 0.07 to 0.14 gal (250 to 500 ml). Drenching with magnesium oxide, a calcined magnesite, at the rate of 2 ounces (0.06 kg) per animal per day for a few days will help to prevent a relapse. Drenching with magnesium sulfate, as Epsom salts, is usually not recommended because, if given in quantities sufficient to provide the magnesium needed, its laxative effect becomes undesirable. Forage magnesium values of less than 0.2 percent magnesium have been considered critical and indicative of possible hypomagnesemia or grass tetany.

Magnesium has been known to be essential for plant growth and development for over 100 years, but its complete role in plant metabolism is still not clearly defined. However, it is known that it is the only metallic constituent in chlorophyll and that it occupies a place in the structure of chlorophyll similar to that occupied by iron in the structure of hemoglobin. Usually, not more than one-fourth of the total leaf magnesium is present in chlorophyll. This suggests that magnesium performs functions other than being a constituent of chlorophyll. Several scientists have suggested that magnesium plays a part in phosphate nutrition and acts as a carrier for phosphorus, particularly into the seeds. Magnesium also appears to be a specific activator of a number of enzymes including certain of the transphosphorylases, dehydrogenases and carboxylases. Magnesium with phosphorus appears to participate in the respiratory mechanism.

The above-ground portion of most mature grain crops and grasses contains from 0.1 to 0.4 percent magnesium. The above-ground portion of cotton, soybean and alfalfa plants averages 0.3 to 0.6 percent magnesium. Tobacco leaves contain up to 1.0 percent magnesium. It is pointed out here that the percentage of calcium in plants, like that of calcium adsorbed on soil colloids, is usually much greater than the magnesium content. Cotton and peanut seeds contain about 0.3 percent magnesium, and barley, corn, oats and rice seeds contain about 0.1 percent magnesium. It has been suggested that magnesium may play an important role in the formation of oils because cotton and other oil seeds contain relatively more magnesium than starchy cereal seeds.[21]

Magnesium performs the following functions in plants:

1. Is an essential part of chlorophyll, which gives the green color to leaves.

---

[20]Grunes and Mayland, "Controlling Grass Tetany."

[21]J. R. Woodruff, M. C. Blount, and S. R. Wilkerson, "Plant Chemistry of Magnesium," in *Magnesium in the Environment,* ed. J. B. Jones et al. (Reynolds, Ga.: Taylor County Printing Co., 1972), pp. 41–60.

2. Is necessary for the formation of sugar from carbon dioxide and water in sunlight.

3. Regulates the uptake of other plant foods.

4. Acts as carrier of phosphorus in the plant.

5. Promotes the formation of oils and fats.

6. Plays a part in the translocation of starch.

Prediction of Magnesium Needs.   Magnesium-deficient cotton and grape have been described as having purplish red leaves with green veins. As the leaf becomes older, it appears to take on a bronze hue. Lower leaves are affected first; as they die and shed prematurely, the next leaves up are affected. On corn, magnesium deficiency results in whitish strips along the veins and often a purplish color on the underside of lower leaves. Magnesium deficiency of tobacco has been observed in North Carolina since 1922. The name "sand drown" was used to describe this symptom because it is found in sandy soils under conditions where excessive leaching occurs. The symptom begins with a loss of green color at the tips of the lower leaves between the veins. As the deficiency develops, larger portions of the lower leaves are affected, and the leaves further up the plant become bleached out and whitish in color. Magnesium deficiency causes leaves to cure out thin, paperish, and lifeless in texture with a dull, dingy, uneven color.

Soil analysis is widely used to diagnose deficiencies and to predict the needs of crops for magnesium. Several workers, using different soils and different crops in widely separated parts of the United States, have concluded that the best level of soil magnesium for plant growth is reached when about 10 percent of the cation exchange capacity of the soil is saturated with magnesium. When the saturation percentage drops below 6.0, magnesium deficiency can be expected.[22]

Lancaster[23] studied the use of selected leaf analysis for the prediction of magnesium needs. It was found that typical deficiency symptoms develop with a leaf analysis of less than 0.025 percent magnesium in cotton.

Crop Response.   Magnesium is generally thought to be deficient in coastal plain soils from Maine to Texas that have not been limed with a magnesium-bearing limestone. Since soils for growing potatoes, tobacco and some tree and bush fruits are not generally limed, fertilizers for these crops in the Atlantic and Gulf Coastal Regions of the United States often include guaranteed amounts of magnesium. Cotton grown on black-belt soils derived from Selma chalk and sugar beets growing on acid mineral soils in Michigan also respond to applications of magnesium.

[22]F. E. Bear, A. L. Prince, S. Y. Toth, and E. R. Purvis, "Magnesium in Plants and Soils," New Jersey Experimental Station Bulletin 460, 1951.

[23]Lancaster, "Magnesium Status."

Sweet corn and potatoes have responded to magnesium in Columbiana County, Ohio. Potato farmers use soluble sources, but others use dolomitic limestone. In Michigan, magnesium-deficient potatoes have been noted on acid mineral soils. On soils that are below a pH of 5.5 and have less than 75 lb/acre (84 kg/ha) of exchangeable magnesium, dolomitic lime increased the growth and yield of some varieties. On soils that are not acid, soluble fertilizer magnesium at the rate of about 30 lb/acre (34 kg/ha) is suggested.

The northeastern, southwestern and central parts of Pennsylvania have low magnesium soils. On tomatoes in these areas and in South Carolina, 15 to 30 lb/acre (16.8 to 33.6 kg/ha) is suggested.

Magnesium at the rate of about 50 lb/acre (56 kg/ha) is generally recommended for flue-cured tobacco. Clemson University recommendations suggest a tobacco fertilizer containing 2 percent magnesium, at least one-half of which should be water soluble.[24] A similar rate is suggested for grapes. Deficiencies have been noted on pearl millet and vegetable green crops. Where the soil pH is above 7.0 or where it is too late to add lime, 25 lb/acre (28 kg/ha) of a soluble source is suggested as a spray or as a soil side-dressing.

As of this writing, insufficient data are available to define low soil-test values for many plants and crops. However, evidence has accumulated indicating that magnesium is a limiting factor in plant growth on some soils. Where soil-test values indicate low amounts and in the presence of observed or suspected magnesium deficiencies, 25 lb/acre (28 kg/ha) of magnesium is suggested.

Cotton has been studied in terms of magnesium nutrition about as much as any plant, and independent scientists, working on soil types varying from acid, gray, sandy soils to alkaline, black, fine-textured clay soils, have concluded that 20 pounds of magnesium from either dolomitic limestone or potassium-magnesium sulfate added annually is about the correct rate.

Fifteen hundred pounds of dolomitic limestone per acre every five years has been found to be sufficient to correct magnesium deficiency and soil acidity of Lakeland sand growing field crops and peaches at the Sandhill Experiment Station in South Carolina.

Sources of Magnesium. Most states in the United States recommend the use of dolomitic limestone as the least expensive magnesium source. A tremendous amount of data, discussed in the section "Liming Acid Soils," indicates that finely ground dolomitic limestone is an effective material for correcting or preventing magnesium deficiency as well as reducing soil acidity and supplying calcium. Under certain conditions, some workers have shown an advantage of a water-soluble source of magnesium over

[24] "Fertilizer Recommendations."

dolomite. Underwood and Miller[25] reported that total magnesium uptake from fine dolomitic lime was significantly greater than from coarse dolomitic lime but usually less than from magnesium sulfate, hydrated dolomitic lime and burnt dolomitic lime. With the exception of coarse dolomite, applications of magnesium at the rate of 15 to 30 lb/acre (16.8 to 33.6 kg/ha) significantly increased corn dry weight regardless of source. Brupbacher and Sedberry[26] reported that total magnesium uptake by clover from soluble magnesium sources was greater than from dolomitic lime or oyster shell lime. There was no difference in yield of clover between soluble and insoluble sources of magnesium where ordinary superphosphate was the source of phosphorus; there did appear to be differences where concentrated superphosphate was the source of phosphorus, but this was probably due to confusing magnesium response with sulfur response. The chemical characteristics of magnesium in dolomite and some other commonly used magnesium-bearing fertilizer materials are shown in Table 7.7.

Where the pH of the soil is above 6.0, water-soluble magnesium sulfate may be preferable to dolomitic limestone for supplying magnesium. Epsom salts and sulfate of potash-magnesia contain water-soluble magnesium. Such materials may be preferred where a quick crop response is required in order to correct rapidly an acute magnesium deficiency. However, it has been reported that both $MgSO_4 \cdot 7H_2O$ and dolomitic limestone increased the yield of tomatoes equally on mulched, loamy sand soil (Figure 7.7).[27]

Under conditions where close contact of seed and fertilizer is desirable or unavoidable and soil pH is too high for the use of dolomitic limestone, a magnesium material of slight solubility can be helpful. Several

Table 7.7
CHEMICAL CHARACTERISTICS OF MAGNESIUM IN SOME COMMONLY USED FERTILIZER AND LIMING MATERIALS

Source	Common Name	Formula	Mg(%)	H_2O Solubility g/100g
Magnesium oxide	Magnesia	MgO	45.0	0.00062
Magnesium ammonium phosphate	Mag-Amp®	$MgNH_4PO_4 \cdot H_2O$	14.8	0.014
Dolomite	Dolomite	$CaCO_3 \cdot MgCO_3$	12.0	0.032
Kieserite	EMJEO®	$MgSO_4 \cdot H_2O$	18.2	68.4
Langbeinite	K-Mag®; SUL-PO-MAG®	$K_2SO_4 \cdot 2MgSO_4$	11.2	soluble
Magnesium sulfate	Epsom salt	$MgSO_4 \cdot 7H_2O$	10.5	91.0

[25]Underwood and Miller, "Effect of Soil pH."

[26]Brupbacher and Sedberry, "Effects of Magnesium and Sulfur."

[27]Jones and Jones, "Influence of Polyethylene Mulch and Magnesium."

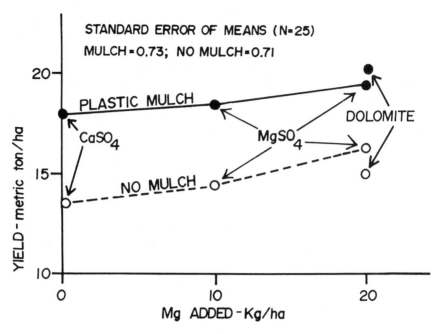

Figure 7.7 Mean yields of three crops of "Walter" tomatoes as influenced by polyethylene mulch and Mg as $MgSO_4 \cdot 7H_2O$ and dolomitic limestone.

years of research with slightly soluble magnesium ammonium phosphate on various soils have shown that the material has nonburning and non-leaching characteristics, which are so important under conditions noted above when it is in contact with seed. Experiments were conducted in which oats, corn, beans, wheat, alfalfa and rye grass were germinated and

Figure 7.8 Pine seedlings growing in $MgNH_4PO_4 \cdot H_2O$ compared with growth in unfertilized soil.

grown satisfactorily for a few weeks in a pot containing only $MgNH_4PO_4 \cdot H_2O$. Pine seedlings were transplanted in the compound, and they grew better for one month in the slightly soluble salt than in unfertilized soil (Figure 7.8). The manufacturer suggests that all sorts of tree seedlings and nursery crops can be fertilized directly under the roots with this material without burning the plant. Chemical characteristics of the material are given in Table 7.7. In addition to the magnesium content, a pelleted magnesium ammonium phosphate produced on a semi-commercial basis in 1960 contained 8 percent nitrogen and 40 percent P_2O_5.

SUMMARY

1. Determination of lime requirement is one of the most accurate soil tests.

2. The degree of soil acidity is expressed in terms of pH.

3. Soil pH varies from about three to nine; conditions favorable for most plant growth are between pH 5 and 8.

4. If pH of a mineral soil is 5.8 or less, for most field crops, it should be limed to give pH 6.0 to 6.5. Mucks and peat soils should not be limed beyond a pH of about 5.3.

5. Phosphorus availability in humid tropical, subtropical and temperate zone soils is reduced by a decrease in soil pH, which brings about iron and aluminum solubility. The iron and aluminum "fix" or "tie up" phosphorus in insoluble forms.

6. Lime is lost and soil acidity is increased (soil pH is decreased) by crop removal, erosion and water passing through soil. Cultivation, crop residues and fertilization with nitrogen and sulfur contribute to the loss of bases and hence to the increase in acidity, brought about by water leaching nutrients out of the root zone.

7. Effectiveness of liming materials is influenced by neutralizing value, fineness, magnesium content and uniformity of lime applications.

8. A soil pH above 8.5 usually indicates the presence of sodium and an alkaline soil. This is undesirable because excess sodium destroys soil structure and depresses calcium, magnesium and potassium uptake. Alkaline soils high in pH and with excess sodium can be reclaimed by adding calcium sulfate (gypsum) and then leaching the sodium from the soil.

9. Calcium is essential for plant and animal growth.

10. Humans encounter calcium and vitamin D deficiencies in a disease known as rickets.

11. Calcium promotes plant vigor and early root growth. A deficiency in plants can be recognized by a stunting effect.

12. In nutrient solutions low in calcium, soybeans exhibit necrosis or brown spots on the edges of leaves.

13. Calcium deficiencies are known as blossom-end rot of tomatoes and pimiento peppers, cavity spot in carrots, brown heart of escarole, black heart of celery and internal-tip burn of cabbage.

14. Calcium is usually the dominant cation adsorbed on the soil cation exchange materials, and significant amounts are found in the soil solution, except in very acid soils.

15. Since plants require less calcium than several other nutrients and soils commonly contain more calcium than some others, calcium deficiency does not occur frequently, except on very acid soils. Hence, soil pH is used as an indirect measure of available calcium. A calcium deficiency will be corrected by liming.

16. Modern fertilizer technology has replaced the ingredients formerly used in multinutrient fertilizers that contained calcium and magnesium. Because all liquid fertilizers and, perhaps, 50 percent of modern solid multinutrient fertilizers are devoid of calcium and magnesium, it is no longer feasible to assume that these nutrients will be supplied as incidental ingredients. Superphosphates, used in about one-half of multinutrient goods, contain calcium but very little magnesium.

17. Unless magnesium is added to fertilizer as dolomitic limestone or as another fertilizer ingredient such as Epsom salt, potassium-magnesium sulfate or magnesium oxide, the chance that a mixed fertilizer contains any significant amount of magnesium is slim indeed.

18. Magnesium is essential for both plant and animal growth. It is the only metallic constituent of chlorophyll, the "green blood" of plants.

19. Hypomagnesemia is an animal disease caused by low blood magnesium. It usually affects lactating animals grazing on grasses or small grain forage.

20. Magnesium deficiency usually occurs on older leaves of plants causing a purplish bronze color of leaves on grape and cotton.

21. For corn, interveinal chlorosis (yellowing between the veins) causes a striped effect on the lower, older leaves. Magnesium deficiency symptoms become progressively weaker with younger leaves.

22. Available magnesium in soils is present as an exchangeable cation adsorbed on cation exchange material. There is usually one-fourth to one-half as much exchangeable magnesium as exchangeable calcium.

23. Exchangeable magnesium is not held as tightly, and consequently there may be as much magnesium in the soil solution as calcium.

24. Because magnesium is not held as tightly, it is also more subject to leaching than calcium.

25. Because plants require similar quantities of these two nutrients, a deficiency of magnesium is more common than one of calcium.

26. Magnesium deficiency is most likely to occur on acid soils that are low in exchange capacity or on alkaline soils high in calcium.

27. Applying large amounts of potassium will suppress the uptake of magnesium.

28. Supplying large quantities of calcium without magnesium also suppresses magnesium uptake by plants.

True or False:

1. _____Agricultural lime refers to any material containing calcium and/or magnesium in forms capable of neutralizing acid soils.

2. _____Soil tests for lime requirement are about the most accurate tests there are, and the suitable pH for most field and vegetable crops will be about 6.0 to 6.5.

3. _____The proper use of lime may be the backbone of permanent agriculture in a humid region.

4. _____For best results, agricultural lime should be spread evenly on the land and then plowed or disked in until it is intimately mixed with the furrow slice of soil.

5. _____Calcium deficiencies, that are characteristic in several vegetable crops, are blossom-end rot of tomatoes and pimiento peppers, cavity spot in carrots, brown heart of escarole, black heart of celery and internal tip burn of cabbage.

6. _____Most states east of the Lime Line in the United States recommend the use of dolomitic limestone as the least expensive calcium and magnesium source.

7. _____In the past, magnesium was supplied as an incidental ingredient of mixed fertilizers. Today, unless magnesium is specifically added, the chance that a multinutrient fertilizer contains any is slim indeed.

8. _____Forage magnesium values of less than 0.2 percent magnesium have been considered critical and indicative of possible hypomagnesemia or grass tetany.

9. _____On corn, magnesium deficiency results in whitish strips along the veins and often a purplish color on the underside of lower leaves.

10. _____Magnesium at the rate of about 50 pounds per acre (56 kg/ha), one-half of which should be water soluble, is generally recommended for flue-cured tobacco.

11. _____Where the pH of the soil is *below* 6.0, water-soluble magnesium sulfate may be preferable to dolomitic limestone for supplying magnesium.

Multiple Choice:

More than one answer may be correct.

12. Liming acid soils is a desirable practice because it:
    a. corrects soil acidity.
    b. supplies calcium and magnesium.
    c. increases crop yields.
    d. speeds the decay of organic matter and the liberation of plant nutrients.
    e. all of the above.

13. Calcium functions in plants as follows:
    a. promotes early root formation and growth.
    b. is an essential part of chlorophyll.
    c. improves general plant vigor and stiffness of straw.
    d. encourages grain and seed production.
    e. all of the above.
14. Magnesium functions in plants as follows:
    a. promotes early root formation and growth.
    b. is an essential part of chlorophyll.
    c. is necessary for the formation of sugar from carbon dioxide and water in sunlight.
    d. plays a part in the translocation of starch.
    e. all of the above.

Completion:

15. Lime is lost from the soil by:
    a. _____
    b. _____
    c. _____
    d. _____
16. Lime should be fine enough for 90 percent of it to pass a ____-mesh screen, and at least 50 percent should pass a ____-mesh screen.
17. Complete and balance the following reaction:

$$Ca\boxed{Micelle} + 2H_2CO_3 \rightarrow \underline{\quad} + Ca(HCO_3)_2$$

Problems:

18. How much sulfur after oxidation and hydrolysis is required to neutralize 3,000 pounds (1,363 kg) of lime that is 95 percent $CaCO_3$ equivalent? Atomic weight of calcium = 40; sulfur = 32; oxygen = 16; hydrogen = 1; carbon = 12.

19. Pure calcium carbonate has a neutralizing value of 100. Use atomic weights listed below in Question 20 to determine the neutralizing value of:
    a. $MgCO_3$ = _____
    b. $½MgCO_3·½CaCO_3$ = _____
    c. $Ca(OH)_2$ = _____
    d. CaO = _____
20. What is the percent calcium and percent sulfur content of gypsum $(CaSO_4·2H_2O)$? Atomic weights of calcium = 40; sulfur = 32; oxygen = 16; hydrogen = 1; magnesium = 24; carbon = 12.
21. How many pounds of $CaCO_3$ per acre (2,000,000 lb) would be required to supply 2 milliequivalents of Ca^{++} per 100 grams of soil? Use the atomic weights given in Problem 20.

chapter 8

Micronutrients—For Healthy Plant Growth

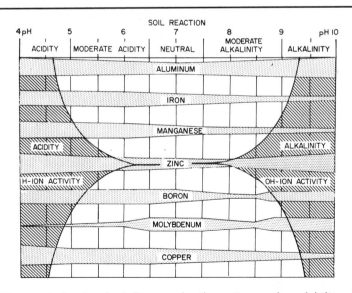

Diagram indicating the influence of soil reaction on the solubility of some metallic elements, boron, and molybdenum. All except aluminum are micronutrient elements essential for plant growth. The width of bands indicates more or less solubility of the element, as influenced by conditions associated with pH, and not necessarily to actual amounts present.

Seven elements essential for plant growth are designated and commonly referred to as micronutrients. They are zinc, iron, manganese, copper, boron, molybdenum and chlorine. The first six micronutrients listed above appear to be required in concentrations as small as one-fourth part per million, while as much as two parts per million in solution may prove damaging, if not lethal, to some plants. Thus, the name micronutrients. Cobalt is an essential element for animal growth and has been found necessary for nodulation of certain legumes. Chlorine is found in plants in concentrations higher than that of either phosphorus or sulfur; however, it is considered as an essential micronutrient. The air around a plant has to be filtered in order to demonstrate that chlorine is essential. Barnyard manure is a good source of most micronutrients. Fertilizer, fortified with all the known essential micronutrients, has been offered by several manufacturers (Table 8.1).

The functions of micronutrients for plant growth are as follows:

1. They may be essential nutrients for life and growth.

2. They may serve as substitutes in part for other plant nutrients.

3. They may help to form enzymes or vitamins.

4. They may antidote one another.

5. One element may aid in keeping another reduced or oxidized.

Table 8.1
IF THE ADDITION OF MICRONUTRIENTS TO REGULAR GRADES OF
FERTILIZER IS DESIRED, THE FOLLOWING MAY BE USED AS A GUIDE:

Element	Minimum Lb per Ton[1]	Percent	Safe Maximum Lb per Acre[2]
Molybdenum	0.10	0.005	0.2
Zinc	5.0	0.25	10.0
Manganese	3.0	0.15	10.0
Iron	5.0	0.25	10.0
Copper	3.0	0.15	10.0
Boron	0.5	0.025	1.0[3]

[1]Lb/ton x 0.5 = kg/m ton

[2]Lb/acre x 1.12 = kg/ha

[3]For crops like cucumbers, melons and snap beans, not over 0.5 lb per acre of boron, if any, should be used. Alfalfa, garden beets and similar crops may use up to 4.0 lb boron per acre without damage. If boron is included in fertilizer, some fertilizer laws require addition of borax in proportions of 5 lb of borax per ton. This amount of borax contains 0.57 lb of boron.

246

6. They may increase resistance to disease.

7. The addition or presence of one element may precipitate others.

8. An indirect effect may occur from the influence of an element on microorganisms.

9. There is considerable evidence of interactions among the micronutrients and with the secondary or major plant nutrients.

It is logical that the primary and secondary plant nutrients discussed before have received most of the attention by agronomists and the fertilizer industry in the past. Micronutrients have been established as essential plant nutrients only after precise chemical methods of analysis were developed with sufficient refinement and accuracy to quantitatively determine the micronutrients and with techniques to purify growth media. It is a tribute to agricultural scientists who have discovered by careful research the essentiality of micronutrients to healthy plant growth and have made it possible to avoid deficiencies and resultant crop losses.

Deficiency symptoms and patterns of several micronutrients have been established and described for many crops of economic importance. Likewise, soil types and conditions most likely to be subject to micronutrient deficiencies have been mapped and catalogued (Table 8.2). A symposium consisting of papers by several authorities and published in the *Journal of Agricultural and Food Chemistry*,[1] may be referred to for details of developments on the needs for micronutrients, the properties of materials supplying these needs and the methods by which these plant nutrient needs are satisfied.

The Soil-Testing Committee of the 1963 Soil Science Society of America surveyed the known micronutrient deficiencies by crop as reported by workers in each state of the United States. Deficiencies were reported on one or more crops as follows: boron in 41 states, copper in 13, manganese in 25, molybdenum in 21, iron in 25, and zinc in 30. Alfalfa was reported deficient in boron in 38 states and corn was deficient in zinc in 30. Some of the results obtained after correction of micronutrient deficiencies have been spectacular and of major economic importance.

Interactions among micronutrients with major elements are known to occur and to influence plant growth. It is known, for example, that phosphorus/zinc, potassium/boron and nitrogen/boron interactions occur in soils and crops. Closely observed field trials by agronomists will be required to obtain desired results from favorable or unfavorable nutrient balance.

Since the necessity and value of micronutrients have been reported and discussed on a world-wide basis, the practical problem of where they are needed and how to supply them arises. Some believe that the fertilizer

[1]Micronutrient Symposium, American Chemical Society, *J. Agric. Food Chem.*, 10 (3) (1962).

Table 8.2
SOIL CONDITIONS AND CROPS MOST SUSCEPTIBLE TO DEFICIENCIES OF
MICRONUTRIENTS

Micronutrients	General Soil Type and Conditions	Crops Most Likely Susceptible
Boron	Acid leached soils, coarse-textured sandy soils, peats and mucks, droughty conditions, overlimed acid soils.	Alfalfa, apples, beets, clovers, citrus, cotton, cauliflower, celery, corn, sweet potatoes, tomatoes, tree crops, sugar beets.
Chlorine	Unknown in field soils.	Beets, tomatoes.
Copper	Sandy soils, peats and mucks, over-limed acid soils.	Small grains, vegetable and tree fruits.
Iron	Alkaline soils, particularly when cold and wet; excess phosphorus.	Beans, soybeans, corn, sorghum, tree fruits, ornamentals.
Manganese	Sands, mucks and peats, alkaline, particularly calcareous soils.	Soybeans, small grains, tree fruits, cotton, leafy vegetables.
Molybdenum	Highly weathered acidic leached soils, acid soils.	Cauliflower, citrus, all legumes.
Zinc	Calcareous soils after leaching and erosion, acid-leached soils, after heavy phosphorus, coarse sands, subsoil exposed by land leveling.	Beans, soybeans, citrus, corn, sorghum, onions, potatoes, tree fruits, flax, sugar beets, rice.

industry should provide a mixture for general use containing all of the necessary plant food elements for the most efficient and profitable crop production. Because relatively small amounts of micronutrients that are needed by some plants may be damaging to others, the formulation of this all-encompassing plant food is not as simple as it first appears.

Insoluble materials have been produced in which certain of the micronutrients have been incorporated. The idea has been to make compounds very slightly water-soluble to furnish growing plants with fool-proof micronutrients in the needed concentration but at the same time not so soluble as to be damaging to plants if added in excess. Particular attention is being given to those micronutrients that are required by plants in small amounts but that produce injurious effects when added in excess.

Some of these slowly soluble compounds of the heavy metals, zinc, iron, manganese and copper, include the metallic frits, metallic oxides, ferrous oxalate, zinc oxysulfate and metallic ammonium phosphates. These materials as well as more soluble metallic sulfates, metallic chelates and compounds of boron and molybdenum will be discussed in terms of their chemical characteristics, method of manufacture, handling characteristics and uses.

The heavy metals essential for plant growth are iron, manganese, copper and zinc. All tend to become less available to plants as the soil pH is raised. As the pH goes up, hydroxides or oxides of low solubility are formed. Since this is a characteristic induced by high pH, deficiencies of these metals are difficult to correct in soils with pH values of 6.3 or above, particularly by soil applications of compounds in which the heavy metals exhibit cationic properties. The metal chelates offer one means of minimizing the chemical activity of the heavy metal ions in the soil. Additions of these compounds to the soil have been used successfully to eliminate plant deficiencies of iron and zinc.[2]

The word "chelate" is derived from the Greek word meaning "claw." In organic chemistry, it refers to a ring structure produced when a metal ion combines with two or more electron donor groups to form a single molecule. Metals bound in chelate rings lose their ability to act as ions and therefore, are less likely to take part in chemical reactions that precipitate them or otherwise render them unavailable to plants.

Iron, for example, has a configuration number of six, i.e., six oxygen-size atoms can surround it, giving it an octahedral configuration like a bird enclosed in two pyramid-shaped cages that are connected at the base. The disodium iron salt of ethylene diamine tetraacetic acid is one of the chelates that has this octahedral configuration.

All polyvalent cations will form chelates, but each metal differs in the ease with which it chelates. The stability of the metal chelates or the replacing power of the elements in decreasing order is: Fe^{3+}, Cu^{++}, Zn^{++}, Fe^{++}, Mn^{++}, Ca^{++} and Mg^{++}. Ferric iron chelate is more stable than any of the other chelates of metals essential for plant growth, and the ferric ion would be expected to replace equal concentrations of any of the other metals from the chelate ring. Chelates can be used in several ways to correct nutrient deficiencies: they can be added as a spray on the foliage, or they can be added to the soil. Different kinds of chelating agents are used. Some are effective on acid soils, some on calcareous soils, and some are effective on both acid and basic soils. A description of the chelating agents and characteristics of the chelates are outlined below.

Necessary Characteristics of a Chelate Added as a Spray on Foliage

1. Easily absorbed by plants.

2. Translocated readily within the plant.

3. Easily decomposed so that the metal becomes available.

[2]W. A. Norvell, "Equilibria of Metal Chelates in Soil Solution," in *Micronutrients in Agriculture*, eds. J. J. Mortvedt et al. (Madison, WI: Soil Science Society of America, 1972), pp. 115–136.

249

4. Nondamaging to plants at concentrations necessary to control deficiencies.

Necessary Characteristics of a Chelate Added as a Soil Amendment

1. Should not be easily replaced by other polyvalent cations in the soil.
2. Must be stable against hydrolysis.
3. Has to be resistant to microbiological decomposition.
4. Should be soluble in water.
5. Not easily precipitated by ions or colloids in soils.
6. Must be available to plants either at root surface or within the plant.
7. Must be nondamaging to plants at concentrations required to prevent deficiencies.

Chelating Agents

1. EDTA—Ethylene diamine tetraacetic acid is most effective on acid soils. In calcareous soils, it is detrimental to plant growth under some conditions and is not sufficiently stable under other conditions. Also, iron EDTA is fixed in insoluble form on soil clays.
2. DTPA—Diethylene triamine pentaacetic acid.
3. HEEDTA—Hydroxyethyl ethylene diamine triacetic acid.
4. CDTA—Cyclohexane trans 1,2-diamino tetraacetic acid.
5. EDDHA—Ethylene bis Alpha-imino-2-hydroxy-phenyl-acetic acid.

Chelating agents 2, 3, and 4 have increasing stability over EDTA in calcareous soils. EDDHA, when chelated with iron, has overcome all of the objections of EDTA on calcareous soils, its chelates are effective in both acid and basic soils.

CHELATED METALS USED IN AGRICULTURE

Iron Chelates. These are the most widely used. Ferric iron forms the most stable chelate. Therefore, it will remain in the soil in an effective form. Besides chelates, the only effective soil treatment for iron chlorosis has been the addition of inorganic iron salts to flooded rice.

Copper Chelates. Next to ferric iron, copper chelates are the most stable and should be an effective source of copper. They have not been used commercially because copper can be utilized by plants from copper salts, even when the pH is high, because of the strong chelating action of soil organic matter for the copper ion supplied as a salt.

Zinc Chelates.   Soil applications of zinc chelates have been more effective than the stability constant would indicate. The success is partly due to the fact that zinc chelates do not become fixed on the soil clay as iron chelates do and partly due to the low concentration of ferrous (0.00027 ppm) and ferric (0.00000006 ppm) ions at pH 6.0. At higher pH values, the concentration of ferrous and ferric ions decreases rapidly. It is at soil pH values of 6.0 and above that zinc deficiency usually occurs. The use of zinc chelates will probably be limited because less expensive zinc sprays on foliage provide satisfactory control of zinc deficiency.

Manganese Chelates.   Because of the low stability constant, soil applications of manganese chelates are not effective. Calcium exists in soil at high concentrations and by mass action will replace some of the manganese from the chelate even though calcium is lower than manganese on the scale of stability and replacing power. Also, effective foliage sprays of manganese salts are available commercially.

Compatability with Commercial Fertilizers

1. Chelated nutrients are compatible with dry fertilizer mixes. From 5 to 20 pounds per ton (2.5 to 10 kg/m ton) of iron chelate has proved effective on some soils.
2. They cannot be mixed with liquid fertilizers that contain:
   (a) Ammonia. The high pH precipitates the metals because the (OH) groups compete with the donor groups of the chelating agent for the metal ions.
   (b) Free phosphoric acid. The hydrogen ions will effectively compete with iron and other metals for the donor groups.

FRITS FOR THE SOIL

Fritted micronutrients have a promising future if experiment station tests continue to show a need for their use. Interest in agricultural frits is growing rapidly, particularly along the Atlantic and Gulf Coasts of the United States and in the West. In humid coastal areas, heavy cropping and/or leaching has depleted these elements from the soil. In the western part of the United States, high pH and the general calcareous nature of the soils render micronutrients unavailable to plants.

Description.   Brown-colored, pulverized, specially compounded glasses or frits that slowly release micronutrients such as boron, copper, iron, manganese, molybdenum and zinc are being used for many crops. They are offered in powdered or granulated form.

Method of Manufacture. Proper proportions of feldspar, soda ash, silica, fluorspar (CaF_2), cryolite ($AlF_3 \cdot 3NaF$), sodium nitrate and borax are thoroughly mixed. These glass components are mixed and fused in a smelting furnace at 1652° to 1832°F (900° to 1000°C). The molten material is quenched in water, dried and milled. The end product is finely ground, moderately soluble fritted glass. Solubility of the metallic salts in frits is controlled by particle size and changes in the composition of the matrix. Developments include the incorporation of fritted material in superphosphate granules.

Chemical Characteristics. A fritted product containing 6 percent boron and 18 percent manganese has been offered in the United States. It reportedly sells for about 18 percent more than equivalent amounts of manganese sulfate and borax. Another frit contains 3.8 percent boron, 7 percent copper, 14 percent iron, 7 percent manganese, 0.07 percent molybdenum and 7 percent zinc.

Handling Characteristics. The physical nature of a frit, usually finely ground and relatively inert, provides handling ease in plant operations. Frits can be added to mixed fertilizers at any stage of their manufacture without causing the product to cake during storage. Recent developments in granular frit production permit the material to be incorporated in granular mixed fertilizer although it must be remembered that the solubility of the metallic frits is controlled by particle size. The amount of soluble micronutrients supplied by frits is relatively small in the early and most sensitive stages of crop growth. Therefore, material crop damage from excess concentrations of the nutrients, especially boron, is much less likely to occur.

Most frits are applied to the soil as ingredients of mixed pulverized fertilizers; very little is added directly to the soil. However, there have been tests of application by injecting a water suspension around the roots of trees. Another possibility is to apply the frit with crop-dusting equipment.

Frits are not yet in wide use by farmers. Many do not know much about them, and those who do, in many cases, are awaiting test results from state experiment stations, which have research on micronutrients in progress.

Crop Response. At Clemson University in South Carolina, field tests on cotton, alfalfa, corn and peaches show the frits now manufactured are as effective on a pound-for-pound (kilogram-for-kilogram) basis as are quickly soluble micronutrient sources. Boron and manganese requirements for cotton can be fulfilled by the use of manganese-boron frits. Requirements of boron for alfalfa can be fulfilled by the use of a boron frit. Under South Carolina conditions, from 30 to 40 pounds per ton (15 to 20 kg/m ton) of mixed fertilizer will meet the micronutrient requirements for most crops.

When applied at the rate of 10 to 15 pounds per acre (11.2 to 16.8 kg/ha), the frit is used up in about a year.

Results from Florida's agricultural experiment station place the micronutrients in frits on par in availability with the equivalent micronutrients applied as soluble compounds. Comparison tests were made on corn, peanuts, oats, vegetable crops and pastures.

In New Jersey, the experiment station at Rutgers explains, "The general use of frits would be primarily a matter of providing 'insurance' quantity additions of micronutrients to commercial fertilizers."

Testing has also been conducted in Virginia, North Carolina (on carnations) and Georgia (on clover).

Most testing completed thus far has been conducted in the coastal states. The midwestern and south central parts of the United States appear to have little need for frits, and there is little practical advantage for frits over soluble products in those areas. But many state experiment stations have conducted tests as follows:

1. In Arizona on deciduous fruits, tomatoes, lettuce and soybeans.
2. In California on peaches, pears, walnut and cherry trees, potatoes and ornamentals.
3. In New Mexico on vegetable crops, peanuts, trees and ornamentals.
4. In Idaho on hops.
5. In Colorado on cherries, peaches and pears.

These widely varied tests point out the fact that research with frits is carried on in areas where trace elements are thought to be deficient in the soil. These are usually sandy and rainy areas. In heavy (fine-textured) soils, micronutrients are not lost so rapidly by leaching from the root zone; thus, the need for fritted trace elements is not so great in these soils as it is in sandy soils. In general, growers on sandy soils in humid regions who raise high-value crops, such as vegetables, and who seek early markets are likely to give serious thought to fertilizers containing frits.

ZINC: FRIT, CHELATE, SULFATE, PHOSPHATE, SULFIDE AND OXIDE

Description. Frits containing zinc and other heavy metals for plant growth are finely ground, brown, powdery materials. The zinc chelate is a finely divided or granulated white material. Zinc sulfate is a white, finely divided crystal having a specific gravity of about 2.0. Zinc ammonium phosphate is a white granular or pulverized salt. Sphalerite[3] or zinc sulfide is a white

[3]K. B. Krauskopf, "Geochemistry of Micronutrients," in *Micronutrients in Agriculture*, ed. J. J. Mortvedt et al. (Madison, WI: 1972), pp. 7–36.

pulverized material in the range of 10 to 60 mesh (1.4 to 0.25 mm) and has a specific gravity of about 4.0. Zinc oxide is a white or yellowish amorphous powder. It is usually the least expensive and contains more actual zinc than other materials. It can be added to ammonium phosphates, ammonium polyphosphates and ammoniated superphosphates. Zinc oxide and zinc sulfate monohydrate are comparable for serving as sources of zinc to the plant. Only the chelate and, to a very limited extent, the sulfate of zinc are soluble enough to be used in most clear liquid fertilizers (Table 8.20). A zinc ammonium nitrate compound has been produced with a trade name of NZN®.

Method of Manufacture. Frit manufacture was described in a previous section based on Riegel's works.[4] By adding the metal salt, zinc sulfate, in proper proportions prior to fusing, the fritted zinc micronutrient is formulated. A frit containing zinc as the only micronutrient is not currently available. Frits usually contain not only zinc, but also boron, copper, iron, manganese and molybdenum.

The chelates, zinc EDTA and disodium zinc EDTA, are made by complexing the metals with ethylene diamine tetraacetic acid. This stable compound is thought to be broken down with difficulty in the soil, yet its zinc appears to be readily available to plants. Zinc HEEDA, EDDHA, and ammonium lignin sulfonate chelated with zinc have been manufactured.

Zinc sulfate is the most commonly used source of zinc for sprays; it is also used for mixing with solid fertilizer. A basic zinc sulfate has been used that is designated as zinc oxysulfate.[5] This is a chemically combined zinc oxide–zinc sulfate compound. It is adjusted to pH 7.4 and is adapted primarily for foliar application. As a foliar application, zinc oxysulfate is also used to control bacterial spot, *Xanthomonas pruni*, of peaches.

A zinc iron ammonium sulfate containing 4 percent zinc has been manufactured using spent sulfuric acid containing zinc and iron impurities.

A zinc compound developed for agricultural use is zinc ammonium phosphate. All of the metal ammonium phosphates are produced in the same manner as described by Bridger et al.[6] They are very slightly water soluble, will not burn roots or foliage and may be used for soil or foliar application.

Sphalerite, ZnS, is produced as "fines" through flotation processing of ground mother dolomite of the Knox formation in Tennessee and similar deposits. The ground agricultural lime from such deposits added to acid

[4]E. R. Riegel, *Industrial Chemistry,* 5th ed. (New York: Reinhold Publishing Co., 1949).

[5]W. D. Warriner and E. H. Conray, Production of Zinc Oxysulfate. U.S. Patent #2772151. Nov. 27, 1956.

[6]G. L. Bridger, M. C. Salutsky, and R. W. Starotska, "Metal Ammonium Phosphates as Fertilizers," *J. Agric. Food Chem.* 10 (1962), pp. 181–188.

soils will supply ample zinc for plant growth. After sphalerite is concentrated by flotation, it is roasted to convert the sulfide into the oxide.

$$2ZnS + 3O_2 \longrightarrow 2ZnO + 2SO_2 \uparrow$$

Handling Characteristics. Because of the small quantities of zinc and other micronutrients required by plants, it is usually necessary that a diluent be employed before attempting to spread amounts such as 10 lb/acre (11.2 kg/ha) or 20 grams per tree. For soil additions, multinutrient or compounded fertilizers are the usual diluent, and water is the usual diluent for foliar additions.

When frits or chelates are added to liquids or slurries containing ammonia or free phosphoric acid and then are processed into finished solid goods, the chemical nature of the micronutrients is changed. Chemical reactions taking place in the factory during processing, particularly of ammoniated superphosphates, nitric phosphates and ammonium phosphates, are likely to influence materially the solubility and other characteristics of the added micronutrients regardless of their original form.

Solid ammonium phosphate and polyphosphate fertilizers containing zinc have been made by the Tennessee Valley Authority using zinc oxide. They both possess excellent physical qualities. On the other hand, zinc sulfate, added to mixed pulverized fertilizer, causes the product to cake during storage.

Chelates of zinc have been successfully added to clear liquid ammonium polyphosphates.

Sphalerite does not cause a chemical reaction when dry mixed with pulverized fertilizers; therefore, it does not cause caking that results from the inclusion of soluble zinc salts.

Chemical Characteristics. The zinc in a frit is very difficultly soluble in water but is soluble in weak acids. The water solubility of the metal cation in such frits is controlled by particle size. Frits containing 7 to 9 percent zinc are manufactured.

It appears that chelated zinc itself is not taken up by plant roots. A cation exchange takes place between the chelates and the roots providing a mechanism by which the plant roots can feed on the zinc even though the stable chelate itself is not easily decomposed in soil.

Zinc EDTA is made containing about 6 percent zinc. Fertilizer-grade zinc sulfate contains 27.8 percent zinc, 13.6 percent sulfur, 0.06 percent magnesium, and 0.02 percent copper. Zinc sulfate monohydrate is used as fertilizer and contains 35 percent zinc. The formula for the crystallized salt is $ZnSO_4 \cdot 7H_2O$, and it contains 22.6 percent zinc. Zinc oxysulfate contains about 52 percent zinc.

Zinc oxide is insoluble in water. It is relatively inexpensive and contains about 80 percent zinc. When it is added to polyphosphates, it is thought that the polyphosphates complex the zinc in accordance with the equation postulated below.

$$ZnO + (NH_4)_4P_2O_7 \rightarrow Zn(NH_4)_4(PO_4)_2$$

Zinc ammonium phosphate contains 33.5 percent zinc and is very slightly soluble in water. It is used for soil or foliar applications.

Zinc sulfide or sphalerite contains 61 percent zinc and 30 percent sulfur and is very slightly soluble in water. No trace of H_2S was found by nasal test when this material was incorporated for 35 days in a 12-3-5 (12-6-6)[7] pulverized mixed fertilizer.

Soluble copper, iron and zinc will form insoluble phosphate compounds when added to superphosphate.

Chemistry of Zinc in Soils.   Like several of the other micronutrients in acid soils, zinc can become soluble enough to be damaging to plants. The availability of zinc is reduced by increasing soil pH. This micronutrient is frequently found deficient in neutral to alkaline sandy soils. Zinc is subject to precipitation or fixation in the crystal structure of soil clays and/or just inside the cell wall of roots, in which form it is difficultly available for plant metabolism.

Because polyphosphates are able to sequester heavy metals, there is a feeling among some that ammonium polyphosphate fertilizer is able to complex zinc and then to release it gradually into the soil solution. In doing this, the heavy metal would avoid being fixed in large quantities before the plant root would have an opportunity to take it up.

A value of 2 ppm zinc using 0.1 N HCl extraction of soil was the critical value below which a response to zinc might be obtained on neutral or slightly acid soils.

The liming of acid soils definitely increases the concentration at which zinc becomes injurious to crops. The relation between free lime in soil and the acid-extractable zinc is the basis of soil-test recommendations for corn and sorghum (Figure 8.1). Phosphorus and zinc are mutually antagonistic[8] because they form an insoluble precipitate of zinc phosphate. Figure 8.2 shows the zinc concentration of the corn plant and its relationship to available soil phosphorus. For example, corn is more seriously affected by zinc deficiency in the high phosphate soils of the Outer Basin of middle Tennessee than in other parts of the state.

Agricultural limestone mined from certain formations and made available to farmers in parts of the midwestern and southeastern United States contains zinc as an impurity. This zinc is in the form of sphalerite,

[7]Numbers in parentheses refer to nitrogen, P_2O_5 and K_2O.

[8]R. A. Olson, D. D. Stukenholtz, and C. A. Hooker, "Phosphorus-Zinc Relations in Corn and Sorghum Production," *Better Crops Plant Food* 49 (1) (1965), pp. 19–24.

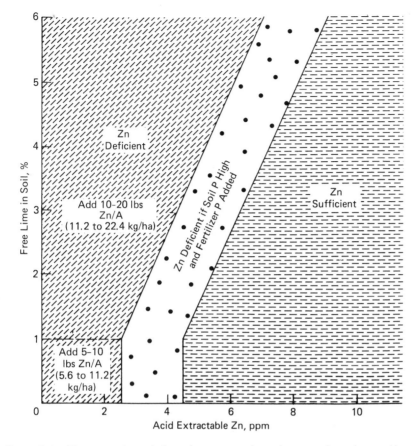

Figure 8.1 Zinc recommendations for corn and sorghum are based on soil tests for free lime and zinc.

which reportedly undergoes progressive conversion to soluble salts in the soil as the plant roots require it. The portion not used by plants remains in the form of sphalerite and is not lost by leaching.

At present, zinc deficiency in certain citrus and corn growing areas appears to be more widespread than the deficiency of any other element, with the exception of nitrogen. It has been reported that there is more zinc deficiency than even nitrogen deficiency in the corn- and sorghum-growing areas of Nebraska and the adjoining states with similar soil and climatic conditions.

CROP RESPONSE TO ZINC

Symptoms. "White bud" disease of corn is a zinc deficiency. "Rosette" of pecans results from a deficiency of zinc. "Bronzing" of tung nut trees, "yellows" of walnut, "mottle leaf" of citrus, "rosette" of apple and "little leaf" of stone fruits and grapes are symptoms reported as zinc

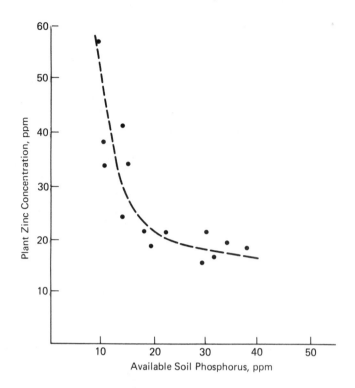

Figure 8.2   Zinc concentration of non-fertilized corn grown on fourteen slightly acid to alkaline soils in the greenhouse.

deficiency. Zinc deficiency of tobacco is usually indicated by the development of dead spots all over the leaf and not specifically at the tips and margins.

Zinc deficiency, which causes "fern leaf" growth disorders, has been found on Irish potatoes in a Washington experimental field where considerable surface soil had been removed. The deficiency was corrected by adding zinc sulfate at a rate of 10 pounds per acre (11.2 kg/ha). Zinc deficiency commonly occurs in corn, beans and rice grown on land that has been recently leveled for irrigation. Deficiency symptoms in plants are most prominent the first year after land is leveled. Weathering of the subsoil, exposed by land leveling, seems to promote the availability of native soil zinc.

Rates.   On slightly acid soils, 10 to 20 pounds per acre (11.2 to 22.4 kg/ha) of $ZnSO_4 \cdot 7H_2O$, or its equivalent of other zinc compounds, is the usual amount added to field crops; 0.5 pounds (0.23 kg) per tree, in small concentrated areas around the trunk, is applied for tree crops. At a rate of 75 pounds per acre (84 kg/ha), $ZnSO_4 \cdot 7H_2O$ was harmful to soybeans grown on Lakeland sand of pH 5.5. Sprays containing 1 to 5 pounds (0.45 to 2.27 kg) of $ZnSO_4 \cdot 7H_2O$ or its equivalent per 50 gallons (189.5 l) of water have been used successfully on trees, shrubs and field crops in humid regions on acid soils.

A rate of 40 to 70 pounds per acre (44.8 to 78.4 kg/ha) of $ZnSO_4 \cdot 7H_2O$ is a practical application for annual crops and fruit trees on alkaline, fine sandy loam soils. A spray consisting of 25 pounds (11.4 kg) of $ZnSO_4 \cdot 7H_2O$ per 100 gallons (379 l) of water, applied before the buds open, was a practical treatment for fruit trees on alkaline soils.

For dormant sprays of fruit trees, $ZnSO_4 \cdot 7H_2O$ at rates of 5 to 25 pounds (2.7 to 11.4 kg) per 100 gallons (379 l) of water is effective. In sprays on leaves, lime hydrate is added to prevent leaf burn.

Adding $ZnSO_4 \cdot 7H_2O$ at a rate of 0.1 pound (0.05 kg) per 5 gallons (18.9 l) of transplant solution was the best treatment for tomatoes. In general, sprays may be more reliable for correcting deficiency than soil treatment.

Zinc is recommended for corn on sandy soils after liming and when the pH is 6.0 to 6.9 at the rate of 3 pounds per acre (3.36 kg/ha). For pecans on the same kind of soil, zinc is recommended at a rate of ¼ pound (0.11 kg) per tree per year of age up to 3 pounds (1.36 kg) per tree in order to correct "rosette"; then apply ¾ pound (0.34 kg) of zinc per tree each year or 3 pounds (1.36 kg) of zinc per tree every 5 years. For apples, a corrective treatment of 15 lb/acre (16.8 kg/ha) of zinc is recommended.

Diagnosis. It is possible and very economical to analyze soils for zinc. However, data indicate that the soil pH and amount of soil phosphorus is as much of a determining factor in zinc needs as is the amount of soil zinc, in practically all cases. This may change as more zinc is applied as manures and pesticides, and it will eventually become necessary to analyze for zinc not only to determine deficiencies but also to diagnose for excessive amounts.

Deficiencies. Lancaster[9] compiled a very comprehensive review of zinc and other micronutrient deficiencies and recommendations for the southeastern United States. According to him, field deficiencies of zinc were first observed in Florida in 1927 on crops growing on organic soils. It was used in the 1930s to correct nutritional disorders in citrus in Florida. Later, deficiencies were encountered in tung and pecans in Florida and other states. Still later, zinc deficiency was observed in corn in a number of states. Field deficiencies of zinc in crops for the different states of the southern region of the United States are shown in Table 8.3. Except for corn and tree crops, occurrence of zinc deficiency in the South has been confined almost entirely to Florida and Texas. In Florida, it is more of a problem on horticultural than on agronomic crops.

[9]J. D. Lancaster, *Micronutrient Deficiencies and Recommendations in the Southeast*. Mississippi State University, 1968.

Table 8.3
ZINC DEFICIENCY IN AGRONOMIC AND HORTICULTURAL CROPS OF THE SOUTHERN REGION

Crop	Va.	N.C.	S.C.	Ga.	Fla.	Ala.	Miss.	La.	Ark.	Tenn.	Ky.	Tex.	Okla.
Avocado					+								
Bean					+							+	
Celery					+								
Clover					+								
Corn	+		+	+	+	+	+			+	+	+	
Cotton												+	
Cow pea					+								
Flax												+	
Guava					+								
Mango					+								
Peach	+	+	+	+	+	+						+	
Peanut					+							+	
Pecan			+	+	+	+	+	+	+	+		+	+
Rice								+					
Sorghum					+					+		+	+
Squash					+								
Tung					+		+	+					

Viets et al.[10] have classified certain crops according to their capacity to utilize native soil zinc. This classification in modified form is presented in Table 8.4. Except in Florida and Texas, zinc deficiency has occurred only in those crops that are very sensitive to the supply of soil zinc. It is interesting to note that soybeans were included by Viets et al. in the very sensitive group.

Recommendations. Because zinc deficiency is affected so much by soil pH, general recommendations for zinc fertilization of corn are almost always restricted to soils having a pH of 6.0 or above. Very light sandy soils are sometimes an exception. In the southern United States, general recommendations are made for zinc on corn in Alabama, Florida, Georgia, Mississippi (Hill Section) and Tennessee but with restrictions on soil pH, phosphate level, or soil texture in every state except Florida. General recommendations are supplemented by soil tests in Alabama and Georgia, but the soil tests are more successful on soils of light to medium texture. In other southern states, recommendations are made primarily on the basis of a visual deficiency.

Rates of zinc commonly recommended for agronomic and some other crops are given in Table 8.5. Both soil and foliar applications are

[10]F. G. Viets, Jr., L. C. Bowan, and C. L. Crawford, "Zinc Content and Deficiency Symptoms of 26 Crops Grown on Zinc-Deficient Soil," *Soil Sci.* 78 (1954), pp. 305–316.

260

Table 8.4
CLASSIFICATION OF CROPS ACCORDING TO SENSITIVITY
TO SOIL ZINC SUPPLY[1]

Very Sensitive	Mildly Sensitive	Insensitive
Beans	Alfalfa	Asparagus
Corn	Beets	Barley
Flax	Clover	Carrots
Grapes	Cotton[2]	Cruciferae[2]
Peach[2]	Onion	Grasses (forage)
Pecan[2]	Potato (white)	Mustard
Soybean	Peanut[2]	Oats
Tung[2]	Sorghum	Peas
	Sudan grass	Wheat
	Tomato	

[1]Viets, Bowan, and Crawford, "Zinc Symptoms."
[2]Added by Lancaster, Micronutrient Deficiencies.

recommended, but soil application predominates for agronomic crops. Poor results frequently are obtained from foliar application to agronomic crops, particularly under conditions of moderate to severe deficiency. Although it may be added separately, zinc usually is applied as a constituent of another fertilizer.

In Florida, fertilizers containing micronutrient mixtures are recommended to supply zinc for agronomic crops.

Table 8.5
RATES OF ZINC MOST COMMONLY RECOMMENDED FOR ZINC-DEFICIENT AGRONOMIC AND SOME OTHER CROPS IN SOUTHERN REGION OF THE UNITED STATES

Crop	Recommendation as $ZnSO_4 \cdot 7H_2O$ Containing 23% Zinc[1]		
	Soil Lb/Acre[2]	Foliar Lb/100 Gal[3] H_2O	Remarks
Corn	10–20	3	Foliar treatment not always satisfactory
Cotton	15–25	3	Only in Texas
Flax	15–25	3	Only in Texas
Rice	25–45	3	Only in Texas
Sorghum (Gr.)	15–25	3	Only in Texas
Citrus		5	Soil application not fully effective
Pecan	40–70	5	Corrective-soil Maintenance-soil

[1]Other sources are effective. In alkaline soils a lower rate of zinc chelate is usually recommended.
[2]Lb/acre × 1.12 = kg/ha.
[3]Lb/100 gal H_2O × 0.0172 = kg/l.

Zinc added to soils through plant residues and in fertilizers accumulates in the surface soil where it is placed; under some conditions, a buildup develops to the extent that toxicity occurs.[11] Therefore, as a precautionary measure, fertilization with zinc at rates much in excess of crop removal should be avoided, except where a deficiency is known to exist or is highly probable. The grain of a 100-bushel (2,727-kg) crop of corn contains approximately 0.12 pound (0.05 kg) of zinc, 40 bushels (1,090 kg) of soybeans contains about 0.06 pound (0.03 kg) and 3,000 pounds (1,363 kg) of seed cotton contains around 0.04 pound (0.02 kg). Unless zinc fixation in highly unavailable forms is a factor, a total application of 50 lb/acre (56 kg/ha) of zinc sulfate should correct zinc deficiency for perhaps decades. Rates and frequency of application of zinc for crop production merit careful consideration.

MANGANESE: FRIT, CHELATE, SULFATE, PHOSPHATE AND OXIDE

Description: Manganese frit[12] is similar to the zinc frit described previously. Basic slag manufactured by U.S. Steel and described previously contains manganese probably in the form of manganese dioxide, MnO_2. U.S. Steel also offers a product containing 40 percent manganous oxide, MnO, Agmanox®, a flue-dust by-product of the manufacture of ferromanganese metal. Manganese sulfate is a purple, crystalline material sold for fertilizer in several forms and trade names, one of which is Tecmangam® manufactured by Eastman Chemicals. Manganese ammonium phosphate is a pink, pelleted material.[13] Manganous oxide is a dark purple, almost black, material suitable for incorporating into fertilizer or for being finely ground for liquid sprays. A manganese chelate absorbed on vermiculite has a greenish brown color and a volume weight about the same as a sandy loam soil. One such product is produced by Dow Chemical Company.

Method of Manufacture. Manganese frit, phosphate and sulfate are manufactured similarly to the zinc materials described previously. The production of basic slag containing 1 to 3 percent manganese has also been described before.

A promising manganese compound recently offered for plant nutrition is manganous oxide, MnO. It is made by reducing pulverized ore consisting mainly of manganese dioxide in a rotary kiln. The reducing medium of

[11]F. R. Cox and J. I. Wear, eds., "Diagnosis and Correction of Zinc Problems in Corn and Rice Production," Southern Cooperative Series Bulletin 222, 1977.

[12]Riegel, *Industrial Chemistry*.

[13]Bridger, Salutsky, and Starotska, "Metal Phosphates."

carbon monoxide is produced by the incomplete combustion of powdered coke or coal.

One manganese chelate is made by absorbing manganese disodium ethylene diamine tetraacetate on vermiculite. Another chelated manganese material absorbed on lignin from hemlock bark is manufactured by Rayonier, Inc. Ammonium phosphate and polyphosphate fertilizers containing 3 to 5 percent manganese have been made by the Tennessee Valley Authority and other manufacturers by incorporating either manganous oxide, manganese sulfate, manganese dioxide or manganese carbonate.

Handling Characteristics.   The addition of small quantities per acre or per tree requires special handling considerations. These handling considerations include choice of diluent, as well as the uniformity of mixing or blending with fertilizer into which manganese is incorporated as a frit, sulfate, phosphate or oxide. Ammonium phosphates and polyphosphates containing manganese possess good physical and handling characteristics. Manganese in some forms may be sprayed or dusted on the foliage, it may be added to the soil directly, or it may be incorporated in the fertilizer. A chelate containing 5 percent manganese may be added directly to the soil at a rate of 200 lb/acre (224 kg/ha) when an amount of 10 lb/acre (11.2 kg/ha) of manganese is required.

Chemical Characteristics.   Manganous oxide has a pH of about 7.0; consequently there is little danger of burning when sprayed on foliage. Two grades have been offered, one containing 48 percent manganese and the other 65 percent manganese. The former is suitable for incorporating in fertilizer, and the latter is a finely ground grade suitable for spray or foliar application. They are only very slightly water soluble but are soluble in weak acids. Availability for plant nutrition compares favorably with manganese sulfate. Manganous oxide is much more effective than manganese dioxide when compared on an equal manganese basis.

Because polyphosphates are able to sequester heavy metals, it is thought that the polyphosphates will be able to complex about 3 to 4 percent manganese and then gradually release it into the soil solution.

Frits contain around 7 percent manganese along with varying quantities of other micronutrients. A popular frit sold along the Atlantic Coast of the United States contains 6 percent boron and 18 percent manganese.

A chelate produced by the Dow Chemical Company contains 6 percent manganese, and one produced by Rayonier, Inc. contains 9 percent manganese.

Manganese ammonium phosphate contains 27.6 percent manganese, and its chemical characteristics are similar to those of the zinc compound.[14]

An average analysis of 48 samples of fertilizer-grade manganese

[14]Bridger, Salutsky, and Starotska, "Metal Phosphates."

sulfates contains 25.1 percent manganese, 14.5 percent sulfur, 1.9 percent magnesium, 6.6 percent calcium, 0.05 percent copper, 0.08 percent zinc and 0.3 percent boron. Manganese sulfate is probably the least expensive of all carriers of manganese used in fertilizers. Pure anhydrous manganous sulfate, $MnSO_4$, contains 36 percent manganese.

Chemistry of Manganese in Plants and Soils. The manganese content of the same crop will vary widely; in this respect, it is probably the most variable of the mineral nutrients. As an example, manganese content of cotton leaves can vary from 15 to 1,500 ppm. Even so, foliar composition of plant tissue is helpful. Values less than 25 ppm are usually deficient; those of 50 to 200 ppm are normal; and those over 400 ppm are considered excessive. Plants may selectively absorb manganese ahead of iron, and manganese to some extent is capable of replacing iron.

Excess quantities of soluble manganese, such as may be present in very acid soils, may be injurious to plants, and this explains in part the ill effects of soil acidity on crops. Above a soil pH of 7.5, the availability of manganese is very low because of the formation of insoluble basic hydroxides and carbonates. The pH range of low solubility of manganese in soils is about 6.5 to 8.5. The pH level that is conducive to medium or high solubility of manganese is below 6.0. In poorly drained soils, oxidation is curtailed, thus encouraging the accumulation of organic matter and the formation of reduced manganese, such as MnO, that is readily available to plants.[15] Well-drained, aerated soils encourage the formation of oxidized manganese, such as MnO_2, that is only one-fourth to one-fifth as available to plants as MnO. The solubility of manganese increases with increasing organic matter content; this results in loss by leaching and crop removal, especially in cultivated soils that are classified as peats and mucks. Coarse-textured soils are usually more deficient than fine-textured soils. Thus, soil pH, organic matter content, redox potential, moisture, texture and parent material all influence the response of crops to manganese.

CROP RESPONSE TO MANGANESE

Manganese is added to many Florida soils for many crops, to South Carolina upper coastal plain soils for cotton, and to soils in a few other areas for cotton, peanuts, soybeans, fruits, beans, vegetables and small grains. Deficiencies have been reported in 25 states. Legumes appear to be more sensitive to a deficiency of manganese than nonlegumes, but legumes usually contain the least amount of manganese whereas the grasses contain the greatest amount. A ton (0.91 m ton) of alfalfa, however, may contain as much as a pound (0.45 kg) of manganese.

[15]F. N. Ponnamperuma, "Behavior of Minor Elements in Paddy Soils," International Rice Research Institute Research Series, No. 8, 1977.

Rates. When manganese deficiency occurs in light, sandy soils, 5 to 10 pounds per acre (5.6 to 11.2 kg/ha) of manganese is generally suggested. Higher rates may be used on alkaline and organic soils. Sprays for foliage containing 0.05 to 0.1 percent manganese, an amount equivalent to about 5 pounds (2.27 kg) of manganese sulfate dissolved in 200 gallons (758 l) of water and sprayed on one acre (0.4 ha), have corrected manganese deficiency (Table 8.6).

Table 8.6
RATES OF MANGANESE COMMONLY RECOMMENDED FOR AGRONOMIC AND SOME OTHER CROPS

Crop	Recommendation—Pounds of Manganese[1] per Acre[2]		
	Soil	Foliar	Remarks
Citrus	8–12	0.75[3]	Apply when visual symptoms are evident. Make only foliar application to calcareous soils.
Cotton	1–2.5		
Soybean	10–15	3–5	In 3 to 5 applications of 1 lb (0.45 kg) each.
Small grain	5–10		
Peanut	10–15	3–5	In 3 to 5 applications of 1 lb (0.45 kg) each.
Vegetables	5–15	1–3	

[1]As manganese sulfate or manganous oxide. Lower or higher rates may be required with other sources.
[2]Lb/acre × 1.12 = kg/ha.
[3]In 100 gallons (379 l) of water.

Symptoms. Harmful amounts of soluble manganese are not uncommon in very acid mineral soils. Malnutrition of cotton known as "crinkly leaf" or "curly top" is caused by an excess of manganese in the plant. Lime will correct the symptom. The same symptom and correction have been noted on leafy green crops, such as turnip greens, collards and mustard.[16]

Lucas[17] reports that crops showing high response to manganese are beans, lettuce, oats, onions, peas, potatoes, spinach, soybeans, Sudan grass, sugar beets, table beets and wheat; those showing low to medium response are alfalfa, asparagus, barley, cabbage, carrots, cauliflower, clover, celery, corn, cucumbers, grass and mint. The amount of manganese recommended varies depending on soil pH, texture and organic content

[16]F. B. Stewart, "Manganese Fertilization of Turnip and Mustard Greens," *The Veg. Grow. News* 26 (1971), pp. 3–4.

[17]R. E. Lucas, "Micronutrients for Vegetables and Field Crop," Extension Bulletin E-486, Michigan State University, 1964.

(Table 8.7). Manganese deficiency in crops can be prevented by adding manganese fertilizer to the soil or on the foliage or by making the soil more acidic with the addition of sulfur or aluminum sulfate or the use of acid-forming fertilizers, especially if banded near the plant. Steam or chemical fumigation will also give temporary correction to deficiencies of manganese.

Table 8.7
MANGANESE RECOMMENDATIONS WHEN APPLIED IN A BAND NEAR THE SEED OR PLANT[1]

| | | Pounds of Manganese[2] per Acre[3] | |
| | | High Responsive Crops | Low to Medium Responsive Crops |
Soil	pH Range		
Peats, mucks and	5.8 to 6.4	10	5
dark colored	6.5 to 7.2	15	8
	7.3 to 8.5	20	10
Coarse textured	6.0 to 6.4	5	–
All mineral soil	6.5 to 7.2	8	4
All mineral soil	7.3 to 8.5	10	5

[1]Does not include chelated manganese.
[2]Fertilizer control regulations permit manganese to be mixed in nitrogen-phosphorus-potassium fertilizers. Thus, an application of 500 lb/acre (560 kg/ha) of fertilizer containing 2 percent manganese will supply 10 lb/acre (11.2 kg/ha) of the element.
[3]Lb/acre $\times$ 1.12 = kg/ha.

Deficiencies. Manganese deficiency in the southern United States has been limited almost exclusively to soils in the Atlantic Coastal Plain, particularly in Virginia and the Carolinas with respect to agronomic crops. Recommendations for crop fertilization with manganese in this area have been issued for many years. The distribution of manganese deficiency by crops in the region is shown in Table 8.8. Manganese toxicity has occurred on strongly acid soils in a number of states.

With regard to the effect of soil reaction on the availability of manganese, it is interesting to note that manganese deficiency seldom occurs on the neutral and alkaline soils of the western United States, except on citrus.[18]

Diagnosis. The ease with which soil manganese undergoes reduction is a measure of its reactivity and availability to crops. In contrast to soils in other areas of the Southern Region, sandy soils of the Atlantic Coastal Plain are unusually low in easily reducible manganese and probably in total content as well.

[18]*Western Fertilizer Handbook* (Danville, Illinois: Interstate Printers and Publishers, 1975).

Table 8.8
MANGANESE DEFICIENCY IN AGRONOMIC AND HORTICULTURAL CROPS OF THE SOUTHERN REGION

Crop	Va.	N.C.	S.C.	Ga.	Fla.	Ala.	Miss.	La.	Ark.	Tenn.	Ky.	Tex.	Okla.
Bean			+		+					+			
Cabbage					+								
Citrus					+							+	
Clover					+								
Corn	+				+							+	
Cotton			+	+									
Peach	+				+								
Peanut	+	+											
Pecan					+								
Potato					+								
Small grain	+	+	+										
Soybean	+	+	+				+				+		
Spinach	+												
Sweet potato				+									
Tomato					+								

Of the agronomic crops, cotton, peanuts, small grain and soybean are most severely affected by low manganese levels. Manganese is recommended for peanuts and soybeans in North Carolina and Virginia and for soybeans in Kentucky. Visual observations, soil tests (manganese and/or pH) and past history serve as bases for recommendations. Both soil and foliar applications are recommended, but more than one foliar treatment is usually required for optimum response. On alkaline, peaty soils, soil application is rather ineffective. Banding apparently is greatly superior to broadcasting manganese sulfate.

General recommendations for manganese on cotton are made in Georgia and South Carolina. In Georgia, the recommendations are restricted to soils having a pH above 6.2. Manganese is recommended for cotton in North Carolina on deep sands and on other soils having a pH above 6.5 that are low in manganese.

An interesting aspect of the response of cotton to manganese in Georgia has been the apparent lack of association of yield response with leaf-petiole content of manganese at average manganese concentrations within the range of 100 to 300 ppm. Deficiency symptons in plants usually do not appear until plant concentrations are below 25 ppm. Much of the response in Georgia seems to have been associated with increased earliness of the crop. In Mississippi, cotton has shown no response to manganese with leaf concentrations as low as 40 ppm. No response has been obtained in Alabama, and results show that the content of manganese in Alabama and Mississippi soils is higher than in the Carolinas.

267

Recommendations. Fertilizers containing a micronutrient mixture of boron, copper, iron, manganese, molybdenum and zinc are suggested as insurance against micronutrient deficiency in agronomic crops in Florida. Rates of manganese commonly recommended elsewhere are given in Table 8.6. Higher rates for soil application have been indicated for both organic soils and sandy soils (high in organic matter) that are alkaline in reaction. It is interesting to compare the Lucas rates for soils in Michigan in Table 8.7 with Lancaster's rates for soils in Southeast United States in Table 8.6. The former suggests from 4 to 20 lb/acre (4.5 to 22.4 kg/ha) while the latter suggests slightly less, from 2.5 to 15 lb/acre (2.8 to 16.8 kg/ha).

In relation to the amount of manganese in the soil and to that applied in fertilizers, crop removal of this micronutrient is rather insignificant, particularly where only grain is harvested. Therefore, where manganese is added, a buildup of manganese will occur; after some time, the requirement for additional manganese should be greatly reduced, or eliminated, except where very severe soil fixation is a factor. Apparently, considerable accumulation has already occurred in Florida in areas where fertilization with manganese has long been practiced.

It is probable that fertilization with manganese will not be required to any great extent except in soil areas and on crops where deficiencies have already been noted in the United States.

IRON: FRIT, CHELATE, SULFATE, OXALATE AND PHOSPHATE

Description. Iron frit and chelate are similar to the zinc ones described previously. Iron imparts a dark brown color to the glassy frits, giving them an appearance similar to cocoa. Basic slag contains iron, and this material has previously been described. Iron chelate is a light to dark brown, powdery material. Ferric sulfate, $Fe_2(SO_4)_3 \cdot H_2O$, is 20 percent iron and is a brownish red salt made by several corporations. Ferrous sulfate, $FeSO_4 \cdot 7H_2O$, sometimes called copperas or green vitriol, is a blue green salt. Ferrous oxalate, $FeC_2O_4 \cdot 2H_2O$, is a pale yellow crystal of fine particle size and is marketed under trade names, including Nu-Iron®; it is another type of chelated iron. Iron ammonium phosphate is a green salt.[19]

Method of Manufacture. Iron frits are manufactured similarly to the other metallic frits. Iron chelates include monosodium ferric ethylene diamine tetraacetate (NaFe EDTA), disodium ferric ethylene diamine tetraacetate (Na₂Fe EDTA), monosodium ferric diethylene triamine pentaacetate (NaFe DTPA) and monosodium ferric hyroxyethyl ethylene diamine triacetate (NaFe HEEDTA). Iron chelates are the most widely used chelates. EDDHA chelated with iron has overcome all of the objections to EDTA

---

[19]Bridger, Salutsky, and Starotska, ''Metal Phosphates.''

on calcareous soils. Its chelates are effective in both acid and alkaline soils.

The most common iron salt used in plant nutrition is ferrous sulfate, a by-product of the pickling operation in steel manufacture. Another salt, ferrous oxalate, is produced by reacting a soluble iron salt with oxalic acid. Ferric sulfate is made by treating flue dust, from a Cottrell precipitator in smelting operations, with sulfuric acid. The ferrous iron salts, in contrast to the ferric iron salts, are more soluble and more readily available to plants, but some of them are unstable. For example, ferrous sulfate, $FeSO_4 \cdot 7H_2O$, readily loses water, and the salt undergoes oxidation, changing from a blue green to a brown color, indicating a conversion from ferrous sulfate to ferric sulfate.

Handling Characteristics. Frequently, rates of as little as 10 lb/acre (11.2 kg/ha) of ferrous sulfate or its equivalent in iron are ample to supply the iron needs of crops grown on soils with extremely low available-iron content. Thus, the need for a diluent. Pineapple growers in Hawaii, on acid manganiferous soils, have found it profitable to spray a 5-percent solution of ferrous sulfate (copperas) at the rate of 50 to 100 gallons per acre (471 to 942 l/ha) every 7 to 10 days during the growing season. Soluble iron compounds, after dilution with water and sprayed on foliage has been superior to soil applications.

Ferrous oxalate is nonhygroscopic and remains chemically stable upon exposure to the atmosphere. It retains its chemical stability and fine particle size in storage.

Basic slag is a by-product of the steel industry. Although its principal ingredients are phosphorus, calcium and magnesium oxides, it also contains slightly soluble oxides of iron and manganese. Basic slag is added directly to the soil at convenient rates of 500 to 1,000 pounds per acre (560 to 1,120 kg/ha).

Ferric sulfate broadcast at a rate of 716 pounds per acre (802 kg/ha) [200 pounds (90 kg) of iron] on alkaline soils increased rough rice yields in Arkansas from 4,045 to 5,395 pounds per acre (4,530 to 6,042 kg/ha).[20] Similar results were obtained on alkaline soils in California.[21] Subsequent research by Place[22] showed that the rice response to high rates of ferric sulfate was the result of soil acidity produced or of zinc impurities in the iron salt.

[20]G. A. Place, "Relationship of Iron, Manganese, and Bicarbonate to Chlorosis of Rice Grown in Calcareous Soil," Agricultural Experiment Station Report Series 175, University of Arkansas, 1969.

[21]D. S. Mikkelsen, W. G. Golden, and K. Ingrebretsen, "Response of Rice to Iron Compounds on Alkaline Soils," Rice J. 65(7) (1964), pp. 64–65.

[22]B. R. Wells, L. Thompson, G. A. Place, and P. A. Shockley, "Effect of Zinc Chlorosis and Yield of Rice Grown on Alkaline Soil," Agricultural Experiment Station Report Series 208, University of Arkansas, 1973.

Chemical Characteristics. Frits contain 14 to 19 percent iron. Basic slag consists of 10 to 13 percent iron. Copperas ($FeSO_4 \cdot 7H_2O$) contains 20 percent iron. Ferric sulfate, $Fe_2(SO_4)_3$, is 28 percent iron. Ferrous oxalate carries 30 percent iron, about half of which has chelating properties; it is only slightly water soluble and is neutral in reaction to litmus. Iron ammonium phosphate contains 37 percent iron. A chelated iron, sodium ferric diethylene triamine pentaacetate, containing 10 percent iron is offered by Geigy Corporation.

Chemistry of Iron in Soils. An excess of zinc, copper, phosphorus manganese, or aluminum in the soil has been known to bring about iron deficiency. Conversely, iron salts can antidote an excess of aluminum manganese or zinc.[23] The use of ferrous sulfate sprays on pineapples growing in manganiferous soils for correction of iron deficiency has been mentioned previously. Manganese excess may occur where iron is deficient, and iron excess may occur in the absence of sufficient manganese. The ratio of iron to manganese in a soil may be more important than their actual total amounts.

Nearly all mineral soils contain an abundance of iron, and some contain 20 tons or more of iron per acre furrow slice (44.8 m tons/ha). The amount varies from 200 ppm up to 10 percent. Nevertheless, there are a few acid soils and many alkaline soils that have been found deficient in this element. Too much lime, say 20 tons per acre (44.8 m tons/ha), on acid soils has induced chlorosis of corn, cotton, rice and other crops. When iron deficiency is found on calcareous soils, it may be corrected by the continued use of acid-forming fertilizers, such as sulfate of ammonia or ammonium phosphate sulfate. Very little success has been observed in correcting iron deficiency by additions of soluble iron salts to soils. When they are added to soils, the iron of the soluble salts rapidly converts to insoluble hydroxides, oxides and phosphates.

The pH of a soil is an important factor in determining the solubility of iron in soils and, therefore, the ability of a soil to supply iron to the plant. In the acid range, pH 4 to 5, iron is much more soluble than it is in the slightly acid to alkaline range of pH 6 to 9.

CROP RESPONSE TO IRON

Iron deficiency occurs in dry, subhumid climates like western Nebraska, California and Kansas where soils contain considerable calcium and other bases.

Symptoms. The deficiency symptoms of different species of plants are quite uniform. Fruit trees, ornamentals and grain sorghum are affected more generally than some other crops. The leaf characteristics of the

[23]K. P. P. Nair and G. R. Babu, "Zinc-Phosphorus-Iron Interaction Studies with Maize," *Plant Soil* 42(3) (1975), pp. 517–36.

deficiency symptoms are a distinctive green vein color and yellow inter-veinal areas.[24] Leaf iron content of 10 to 80 ppm is associated with deficiency symptoms. Pineapples respond to iron on soils having a high content of calcium and manganese in Hawaii and Puerto Rico. Deficiencies have been reported in 26 states.

Rates.   Iron chelates have been improved so that the problem of translocation of iron throughout the plant tissue has been solved to a considerable degree. In early stages of development, chelates were only effective for soil application at a pH below 7.0. Developments have now provided materials that may be sprayed on the foliage without burning. Also, chelates of the EDDHA type have been developed that are satisfactory for both acid and alkaline soils. Acceptable results on peaches have been obtained with 0.044 lb (20 grams) of iron per tree as iron chelate (NaFe EDTA). The material was added to acid soil and was effective as a solution or after incorporation in attaclay or vermiculite. The EDTA iron was not sufficiently stable in calcareous soils. Although it is relatively expensive, chelated iron efficiency at low rates of application has resulted in extensive use, primarily in specialty fertilizers for lawns, gardens, shrubs and ornamentals, such as azaleas, roses, camellias and gardenias.

Ferrous oxalate, although only difficultly water soluble, is readily available to the plant either through its foliage or roots and is generally added as a dust or spray by direct foliar application. The spray is made by mixing four heaping teaspoons per gallon (3.8 l) of water or one pound (0.45 kg) per 50 gallons (189.5 l). The dust is made by mixing four parts clay with one part ferrous oxalate. Since it is a neutral salt, there is little danger of burning. For soil application, 2 to 4 pounds per acre (2.24 to 4.48 kg/ha) is suggested. Obviously, this quantity must be added to fertilizer or other diluent for proper distribution. Nu-iron® is recommended for soil mixes for bedding plants such as petunias, snapdragons and tomatoes at the rate of 4 to 8 ounces (0.11 to 0.22 kg) per cubic yard (0.77 m³) of soil.

Basic slag and ferrous sulfate (copperas) are probably the most economical sources of iron for plant nutrition. However, applications to soil of pH 6.5 or higher are often ineffective. Soil applications of iron in the form of slag, frits and copperas around citrus trees, pineapples and grain sorghums seldom produce corrective effects. Sprays on foliage are usually more effective. Foliar sprays of 0.5 percent iron are necessary. Ferrous oxalate is one of the chelates that has been found efficient for foliar application. Iron EDTA is also used on foliage but is a little more expensive to manufacture than is ferrous oxalate.

Ferric sulfate is being used at rates varying from 200 to 1,000 pounds per acre (224 to 1,120 kg/ha) on rice in Texas, California and Arkansas where the soil pH approaches 7.0 or higher and the iron deficiency is caused by calcareous soil conditions. Chlorosis similar to iron deficiency

[24]*Western Fertilizer Handbook.*

is noted in rice grown on saline (soluble salts) or alkali (sodium) conditions, but ferric sulfate is not beneficial when these soil conditions occur.[24]

Recommendations. Some acid-loving plants, such as azalea, gardenia and camellia, are very sensitive to iron availability, and deficiency may occur even on moderately acid soils. Apparently, none of the agronomic and vegetable crops is so sensitive to iron availability in acid soils.

Recommendations for iron on agronomic and/or vegetable crops in the southern United States are given in Table 8.9. Five states, Arkansas, Florida, Oklahoma, South Carolina and Texas, reported deficiencies. Where the pH is high, soil application of soluble salts of iron usually is not very effective because the conditions that render the native iron of soils unavailable also convert the soluble added iron to an unavailable form. Under such conditions, foliar spraying is suggested, but repeated applications are usually necessary. Though other compounds give satisfactory results, ferrous sulfate (copperas) is the most commonly used. A 2.5-percent solution with a wetting agent is a common formulation. Soil application of certain iron chelates and iron polyflavonoid may correct iron deficiency partially or wholly, but the cost often is prohibitive except on very high-value crops.

Deficiencies. Wherever iron deficiency occurs naturally, the problem will no doubt continue to exist. Elsewhere, it seems likely to occur only when induced by improper liming and fertilization practices or by the accumu-

Table 8.9
IRON DEFICIENCY IN THE SOUTHERN REGION

Crop or Plant	Va.	N.C.	S.C.	Ga.	Fla.	Ala.	Miss.	La.	Ark.	Tenn.	Ky.	Tex.	Okla.
Bahia grass					+								
Cabbage					+								
Citrus					+								
Corn					+							+	
Crimson clover													
Fescue													+
Grape													+
Ornamentals	+	+	+	+	+	+	+	+			+	+	+
Peach												+	
Peanut												+	
Rice									+				
Sorghum						+						+	+
Soybean													
Squash					+								
St. Augustine grass					+								
Tobacco		+											
Turnip													

lation of calcareous material, such as is in rice fields that have been irrigated many years with hard water.

In view of the general lack of deficiency in agronomic crops and the lack of response and/or the uneconomical returns to soil application where deficiencies do occur, the addition of iron to fertilizers for crops in most of the southern region of the United States would seem to be of questionable importance, except under specific and limited conditions.

Lucas[25] noted that iron deficiency in field and vegetable crops is not common in the eastern states. It is a problem in the western states where the soils contain considerable sodium and calcium. In Michigan, woody plants, such as pines, pin oaks, roses, certain ornamentals and acid-demanding plants such as blueberries, azaleas and rhododendrons, may need iron. Lawns and particularly putting greens of golf courses sometimes show an insufficient amount of iron for grass because of high pH and high levels of phosphorus. Putting greens made from an excess of coarse sand and acid peat often develop iron deficiencies.

Iron deficiency in many woody plants appears when they are grown in soils low in organic matter and high in pH. Mixing in organic materials, such as manure or acid peat, will help to increase the availability of the iron.

The use of sphagnum moss peat in mixtures with sand, perlite or vermiculite intensifies the need for iron fertilizers for the production of petunias, snapdragons, tomatoes and other bedding plants. This has been widely experienced during recent years.

COPPER: FRIT, CHELATE, PHOSPHATE, SULFATE AND OXIDE

Description. Copper frit is similar to the zinc one described previously. As of this writing, copper is not offered as a separate ingredient in the frit form. However, it is included along with other metals in the frit. Copper is included as a component of the lignin sulfonate chelates. Copper ammonium phosphate is a blue salt offered in granular or powdered form. Copper sulfate, $CuSO_4 \cdot 5H_2O$, known also as blue vitriol or bluestone, in the pure form is an asymmetric crystal and is offered as a solid or spray material; it is the most commonly used copper fertilizer and is easily discernible in a mixed granular or pulverized fertilizer. Cupric oxide, CuO, which is sometimes used has a dark brown color.

Method of Manufacture. Copper frits, phosphates and sulfates are manufactured similarly to the zinc materials described earlier. A copper chelate has been offered in the EDTA form. Cupric oxide, CuO, is formed when copper is heated in air or when the sulfate, nitrate, carbonate or hydroxide of the metal is ignited.

[25]Lucas, "Micronutrients."

Handling Characteristics. Like the other micronutrients, but perhaps to a more pronounced degree, the amount of copper added to crops must be carefully regulated. A damaging effect on soybeans was evident when 75 lb/acre (84 kg/ha) of copper sulfate was added to an acid, Norfolk loamy sand.[26] It has been shown in New Jersey that ¼ ppm of copper in nutrient solution culture seriously damaged flax and barley.[27] In Florida, many fertilizer mixtures have copper sulfate added to them. A common grade for many years was 4-6-8-3-½; the last two figures represent the percentage of magnesium oxide and copper oxide equivalent, respectively. In Michigan, fertilizer control regulations permit the use of 0.5, 1 and 2 percent copper in mixed fertilizer. In many situations, it is best to add copper in a spray directly to deficient plants.

Chemical Characteristics. Copper frits contain 3 to 7 percent copper. Copper ammonium phosphate contains 30 percent copper; its properties are similar to those of the zinc ammonium phosphate compound. Fertilizer-grade copper sulfate is made up of 24.9 percent copper, 12.8 percent sulfur and 0.55 percent zinc. Copper oxide has a copper content of 60 to 80 percent, and in field tests it is as effective a source of copper as copper sulfate.

Chemistry of Copper in Soils and Plants. Soils with 5 percent or higher organic matter content and peaty soils low in ash in Michigan, Wisconsin, Illinois, Indiana, Florida, New Jersey, the Carolinas, Holland and Australia have been reported to be deficient in copper. Where the problem appears on mineral soils, it will most likely be on acid, sandy soils which have been heavily cropped and liberally supplied with mixed fertilizers. In the southeastern United States, copper deficiency in vegetables and wheat is found on Portsmouth and similar types of soils that are acidic and contain 2 percent or more organic matter. In Florida, copper deficiency frequently occurs along with magnesium deficiency on acid Lakeland sand. Copper is readily absorbed by soil colloids, is not easily leached, nor is much used by the crop. For this reason, no further additions of copper are needed on peats and mucks if a total of 20 to 40 pounds per acre (22.4 to 44.8 kg/ha) has been added.

Normal plants contain 8 to 20 ppm, and deficient plants usually have less than 6 ppm. In each ton (0.91 m ton) of dry hay, there is only about 0.002 pound (0.001 kg) of copper. The chief function of copper in plants, soils and warm-blooded animals may be that of an oxidizing agent. It seems to play a definite role in the utilization of iron in warm-blooded

[26]N. McKaig, W. A. Carns, and A. B. Bowen, "Fertilizer Ratio Experiment with Soybeans," South Carolina Experiment Station Annual Report, 1937.

[27]J. G. Lipman and G. MacKinney, "Proof of the Essential Nature of Copper for Higher Green Plants," *Plant Physiol.* 6 (1931), p. 593.

animals. Copper is thought to be involved in oxidizing harmful quantities of soluble sulfide to more insoluble sulfates, soluble ferrous to ferric and manganous to manganic salts.[28]

Copper deficiency in many plants is indicated by wilting of the leaf tips. In grain, the leaves are yellowish, and the leaf tips show a disorder similar to frost damage. Carrot roots, wheat grain and onion bulbs show poor pigmentation.

Crop Response. Deficiencies have been reported in 13 states, mainly in peat, muck and other highly organic soils. Rates of copper commonly used in highly responsive crops are 3 to 6 pounds per acre (3.36 to 6.72 kg/ha), depending on soil pH. The response to copper depends not only on the soil, but also on the crop. Alfalfa, dill, lettuce, onions, spinach, Sudan grass and wheat are more responsive to copper than are other crops. On highly organic soils, barley, carrots, cauliflower, Hungarian millet, oats, radishes, sunflower and table beets have shown medium to large response to copper. In Wisconsin on a Carrington silt loam, a soil with up to 5 percent organic matter, sweet corn yields have been increased 5 to 40 percent by additions of 5 pounds (2.27 kg) of copper sulfate, which was mixed thoroughly with 200 pounds per acre (224 kg/ha) of 3-18-9 fertilizer. Recommendations and rates of copper application from states in the southern region of the United States are given in Tables 8.10 and 8.11.

In the western United States the native copper supply has only rarely been recognized to need supplementation. These few instances have been with tree crops on organic soils and sands.[29]

Bordeaux mixture, a fungicide consisting of copper sulfate (blue vitriol), lime and water, supplies copper to plants that are sprayed with it for the control of diseases. Copper-sulfur dust is used to control *Cercospora* leaf spot on peanuts and grapes but also supplies copper as a micronutrient. This material is a mixture of flowers of sulfur and tribasic copper sulfate, containing 53 percent copper.

Very low levels of copper are used to control algae in fish ponds. The concentration of copper must be carefully controlled to avoid killing the fish. Copper is used to control diseases and is highly toxic; therefore it is not recommended except where a need has been established.

Copper has been used so extensively in fertilizers and fungicidal sprays in Florida that a deficiency seldom occurs now except on new lands or on other soils that have never received copper additions. For most crops, 20 or more years of cropping would be required in order to remove 1 pound (0.45 kg) of copper.

[28]J. E. McMurtrey, "Distinctive Plant Symptom Caused by Deficiency of Any One of the Chemical Elements Essential for Normal Development," *Bot. Rev.* 4 (1938), p. 183.

[29]H. D. Chapman, *Diagnostic Criteria for Plants and Soils* (Riverside: Division of Agricultural Science, University of California, 1966).

Table 8.10
AGRONOMIC AND HORTICULTURAL CROPS OF THE SOUTHERN REGION THAT HAVE SHOWN COPPER DEFICIENCY

Crop	Va.	N.C.[1]	S.C.	Ga.	Fla.	Ala.	Miss.	La.	Ark.	Tenn.	Ky.	Tex.	Okla.
Barley		+											
Bean					+								
Cabbage					+								
Carrot					+								
Celery					+								
Citrus					+								
Corn		+			+								
Eggplant					+								
Lettuce					+								
Oats		+			+								
Onion												+	
Pangola grass					+								
Peach					+								
Peanut					+								
Pepper					+								
Soybean		+											
Sugar cane					+								
Tomato					+								
Watermelon					+								
Wheat		+											
White clover					+								

[1]Copper deficiency has been noted in some horticultural crops.

BORON: FRIT, SILICATE, BORATES AND BORIC ACID

Description. Boron frits are similar in appearance and composition to the zinc frits described previously. Sodium borosilicate is a green, glassy material. Calcium borate, $Ca_2B_6O_{11} \cdot 5H_2O$, is a white, crystalline salt known as colemanite. Borax, sodium tetraborate decahydrate, $Na_2B_4O_7 \cdot 10H_2O$, is a white salt rather finely ground for fertilizer use. It and its anhydrous derivatives are the most commonly used boron fertilizers. Varying degrees of dehydration of borax produce products such as pyrobor and anhydrous borax. Two such products have trade names, Fertilizer Borate 46 (46 percent B_2O_3, 14 percent boron) and Fertilizer Borate 65 (65 percent B_2O_3, 20.2 percent boron). The latter is a canary yellow, glassy, crystalline appearing material with particles about the size and appearance of beach sand; because it is the most concentrated form of boron fertilizer, it is usually the least expensive. Fertilizer Borate 46 is offered in coarse and fine size products and is light gray in color, probably as a result of coating with clay in order to maintain good physical condition. Another product, Solubor® is made by refining Fertilizer Borate

276

Table 8.11

SUGGESTED SOIL TREATMENT FOR CORRECTING COPPER DEFICIENCY IN FLORIDA AND NORTH CAROLINA[1]

State	Crop	$CuSO_4 \cdot 5H_2O$[2] Lb/Acre[3]	Frequency
Florida	Citrus	0.1 to 10	New plantings—annually in fertilizer (0.25% CuO in 8-2-8 grade) for first 10 years, none in groves over 10 years old except where deficiency symptoms occur.
	Tung	25	At five-year intervals.
	Vegetables	50	Initially.
		10	Annually.
	Pangola grass	25 to 50	Initially.
		15	At 5-year intervals thereafter.
	Clover, oats	15	At 5-year intervals.
North Carolina	Small grain	20	At 3-year intervals.
	Corn	20	Initially, as if in rotation with small grain.
	Soybean	10 to 20	Initially, none if in rotation with small grain.
	Vegetables	Not available but probably similar to Florida.	

[1]In part from J. G. Fiskell and S. E. Younts, "Copper Status and Needs in the Southern Region," *Plant Food Rev. 9(1963), 6–10.*

[2]Containing about 25 percent copper as metallic.

[3]Lb/acre x 1.12=kg/ha

65. Solubor® is a finely ground white product that contains 66 percent B_2O_3 (20.5 percent boron). Magnesium borate found in the boracite mineral, $2Mg_3B_8O_{15} \cdot MgCl_2$, is a white to grayish crystalline material. Boric acid, H_3BO_3 is offered as a pharmaceutical in the form of a white powder or in a colorless, aqueous solution.

Method of Manufacture. Boron frits are made carrying various metals, including molybdenum, manganese, zinc, copper and iron. Sodium borosilicate is made by fusing kaolin clay and borax at around 532°F (900°C). The resulting green melt is shredded through fine pores, and a textile product known as fiberglass is produced. Sodium, calcium and magnesium borates are borax, colemanite and boracite, respectively. They are all constituents of a mineral mined as borax in the western United States, Ceylon, India and Tibet. From Searles Lake brine, the mother liquor from the potassium chloride crystallization is diluted and cooled further in vacuum-type coolers to separate crystals of crude borax, $Na_2B_4O_7 \cdot 10H_2O$. From this material, other products, such as refined borax, anhydrous borax, pyrobor and boric acid, are obtained by further processing. Boric acid is also obtained from the volcanic hot springs of

Tuscany, Italy. The hot liquid is evaporated by the heat of the steam jets, and boric acid crystallizes out in thin white plates. It is also prepared by the action of sulfuric acid upon concentrated solutions of borax:

$$Na_2B_4O_7 + H_2SO_4 + 5H_2O \rightleftarrows Na_2SO_4 + 4H_3BO_3$$

Colemanite containing 10 percent boron was offered for a while but had to be withdrawn because of short supply.

Handling Characteristics.   All of the boron materials used for fertilizer are stable chemicals that do not deteriorate in storage. Boric acid and Solubor® have been used as a very dilute spray at the rate of 0.1 to 0.2 lb/acre (0.11 to 0.22 kg/ha) to correct boron deficiency on cotton. Sprays require very fine soluble materials in order to prevent clogging of nozzles. Ordinary agricultural borax is soluble but usually contains a few large crystals that are not immediately dissolved and these clog the nozzles. Solubor® and similar products have been designed especially for spray applications on foliage. Borax products, such as fertilizer borates, colemanite and boracite, as well as boron frits are added to fertilizer for subsequent addition to the soil. When boron is added to fertilizer, it is usually added in amounts of 1 to 5 lb/ton (0.5 to 2.5 kg/m ton). The bags of mixed fertilizer containing boron must be labelled so that the boron content is prominently displayed.

Results of drilling, broadcasting and spraying boron for cotton are compared in Table 8.12. There is no difference in methods of application so far as yield is concerned. Because of the danger involved in using too much boron on young plants, extreme care should be taken in formulation and application. Colemanite and boracite cannot be used as a spray because of their insolubility.

As a result of changes in the composition of various solid materials

Table 8.12
RATES AND METHODS OF APPLICATION OF BORON TO
COTTON IN MISSISSIPPI[1]

Boron Treatment	Relative Yield[2]
No Boron	87
0.25 lb drilled	103
0.5 lb drilled	100
1.0 lb drilled	101
1.0 lb broadcast	100
0.5 lb drilled + foliar spray[3]	101
Foliar spray[3]	104

[1]Data by Lancaster, *Micronutrient Deficiencies.*
[2]Relative to 0.5 lb of boron drilled as 100 percent.
[3]Five to six applications of 0.1 lb each.

that occur in several ingredients of mixed or compound fertilizers, the boron in the final processed product is not the same calcium, magnesium or sodium borate that was added to the initial mix. Boron frit, because of its glassy, silicate nature, is the boron material most likely to remain unaltered by the processing associated with mixed or compound fertilizer manufacture and is the material presently relied upon to supply a relatively insoluble, slowly available boron.

Chemical Characteristics.   The boron frits on the market contain up to 6 percent boron. They are very difficultly soluble in water and weak acid.

Boron frit is a manufactured borosilicate. Its solubility may be altered by:

1. Increasing the percent of boron.
2. Changing the composition of the matrix.
3. Heating and cooling procedures.
4. Altering the fineness of grind.

Silicate or slag-type borate contains 4 to 5 percent boron, and it too is relatively insoluble in water. Fertilizer sodium borate, known as borax, contains 11 percent boron. It is soluble in water and readily available to crops, but is subject to leaching from the soil. The same can be said for Fertilizer Borates and Solubor®. Solubor® contains 20.5 percent boron and is primarily designed for liquids, dusts or spray applications. It has been suggested for use in nitrogen solutions at 0.1 to 0.3 percent boron. Fertilizer Borate 46 contains 14.3 percent boron, and Fertilizer Borate 65 is 20.2 percent boron. Colemanite (calcium borate) and boracite (magnesium borate) are difficultly soluble in water, but they are soluble in weak acids; colemanite contains 10 percent boron and boracite is 21 percent boron. They are available to plants but will not leach excessively from sandy soils. Sodium, calcium and magnesium borates are slightly alkaline salts. Boric acid is soluble in water and is so slightly ionized that its solution hardly affects litmus. It contains 17 percent boron.

Chemistry of Boron in Soils.   Boron is not usually needed on soils that need lime. After acid soils are limed as needed, many of them will respond to boron, particularly if alfalfa or other legumes are grown. There is some evidence that very high soil-phosphate levels accentuate the need for boron. Boron deficiency of crops is correlated not only with the amount of boron in the soil, but is also influenced by soil reaction, species of crop, fertilizer practices and soil organic matter content. Because of the high solubility of potassium tetraborate, heavy addition of potassium to the soil should increase the availability of boron compounds in the soil. The percentage and total amount of boron in plants decrease as the pH of the soil increases. The level of available soil boron required for some field and vegetable crops is shown in Table 8.13.

Table 8.13

BORON REQUIREMENT OF SOME FIELD AND VEGETABLE CROPS
AND PROBABLE LEVEL OF AVAILABLE SOIL BORON FOR OPTIMUM
YIELD[1]

High (>0.5 ppm)	Medium (0.1 to 0.15 ppm)	Low (<0.1 ppm)
Apple	Carrot	Barley
Alfalfa	Cotton	Beans
Asparagus	Lettuce	Citrus
Beets	Clover (seed)	Clover (forage)
Celery	Peanut	Corn
Cruciferae	Peach	Forage grasses
Mangold	Pecan	Small grain
Sunflower	Sweet potato	Soybean
	Tobacco	Strawberry
	Tomato	White potato

[1]Adapted from K. C. Berger, "Boron in Soils and Crops," in *Advances in Agronomy*, eds. A. G. Norman et al., vol 1, (Academic Press, 1946), p. 345.

Soluble boron in soils of humid regions varies from 0.05 to 2.5 ppm; soils from arid regions may contain more than 5 ppm of soluble boron. Soil texture apparently has considerable influence on boron status of soils in Alabama and Mississippi (Table 8.14). Laboratory and greenhouse studies indicate that a soil should contain not less than 0.35 ppm of boron soluble in hot water. Many of the soils of the southeastern United States contain less than 0.35 ppm of hot-water-soluble boron in the top 6 inches (15.2 cm), and most of them contain less than this in the subsoil.

Table 8.14

BORON STATUS OF SOILS IN ALABAMA AND
MISSISSIPPI

Soil Texture	Hot-Water-Soluble Boron—ppm		
	Alabama[1]	Mississippi[2]	
		Hills	Delta
Clays	0.155	0.130	0.76
Silt loams	0.130	0.095	0.48
Sandy loams	0.085	0.086	0.36
Sands	0.032	—	—

[1]J. I. Wear and R. M. Patterson, "Effect of Soil pH and Texture on the Availability of Water-Soluble Boron in the Soil," Soil Sci. Soc. Amer. Proc. 26,1962,144.

[2]Data by J. D. Lancaster, Mississippi State University.

The availability of boron is apparently correlated with the amount of organic matter in the soil. Atlantic coastal terrace soils such as Portsmouth and Bladen are somewhat poorly drained and consequently have more organic matter and more boron than other soils of the coastal plains.

Three regions of low boron content are recognized: the Atlantic and Gulf Coastal Plain; a region across Michigan, Wisconsin and Minnesota; and the Pacific Coastal Region of the Northwest. The soils of the southern Mississippi River Flood Plain have high boron content, but deficiencies are found in much of the southern region outside the Mississippi, Arkansas, and Red River Valleys and the western parts of Oklahoma and Texas (Table 8.15).

CROP RESPONSE TO BORON

Crops that are capable of substituting sodium for a large percentage of their potassium needs are among the first to exhibit boron deficiency. These crops include cruciferae (cabbage, cauliflower, brussel sprouts), tuber crops (beets, potatoes, carrots, turnips) and, to a lesser degree, cotton.

Symptoms. Boron deficiency shows up as a breakdown of the growing-tip tissue or as a shortening of the terminal growth, resulting in a rosetting effect on some plants like alfalfa. Internal tissues of cabbage, beets,

Table 8.15
BORON DEFICIENCY IN AGRONOMIC AND HORTICULTURAL CROPS OF THE SOUTHERN REGION

Crop	Va.	N.C.	S.C.	Ga.	Fla.	Ala.	Miss.	La.	Ark.	Tenn.	Ky.	Tex.	Okla.
Alfalfa	+	+	+	+	+	+	+		+	+	+		+
Apple	+	+	+	+		+			+	+	+		
Beets						+			+			+	
Carrot				+	+								
Celery					+								
Citrus					+								
Clover	+	+	+	+	+	+	+				+		
Corn	+		+		+								
Cotton		+	+	+		+	+	+	+	+			
Cruciferae	+	+	+	+	+	+	+		+	+			
Gladiola		+											
Lettuce					+		+						
Peach					+								
Peanut	+	+		+	+	+							+
Soybean					+				+				
Sweet potato			+	+									
Tobacco	+	+											
Tomato		+	+		+								

turnips and cauliflower show breakdown and corky, dark discoloration. Boron deficiency and leaf-hopper damage in alfalfa are often confused. The deficiency exhibits short nodes, few flowers and yellow to reddish yellow discoloration of the upper leaves. Growing tips of the alfalfa sometimes die and regrowth comes from a new shoot at a lower axil. In contrast, leaf-hopper damage may appear on all parts of the plant and results in a V-shaped yellowing of the affected parts. Unlike boron deficiency, the leaf-hopper disease does not affect the growing tip, and abundant flowering may prevail.

Rates.   Boron has increased the yield or improved the quality of alfalfa, apples and vegetables in many areas. Deficiencies have been reported in 41 states, with many different crops being affected. The quantities that may prove beneficial to alfalfa, beets and celery (Table 8.16) can cause serious damage to small grains, beans and cucumbers. Boron should not be used in combination seedings that contain legumes and grass or small grain because of injury to the grass or grain. It should be added as a top-dressing after the grass has become well established or the grain crop has been harvested.

Where boron is low in the soil, 10 to 50 pounds per ton (5 to 25 kg/m ton) of borax is suggested for use with mixed fertilizer. Only ½ pound per acre (0.56 kg/ha) of boron may be needed for tobacco, but 2 to 3 pounds per acre (2.24 to 3.36 kg/ha) is required for alfalfa, tuber crops and cruciferae, which are classified as high responsive crops by Lucas.[30] The 3 pounds per acre (3.36 kg/ha) required for alfalfa is six times the amount required for tobacco. Severe injury resulted from the addition of about 1.5 pounds (0.68 kg) of boron in the form of 15 pounds per acre (16.8 kg/ha) of borax at seeding time to Austrian winter peas, crimson clover, red clover and white clover on a light, poorly buffered, sandy soil in Alabama. For crops like cucumbers, melons and snap beans, not over 0.5 pounds per acre (0.56 kg/ha) of boron, if any, should be used. Generalized boron recommendations for crops grown on certain types of soils in the humid region of the United States are given in Table 8.16.

The soybean is probably as sensitive to boron as any crop. Visual damage to soybeans has been noticed in some soils when boron is applied at rates as low as 5 pounds per acre (5.6 kg/ha). Tests were conducted for two years in South Carolina with soybeans to evaluate the influence of boron on soybean production. The treatments included soil applications of slowly soluble sources, foliar applications of water-soluble sources, and combinations of the two materials and methods of application. Over a two-year period at four locations, response of soybeans to boron was not evident. Since boron is readily leached from the soil (it is an anion with low soil retention) and soybean seedlings have low boron needs, it is questionable that a boron seed treatment would be beneficial. Boron

[30]Lucas, "Micronutrients."

Table 8.16
GENERALIZED BORON RECOMMENDATIONS FOR CROPS GROWN ON
SLIGHTLY ACID TO ALKALINE SANDY LOAMS, DARK-COLORED SANDY LOAMS,
SANDS, PEATS AND MUCKS IN THE HUMID REGIONS OF THE UNITED STATES

Crop	Recommendations		
	Soil Addition Lb/Acre[1]	Foliar Application	Remarks
Alfalfa	1–3		
Apple	1–1.5	0.2 lb of boron per 100 gal of H_2O[2] at petal fall and 2 weeks later	
Beet	1–3		Higher rate for broadcast application
Carrot	1–2		"
Celery	1–2		"
Citrus	1.0		
Clover (seed)	0.5–1.5		Higher rate for broadcast application
Cotton	0.3–0.5	0.1 to 0.15 lb/acre for 2 to 3 applications at 2-week intervals beginning at early bloom	Up to 1.0 lb/acre[1] broadcast
Corn	0.3–0.5		
Cruciferae	1–3		
Lettuce	1.0		
Peanut	0.5		In Alabama and Georgia where "hollow heart" has been observed
Radish	0.5–1.0		
Rutabagas	1–3		
Spinach	0.5–1.0		
Sweet potato	0.5–1.0		
Tobacco	0.25–0.5		
Tomato	1.0		
Turnips	1–3		

[1]Lb/acre × 1.12 = kg/ha.
[2]Lb/100 gal × 0.0172 = kg/l.

response would more likely be obtained on a coarse, sandy soil low in organic matter at high levels of soil calcium (not necessarily high pH), high levels of phosphorus, and during a dry growing season.

Deficiencies. Since most of the available soil boron is in the topsoil and moisture is essential for plant uptake, a dry year reduces boron uptake and enhances boron deficiency. A boron deficiency in South Dakota sugar beets has been noted by agronomists of the U and I Sugar Company. The condition was scattered throughout the growing area, but not every field was affected. Typical boron deficiencies include small, deformed new leaves that later turn black, chapping on the base of petioles (the branch holding the leaf), twisted petioles and unbalanced leaves, numerous small leaves on the black heart of the crown and ultimate breakdown of the crown and root tissue. The beet had the appearance of having been stepped on. When boron was deficient, both the yield and sugar content were reduced, and tonnage dropped as much as 4 to 5 tons per acre (8.9 to 11.2 m tons/ha). Boron has been known to be essential, but supplies in South Dakota soils were thought to be adequate. Sugar beets do require more of the element than most other crops.

Short, bent cobs of maize with undeveloped tips and poor kernel development at the base are reported to be due to severe boron deficiency in acid, sandy soils in the West Berlin area of Germany. Irregular distribution of kernels and reduced overall growth are considered to be the initial signs of boron deficiency. Soil application of borax at 2 to 5 lb/acre (2.24 to 5.6 kg/ha), is recommended as preventative treatments, except where severe boron deficiency has been confirmed. Under such conditions and where the preceding crop is lucerne or cabbage with a relatively high boron requirement, an initial application of 10 to 20 lb/acre (11.2 to 22.4 kg/ha) of borax is suggested.

In Autumn 1961, following a warm, dry spring, deficiencies in maize observed in fields in West Berlin were recognized as being due to lack of boron. The cob malformations were so serious that they caused one to believe boron deficiency had been present before, but had either not been recognized or had been overlooked.

Recommendations. Boron is suggested on all soils in Alabama at 2 pounds per acre (2.24 kg/ha) for alfalfa and 1 pound per acre (1.12 kg/ha) for apples. Boron is suggested on sandy soils at a rate of 1 pound per acre (1.12 kg/ha) for clover seed production, broccoli, cauliflower and turnips and at a rate of 0.4 pound per acre (0.45 kg/ha) for cotton.

North Carolina fertilizer suggestions call for ½ pound per acre (0.56 kg/ha) of boron on peanuts. It is preferred that the boron be included in a copper-sulfur dust added for peanut-leaf-spot control. Boron can be used as a preplant or foliar application. For soil application, borated gypsum may also be used, in which case 14 pounds (6.36 kg) of agricultural borax is added per ton (0.91 m ton) of gypsum. This formulation is designed to supply 0.5 pound (0.23 kg) of elemental boron equivalent per 600 pounds (272.7 kg) of borated gypsum (land plaster).

Since 1956, boron has been suggested for cotton in South Carolina.[31] Cotton fertilizers in this state should contain a minimum of 0.05 percent

[31]L. S. Murphy and L. M. Walsh, "Correction of Micronutrient Deficiencies with Fertilizers," *Micronutrients in Agriculture*, eds. J. J. Mortvedt et al. (Madison, WI: Soil Science Society of America, 1972), pp. 347–381.

Table 8.17

SUMMARY OF SUGGESTIONS FOR BORON ON COTTON IN THE SOUTHEASTERN UNITED STATES

State	Soil	Boron, Lb/Acre[1]	Method of Application[2]	Expected Increase Lb/Acre[1] Seed Cotton
Alabama	Sandy	0.3 to 0.5	Mixed fertilizer[3] or with preemerge at planting, or Foliar spray at fruiting.	50 to 150
Arkansas	Sandy loams and silt loams	0.3 to 0.5 0.5 to 0.75 1.0 to 1.5 0.2 to 0.4	Band with preemerge or Broadcast before seedbed preparation or Band in mixed fertilizer or nitrogen solution or	100 to 700
	Boron deficient soils	0.3 to 0.6 0.5 to 1.0	3 to 6 foliar sprays of 0.1 lb boron using Solubor.®[4] Soil and foliar.	
Mississippi	Hill section and Delta foothills	0.3 to 0.5 0.5 to 0.7	Borated fertilizers: add 0.3 lb boron to each 48 lb P_2O_5. 5 to 7 foliar applications of 0.1 lb / acre of boron.	100 to 1000
South Carolina	All	0.3 to 1.0 0.6 0.4	Borated fertilizer, preferably with frit or side-dress with fertilizer. Two foliar sprays 2 weeks apart; begin at early blooming.	100 to 300
Tennessee	High pH soils and sandy soils	0.5	Borated fertilizers	100 to 400

[1]Lb × 1.12 = kg/ha.

[2]Caution. apply borated fertilizer 3 inches (7.63 cm) to side of seed zone and 3 inches (7.63 cm) deep.

[3]To obtain 0.5 pound (0.56 kg/ha) of boron in a 500-pound-per-acre (560-kg/ha) application, use a fertilizer containing 0.1 percent boron.

[4]To obtain 0.1 pound (45 g) of boron, spray about 10 gallons (37.9 l) of a 0.1-percent solution of boron.

boron. The law specifies limits for deficiencies and excesses. The fertilizer should be used at a rate that supplies boron at a rate of not less than 0.3 pound/acre (0.34 kg/ha) or more than 1.0 pound/acre (1.12 kg/ha). For foliar spray, 0.4 pound per acre (0.45 kg/ha) is suggested. This should be applied in two applications, 0.2 pound per acre (0.23 kg/ha) at first bloom and 0.2 pound per acre (0.23 kg/ha) ten days to two weeks later. Since Solubor® contains 20.5 percent boron, this amount of boron for each application can be obtained from 1 pound (0.45 kg) of Solubor®.

A summary of boron suggestions for cotton is given for five southeastern states of the United States in Table 8.17 and for other crops in humid regions in Table 8.16. Boron has been toxic as often as deficient in western United States soils. Most boron fertilization has been required by tree crops and legumes in central and western California and in Oregon and Washington.

MOLYBDENUM: FRIT, SODIUM AND AMMONIUM MOLYBDATE

Description. Sodium molybdate, $Na_2MoO_4 \cdot 2H_2O$, and ammonium molybdate, $(NH_4)_6Mo_7O_{24} \cdot 4H_2O$, are white powders. Yellow ammonium phosphomolybdate, $(NH_4)_3P(Mo_3O_{10})_4$, is a reagent-grade experimental material. MOLY-GRO® is the trade name for a water-soluble, crystalline-grade sodium molybdate containing 39 percent molybdenum. It is offered in two forms, one for use in fertilizers and the other for seed-coating or foliar sprays. Superphosphate contains a trace of molybdenum as an impurity and often supplies sufficient quantities for plant growth. A frit containing molybdenum and five other micronutrients has been manufactured.

Method of Manufacture. Sodium molybdate is the sodium salt of molybdic acid. Ammonium molybdate is used in the chemical laboratory as a reagent for quantitatively estimating phosphorus content. It forms with phosphorus a complex, canary yellow compound, ammonium phosphomolybdate, that is easily precipitated. Frits containing molybdenum are manufactured similar to those described previously.

Handling Characteristics. The compounds used for fertilizer are all rather stable, pulverized materials that hold up well in storage. A compound containing a minimum of 38 percent molybdenum is used as a coating on seed. This material, added at a rate of one-fourth ounce per acre (0.02 kg/ha), is sufficient for legumes on molybdenum-deficient soils. Such a small amount is required that seed treatment is the most practical method of applying it to an area. A seed coating of sodium molybdate or similar compound is the best method of application.

In order to treat the seed required to plant four acres (1.6 ha) of any legume, dissolve 2 ounces (0.06 kg) of 38-percent molybdenum compound or its equivalent in one-half pint (250 ml) of water and pour or spray the solution over the amount of seed required for four acres (1.6 ha). Mix the seed thoroughly and let dry. Excess water permits the solution to penetrate the seed embryo and to cause injury.

When inoculating legume seed with nitrogen-fixing bacteria, it is acceptable to mix the inoculant with the molybdenum solution and to apply it to the seed. The inoculant-molybdenum mixture should be added to the seed just before planting in order to avoid harm to the nitrogen-fixing bacteria by the molybdenum. Foliar sprays may also be used by applying 3 to 4 ounces (0.09 to 0.11 kg) of the 38-percent molybdenum compound per acre (0.4 ha). Use wetting agents when adding the material to cauliflower and onions.

Chemical Characteristics.   Sodium molybdate in the anhydrous form contains 46 percent molybdenum and is soluble in water. Ammonium molybdate carries 54 percent molybdenum and is soluble in water. Yellow ammonium phosphomolybdate has about 61 percent molybdenum; it is not soluble in water but is soluble in weak acids. The MOLY-GRO® molybdenum compound that is made for mixing with fertilizer or as a seed coating contains 38 to 41 percent molybdenum and is soluble in water. Molybdenum applied as $(NH_4)_2MoO_4$ was leached through columns of peat and sandy soils but not acid lateritic soils.

Chemistry of Molybdenum in Soils and Plants.   Molybdenum is one of the seven micronutrients recognized as essential to the growth and development of higher plants. It is required for the reduction of nitrates for protein synthesis in all plants and for nitrogen fixation in the root nodules of leguminous plants by *Rhizobia* bacteria. Plant concentrations of this element vary from 0.8 to 5.0 ppm in normal tissue. Some plants have been found to contain as much as 15 ppm. Deficient plants usually have less than 0.5 ppm. Certain nonresponsive crops, such as grass, contain as little as 0.1 ppm. Deficiency results in stunting of plants, causing them to turn yellow. Legume crops and cruciferae suffer where deficiencies exist especially on acid soils. In cruciferae, the deficiency shows up as marginal scorching, curling and crinkling of leaves.

Molybdenum, unlike other micronutrients, is most frequently deficient in soils below pH 5.2; fibrous peats and acid sandy soils are usually the most deficient. Liming such soils will often correct the deficiency. Most fertile soils contain less than 1.0 pound (1.12 kg/ha) of total molybdenum per acre, and only a trace of this is soluble in hot water. The concentration of molybdenum is higher in the surface soil horizons and decreases with depth. Molybdenum appears to be associated with the nitrogen and organic matter content of soils. It acts as a catalyst in the reduction of nitrates to ammonia.

Deficiencies have been reported in 21 states. Cauliflower, citrus, onion, spinach, broccoli, alfalfa, peanuts, clover and soybeans are mostly affected by a shortage. Molybdenum deficiency symptoms are most prevalent on legumes and are identical with those of nitrogen deficiency. This indicates that molybdenum has a basic role in the fixation of atmospheric nitrogen by symbiotic microorganisms associated with legumes rather than a direct nutritional response to the element. The molybdenum content of several legumes grown on acid soil was greatly increased by liming. Typical response of soybeans to lime and molybdenum is shown in Table 8.18 by Anthony.[32] The most severe cases of molybdenum deficiency in nonlegumes have been in cantaloupe, onion, broccoli, citrus, sweet potato and cauliflower (Table 8.19).

Rates. Alfalfa grown in South Carolina on Cecil sandy loam of pH 6.0 responds to one-half ounce per acre (0.035 kg/ha) of molybdenum. One ounce (0.03 kg) of sodium molybdate was dissolved in six gallons (22.7 l) of water and sprayed on one acre (0.4 ha) immediately after the first cutting.

Adding one-fourth ounce per acre (0.02 kg/ha) as a seed treatment has increased yields of soybeans up to 3 bushels per acre (201.6 kg/ha) on many acid soils (Table 8.18). Other crops responding to molybdenum in the southern region of the United States are noted in Table 8.19.

Table 8.18
AVERAGE RESPONSE OF SOYBEANS TO LIME AND MOLYBDENUM AT SIX LOCATIONS IN THE HILL AREAS OF MISSISSIPPI, 1963–1966

Lime Treatment as a Percentage of Lime Requirement	Soil pH, Average	Soybean Yield—Bu/acre[1]	
		No Molybdenum	With Molybdenum
0	5.1	24	30
33	5.4	27	31
67	5.7	29	32
100	6.0	31	32

[1]Bu/acre × 67.2 = kg/ha

Excessive quantities of molybdenum may be absorbed by forage crops, and injury to livestock may result. This injury is known as molybdenosis, a molybdenum-induced copper deficiency of ruminant animals only. This disease can be caused by feeding seed treated with molybdenum or by forage plants grown on soil conditions that cause an

[32]J. L. Anthony, ''Response of Soybeans to Lime and Fertilizer,'' Mississippi Agricultural Experiment Station Bulletin 743, 1967.

Table 8.19
MOLYBDENUM DEFICIENCY IN THE SOUTHERN REGION

Crop	Va.	N.C.	S.C.	Ga.	Fla.	Ala.	Miss.	La.	Ark.	Tenn.	Ky.	Tex.	Okla.
Alfalfa	+	+	+	+		+	+		+	+	+		
Cantaloupe		+											
Cauliflower					+		+	+					
Citrus					+								
Clover				+	+								
Peanut		+		+	+								
Soybean	+	+	+	+	+	+	+	+	+	+	+		
Sweet potato					+								

excessive uptake of molybdenum. When these forages or seed are used as feed, molybdenum intake is sufficient to interfere with the utilization of copper in blood formation and other metabolic processes of ruminant animals. Molybdenosis has been corrected by supplementing the animal's diet with copper.

Research on Montcalm sandy soil showed that molybdenum content of cauliflower was increased 5 times by liming from pH 4.9 to pH 6.7. In a Houghton muck, the content of molybdenum increased 3 fold when the pH was raised from 5.4 to 7.2. Liming severely deficient soils, however, will not completely correct the deficiency.[33]

Recommendations. Molybdenum is suggested as a foliage spray to control "yellow spot" in citrus, but an alternative practice is to maintain the soil pH above 5.5 by liming.

In most instances molybdenum is not recommended on soils having a pH above 6.0 to 6.5. Alluvial soils of the Mississippi Flood Plain generally have not responded to molybdenum even when it is added to legumes on acid soils.[34]

Minute quantities of molybdenum are removed by cropping, viz., 0.0044 to 0.0088 lb (2 to 4 g) in 40 bushels (1,090 kg) of soybeans, 0.0022 to 0.0044 lb (1 to 2 g) in 100 bushels (2,727 kg) of corn, 0.0018 to 0.0026 lb (0.8 to 1.2 g) in 3,000 pounds (1,364 kg) of cotton and 0.0044 to 0.0077 lb (2 to 3.5 g) in 6 tons (5.46 m tons) of alfalfa. Thus, high soil levels of molybdenum, once attained, are not easily reduced by cropping.

Seed treatment of about one-fourth of an ounce per acre (0.02 kg/ha) of molybdenum is the preferred method and rate of application. Do not feed treated seed to humans, livestock or pets, and do not sell treated seed to be processed for oil or meal.

[33]Lucas, "Micronutrients."
[34]Lancaster, "Micronutrient Deficiencies."

The polyphosphate materials pioneered by the Tennessee Valley Authority[35] have been used very beneficially by liquid fertilizer manufacturers to cope with the micronutrient problem (Table 8.20). Polyphosphate materials, such as 11-16-0 (11-37-0),[36] sequester most micronutrient elements, dissolving them and holding in solution substantial quantities of them in accordance with the following type of equation:

$$ZnSO_4 \cdot H_2O + (NH_4)_4P_2O_7 \longrightarrow Zn(NH_4)_2H_2(PO_4)_2 + (NH_4)_2SO_4$$

The amount of P_2O_5 converted to orthophosphate appears to be roughly equivalent to the water of hydration in the zinc sulfate.[37]

Table 8.20
COMPARISONS IN SOLUBILITY AT 75° TO 80°F (24° TO 27°C) OF SOME COMPOUNDS OF THE MOST COMMON MICRONUTRIENTS IN 11-16-0 (11-37-0) SOLUTION MADE FROM ELECTRIC-FURNACE SUPERPHOSPHORIC ACID AND IN 8-10-0 (8-24-0) PRODUCED FROM ORTHOPHOSPHORIC ACID

Material Added	Solubility, Percent by Weight of Element (Zinc, Copper, Iron, Manganese, Boron and Molybdenum)	
	In 11-37-0, base solution	In 8-24-0 base solution
ZnO	3.0	0.05
$ZnSO_4 \cdot H_2O$	3.0[1]	
$ZnCO_3$	3.0	
Mn_3O_4	0.2	<0.02
MnO	0.2[2]	<0.02
$MnSO_4 \cdot H_2O$	0.2[2]	<0.02
CuO	0.7	0.03
$CuSO_4 \cdot 5H_2O$	1.5	0.13
$Fe_2(SO_4)_3 \cdot 9H_2O$	1.0	0.08
$Na_2MoO_4 \cdot 2H_2O$	0.5[3]	0.5[3]
$Na_2B_4O_7 \cdot 10H_2O$	0.9	0.9

[1]A pH of 6 was maintained by addition of ammonia.

[2]Precipitates formed after several days.

[3]Largest amount tested.

[35]T. P. Hignett, Proceedings, Sixteenth Annual Fertilizer Conference, Pacific Northwest, Salt Lake City, Utah, July 13–15, 1965, pp. 5–10.

[36]Numbers in parentheses refer to nitrogen, P_2O_5 and K_2O.

[37]J. R. Lehr, ''Chemical Reations of Micronutrients in Fertilizers,'' in *Micronutrients in Agriculture,* eds. J. J. Mortvedt et al. (Madison, WI: Soil Science Society of America, 1972), pp. 459–502.

The usual urea–ammonium nitrate or ammonium nitrate solutions and aqua ammonia allow reasonably good solubility of micronutrients. An article in *Farm Chemicals*, quoting a National Fertilizer Solutions Association spokesman, gave the following summary based on the micronutrient deficiencies reported and assuming an average 500 pounds per acre (560 kg/ha) of fertilization.[38]

1. The urea–ammonium nitrate, ammonium nitrate and ammonium polyphosphate solutions are capable of containing sufficient individual micronutrients in solution to correct severe molybdenum and boron deficiencies; slight to moderate deficiencies of copper, iron and zinc; and slight manganese deficiencies.

2. Aqua ammonia (30 percent) is capable as a carrier of micronutrients of correcting severe zinc, boron and molybdenum deficiencies.

3. Complete nitrogen-phosphorus-potassium fertilizers in the polyphosphate system with dissolved micronutrients can potentially correct slight to somewhat better than moderate trace-element deficiencies.

4. In orthophosphate liquids with dissolved boron and molybdenum, deficiencies of these two elements can be corrected, but even slight deficiencies in the other trace elements must be corrected by the appropriate organic metal chelate or complex.

To summarize the limitations of the liquids, the way was pointed through use of suspensions to broaden greatly the horizon for micronutrients in fluid fertilizers. Modern fertilizer technology has overcome most of the limitations to adding micronutrients to liquid fertilizers.

SUMMARY

Zinc

1. For the most part, only those crops that are classified as being very sensitive to the soil supply of zinc have exhibited zinc deficiency symptoms in the United States.

2. Corn and pecan are the only crops in which zinc deficiency is somewhat widespread; however, deficiencies have been noted in flax, cotton and grain sorghum on alkaline soils in Texas, in Irish potatoes in Washington, in rice on water-leveled soils of high pH in Arkansas and Louisiana, in sorghum in Kansas and Nebraska, in soybeans on highly limed soil in Florida and in peaches in South Carolina.

3. Zinc deficiency in corn has been associated with soil pH values above 6.0, high phosphate content, or light, sandy soils.

[38]H. E. Ulmer, Liquid Fertilizer Round-up Proceedings, St. Louis, Mo., 1967, pp. 13–16.

4. Proper liming and fertilization of soils for high yields of corn seem likely to increase the need of zinc fertilization.

5. Routine soil testing is being employed by Alabama, Georgia, Nebraska and other states for making recommendations for corn and by Louisiana for rice.

6. A value of 2 ppm zinc using 0.1N HCl extraction of soil was a critical value below which a response to zinc was obtained.

7. A value of 25 ppm zinc in leaf tissue is considered a critical value below which a response to zinc may be expected.

Manganese

1. Manganese may be injurious to some crops in extremely acid soils.

2. Manganese is easily oxidized to unavailable forms in neutral to calcareous soils.

3. The usual sources are $Mn(II)SO_4$ and $Mn(II)O$.

4. Manganese should be applied with acid-forming fertilizers.

5. Amounts are given in text for soil and foliar applications to citrus, cotton, soybean, small grain, peanut and vegetables.

6. Deficiency symptoms seldom appear until plant concentration is below 25 ppm.

Iron

1. Fe(II) is easily oxidized to unavailable Fe(III) in soil.

2. Deficiencies are most prevalent on calcareous soils.

3. Foliar application generally is superior to soil additions.

4. Incorporating iron with phosphorus fertilizers may not be an effective method of application.

5. The main advantage of chelates is improving the availability of iron added to soil rather than as foliar spray.

6. There is no good, calibrated chemical soil test for available iron.

7. With most plants, leaf iron content of 10 to 80 ppm is associated with deficiency symptoms.

8. For soil application, 0.9 to 1.8 pounds (1 to 2 kg/ha) of iron per acre is suggested; for foliar spray, a 0.5-percent iron solution is recommended.

Copper

1. Copper is rather immobile in soil.

2. Copper is strongly sorbed by clay and organic matter.

3. Bordeaux mixture supplies copper.

4. Copper sulfate or copper oxide is effective when added alone or with fertilizer.

5. Copper is found to be critical on *organic soils* above pH 6.0 because of precipitation and chelation with calcium humate and at pH 4.0 because of competition with high concentrations of soluble aluminum.

6. Normal plants contain 8 to 20 ppm; deficient plants have less than 6 ppm; over 20 ppm of copper is excess.

7. For most crops, it requires 20 or more years to remove 1.1 lb (0.5 kg) of copper.

Boron

1. Borax or sodium tetraborate is widely used as a boron fertilizer.

2. Sodium borates at rates that supply 0.45 to 2.67 lb/acre (0.5 to 3.0 kg/ha) of boron to soil take care of most deficiencies.

3. Combination of sodium pentaborate and tetraborate, at rates of 0.09 to 0.27 lb/acre (0.1 to 0.3 kg/ha) of boron, is used as a foliar application because of high solubility and compatibility with other spray materials. Concentration in solution is 0.1 to 0.3 percent boron.

4. Colemanite and a borosilicate glass frit are used on sandy soils because they are slowly soluble and reduce leaching losses.

5. A soil level of 0.1 to 0.5 ppm hot-water-soluble boron is recommended for optimum yield of field and vegetable crops.

6. Some crops can be injured by even low rates of application, particularly on poorly buffered, sandy soils. Examples are clovers, cucumbers, melons, soybeans and snapbeans.

7. Spray applications are the safest and more effective method of application, but they are also the most expensive.

Molybdenum

1. Molybdenum was found to be essential for fixation of atmospheric nitrogen by *Azotobacter* in 1930.

2. Molybdenum was found to be essential for plant growth using tomatoes in 1937.

3. The average molybdenum content of lithosphere is 2.3 ppm. Molybdenum is the heaviest of the essential plant elements, having an atomic weight of 96.

4. Molybdenum is an anion ($MoO_4^=$) not a cation; chemically it acts much like phosphorus ($PO_4^{\equiv}$), another anion.

5. Total molybdenum in soils is 0.24 to 4.45 ppm. Available molybdenum in soils is 0.05 to 0.24 ppm.

6. The $MoO_4(II)$ anion exists in an exchangeable form in soils.

7. The fact that molybdenum availability to plants increases with increasing pH may be explained by an anion exchange:

$$2OH(I) \rightleftarrows MoO_4(II)$$

8. Molybdenum is necessary to reduce NO_3 to NH_2 forms that can be used in legumes or other plants. *Molybdenum is essential for legumes to fix atmospheric nitrogen.*

9. Molybdenum deficiency appears like nitrogen deficiency; actually, that is what it is.

10. Molybdenum deficiency is seldom found at soil pH of 6.5 or above.

11. Molybdenum injury to plants is actually a molybdenum-induced copper deficiency.

12. The compound most commonly used to supply molybdenum as a foliar spray, as a seed treatment or in mixed fertilizers is water-soluble sodium molybdate containing about 39 percent molybdenum.

QUESTIONS

Problems:

1. On slightly acid soils, 20 pounds of $ZnSO_4 \cdot 7H_2O$ per acre is the amount added to field corn. How much zinc is being added to the soil per acre at this rate? Atomic weight of zinc = 65, sulfur = 32, oxygen = 16 and hydrogen = 1.

2. Zinc is recommended for apple at the rate of ¼ lb/tree on sandy soils after liming and when the pH is 6.0 to 6.9. How many pounds of Zn EDTA containing 6 percent zinc would be required to supply the ¼ lb/tree?

3. Nearly all mineral soils contain an abundance of iron, even though some have been found deficient. A Norfolk sandy loam contains 1.75 percent Fe_2O_3. Convert this figure to ppm of iron, percent of iron in the soil and pounds of iron per acre furrow slice (2,000,000 lb). Atomic weight of iron = 56 and oxygen = 16.

4. Manganese toxicity has been observed as "crinkly leaf" of cotton and turnip when the leaf concentration reached about 1,500 ppm of manganese. What is the level of manganese expressed on a percentage basis?

5. On certain types of soil in the humid regions of the United States, it is recommended that 2.0 lb/acre of boron be added to the soil when alfalfa is being grown. How many pounds of borax ($Na_2B_4O_7 \cdot 10H_2O$) would be required to do this? Atomic weight of sodium = 23, boron = 11, oxygen = 16, and hydrogen = 1.

Circle the number(s) you select from the list.

6. Some of the necessary characteristics of a chelate added as a soil amendment are that it must be:

(a) Stable against hydrolysis.
(b) Soluble in water.
(c) Not easily precipitated by ions or colloids in soils.
(d) Nondamaging to plants at concentrations required to prevent deficiencies.
(e) All of the above.

7. Micronutrients may be suspected to be deficient for plant growth if which of the following soil condition(s) prevail(s)?

(a) Acid-leached soils.
(b) Coarse-textured, sandy soils.
(c) Peats and mucks.
(d) Overlimed acid soils.
(e) Any one of the above.

8. Zinc deficiency is somewhat widespread on which of the following agronomic crops?

(a) Oats.
(b) Wheat.
(c) Grasses (forages).
(d) Corn.
(e) All of the above.

Essay:

Briefly answer the following.

9. Given the following graph, explain its meaning and on what soil condition this would occur.

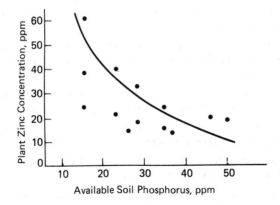

10. Why are foliar applications of iron superior to soil application of iron?

11. Outline the procedure for treating soybean seed with molybdenum prior to planting.

Completion:

12. Oxides of (a) tetravalent manganese (MnO_2) are in equilibrium with (b) divalent manganese (MnO) in the soil solution and with (c) exchangeable manganese ions. The two forms that are considered plant available and increase with decreasing redox potential, associated with poor soil aeration and low pH values, are letters ___and ___.

True or False:

13. _____Micronutrients appear to be required in concentrations as small as ¼ part per million, while as much as 2 parts per million in solution may prove damaging if not lethal, to some plants.

14. _____One micronutrient may serve as substitute in part for another micronutrient.

15. _____There is no evidence of interactions among the micronutrients and with the secondary or major plant nutrients.

16. _____Metals bound in chelate rings lose their ability to act as ions and are therefore *less* likely to take part in chemical reactions that precipitate them or otherwise render them unavailable to plants.

17. _____One of the necessary characteristics of a chelate added as a spray on foliage is that it should be easily absorbed and readily translocated within the plant.

18. _____Frits are brown-colored, pulverized, specially compounded glasses that slowly release micronutrients, such as boron, copper, iron, manganese, molybdenum and zinc, which are offered in powdered or granulated form.

19. _____Zinc is one of the most recommended micronutrients and is generally recommended for *acid* soils.

20. _____In general, manganous oxide (MnO) forms of soil manganese are available to plants whereas highly insoluble manganic forms, such as MnO_2, are not available.

21. _____The pH of a soil is an important factor in determining the solubility of iron in soils and therefore the ability of a soil to supply iron to the plant.

22. _____The addition of soluble iron salts to alkaline soils corrects iron deficiencies and is generally superior to foliar application.

23. _____Copper deficiencies have been reported mainly in areas with peat, muck and other highly organic soils.

24. _____For most crops, approximately one pound per acre (1.1 kg/ha) of copper is removed in 20 years of cropping.

25. _____Boron deficiency of crops is correlated with the amount of boron in the soil but is also influenced by soil pH, species of crop, fertilizer practices and soil organic matter content.

26. _____Molybdenosis, a molybdenum-induced copper deficiency of ruminant animals, can be caused by feeding seed treated with molybdenum or forage plants grown on soil conditions which result in an excessive uptake of molybdenum.

27. _____Molybdenum is generally applied as a broadcast application to soils using a lime spreader.

chapter 9

Protein-Organic
Wastes

The objective of this chapter is to create in the student an awareness of the nature and properties of nitrogen-bearing organic wastes of plant and animal origin and the value of these materials as fertilizers and feedstuffs. Such by-products or waste material will be referred to as protein-organics.

Protein-organic is defined as protein-bearing natural organics derived from plant and animal sources, in contrast to organic chemicals, such as urea and other synthetic amides of carbonic acid. Such organic chemicals are used for feed and fertilizer purposes but are synthesized chemically, rather than biologically, from CO_2 and NH_3. In this context, animal manures and many by-products from food processing are special kinds of solid waste that may be considered as protein-organics. Protein and nitrogen are the principal ingredients of value in most of these materials, although phosphorus is the main ingredient of bones and potassium is the most valuable component of seaweed. The feed and fertilizer use of protein-organic wastes from food processing will be discussed first and afterwards the feed and fertilizer use of protein-organic animal and human manures.

FOOD-PROCESSING WASTE FOR FERTILIZERS

The fertilizer industry was started by meat processors in order to dispose of wastes from slaughterhouses. Today, the fertilizer industry is a major segment of the chemical industry, and no protein-organics are presently used in the manufacture of field fertilizer, with the exception of a few tobacco fertilizer manufacturers on the eastern seaboard of the United States.

In 1900, 91 percent of nitrogen fertilizer was from protein-organic sources; in 1954, less than 3 percent was from this source; present use has declined to about 1 percent. There are two reasons for the decline in use. First, agricultural experiment station research has conclusively shown that inorganic nitrogen is generally superior to organic nitrogen (from plant and animal sources) for making plants grow. Secondly, the costs for hauling, handling, bagging and spreading the bulky organic materials are prohibitive on the competitive market today. Although some organic gardeners and other amateurs in plant nutrition still use products, such as bone meal, cottonseed meal, animal manures and composts, as sources of plant nutrients, there is no basis in the results of scientific research for statements that crops grown with organic fertilizers are nutritionally superior to those fertilized with inorganic forms of nutrients (Table 9.1).

The trend in costs for labor to haul, handle and spread fertilizer has encouraged the use of materials higher and higher in nitrogen content. Thus, in the United States today, NH_3, which contains 82 percent nitrogen, is used far more extensively than any other nitrogen source, whether considered on a basis of tons of nitrogen or tons of material.

Plant utilization of nitrogen occurs in the following order. In the earliest stages of growth, the plant utilizes nitrogen that is stored in the seed; when this reserve is depleted, it is dependent on external sources of nitrogen. Plants tend to use ammonia in very early stages of growth; as plants mature, except for rice and water-grown herbs, they tend to absorb the nitrate form of nitrogen.

Some food-processing wastes formerly used for fertilizer, but also suitable for feeds, are bone meal (steamed bone meal, dissolved bones), oil seed meals (linseed and cottonseed), dried blood, blood meal, fish scrap, offal and tankage. Many scientific man-years were devoted to the research and testing of these materials prior to the 1940s.

Table 9.1
COMPARISON OF PROTEIN-ORGANIC WASTES WITH INORGANIC FERTILIZER AND THE NITROGEN-PHOSPHORUS-POTASSIUM CONTENT

By-product	Rating as Compared with Inorganic Fertilizer Source (%)	Approx. Amount Consumed as Fertilizer (%)	Nitrogen (%)	$P_2 O_5$ (%)	K_2O (%)
Dried blood	80	10.0	12–15	3.0	1.0
Blood meal	75	10.0	15.0	1.3	0.7
Fish scrap, acidulated	75	90.0	7–10	1–2	
Fish scrap, dried	75	5.0	9.0	3.0	
Cottonseed meal	65	5.0	6–7	2.5	1.5
Castor pomace	75	100.0	5.5	1.5	1.2
Linseed meal	65	5.0	6.0	1.0	1.0
Bone meal	68	5.0	4.0	21.0	0.2
Steamed bone meal	70	5.0	1.0	30.0	
Dissolved bones	72	5.0	2.5	8.0	
Poultry manure (dried)	93	95.0	1.7	1.9	1.5
Cattle manure	90	50.0	0.29	0.17	0.35
Horse manure	90	99.0	0.44	0.17	0.35
Swine manure	92	99.0	0.60	0.41	0.13
Sheep manure	93	99.0	0.55	0.31	0.15
Seaweed (dried)		25.0	0.68	0.75	5.0
Tankage	60	5.0	7.0	8.0	

Since early in the twentieth century, an increasing proportion of protein-organic materials formerly used for fertilizer have been moving in feed trade channels because they command a better price there than on the fertilizer market. Because virulent disease organisms, such as salmonella in poultry and hoof-and-mouth disease of cattle, have sometimes been discovered in meat-processing wastes, a certain amount of prejudice developed in some sections against the use of such materials. Now, however, most countries and states require a certificate of sterilization before cargoes of such wastes may be moved into trade channels. Also, the rendering process, melting down to dissolve fats such as lard, destroys most harmful bacteria.

Certain oil-seed wastes and meals also contain substances that are poisonous to animals and thus cannot be fed unless they are further processed to remove the deleterious ingredients. Heavy metals in foods, such as mercury in fish, also present a hazard to animals. A thorough knowledge of the production, handling, and chemical and biological characteristics of protein-organic wastes is needed before they should be used as feeds.

For economic reasons, the use of food-processing by-products for feeds and fertilizer has diminished in importance since the early years of the twentieth century. However, in the past decade and for the same economic reason, there has developed a renewal of interest in the subject because environmental quality has become a watchword of the day and finding uses for wastes, which may no longer be dumped into sewers or on uncovered garbage piles, has become an economic question of no small importance to the food-producing industry.

A brief discussion of the characteristics and possible uses of some protein-organics follow.

DESCRIPTION, PRODUCTION AND USE OF SOME FOOD-PROCESSING WASTES

About one-half million tons of organic waste materials are processed annually in the United States. Such materials include bone meal, castor pomace, coal, dried blood, fish meal, garbage, seaweed, seed hulls and meals, sewage, tankage and fish scraps. Bone meal is primarily a phosphatic material; coal is a rather inert carbonaceous material with a small amount of nitrogen; seaweed a potassium fertilizer; the others may be classified as protein-organics. Processing varies from simple improvements of physical condition to large changes in the chemical composition. Materials may be oxidized, heated, steamed, digested, ammonified, acidulated, nitrated or purified in order to raise the nutrient content and to lower the content of water-insoluble or otherwise intractable materials. The great need for increasing the efficiency of waste disposal will stimulate further development of processed wastes that have value as fertilizers or feeds. However, such materials from an economic standpoint usually must be considered first as end products for waste disposal systems and secondly as competitive feed or fertilizer materials. (Table 9.1).

Dried Blood.   Dried blood is fine sand-sized granular material produced by heating blood from slaughtered animals. It is red or black depending on the amount of heat employed in its preparation. About 20 to 25 pounds (9.1 to 11.4 kg) of dry blood is obtained from 100 pounds (45.5 kg) of liquid. One ton (0.91 m ton) of dried blood is obtained from about 300 slaughtered cattle and has an average composition of about 13 percent nitrogen and less than 1 percent phosphorus and potassium. Over 90 percent of the production is used in hog and poultry feeds. As a fertilizer, none of the protein-organics is as available to crops as inorganic sources of the nutrients. The mineral and inorganic sources of nitrogen are water soluble whereas the protein-organic sources are not. Dried blood, however, is the most available to plants of all the protein-organic materials. Where inorganic sources of nitrogen produce a plant growth increase of 100 percent, blood has a rating of 80 percent.[1]

Tankage.   Tankage is sometimes referred to as meat meal and does resemble lightly cooked hamburger. Animal tankage is the rendered, dried and ground by-product, largely meat and bone, from slaughtered animals. A common grade contains about 9 percent nitrogen and 4 percent phosphorus; an average composition of tankage is about 7 percent nitrogen and 1 to 5 percent phosphorus. It also contains about 3 to 10 percent calcium. The amount of phosphorus and calcium present in the material depends on the amount of bones. About 95 percent of all tankage is consumed as animal feed. Where inorganic sources of nitrogen produce a plant growth increase of 100 percent, tankage has a rating of 60 percent.

Process Tankage.   Process tankage is dark brown, fluffy material produced from materials containing nitrogen that is unavailable to plants but is made available by treatment with steam under pressure of 30 to 100 pounds per square inch (2.1 to 7.0 kg/cm^2) for a period of 20 to 90 minutes. Addition of 0.5 percent sulfuric acid facilitates the hydrolysis at low temperatures. Materials used to make tankage are leather scrap, feathers, hog hair, wool waste, silk, woolen rags, fur garments, felt hats and similar wastes. It contains about 8 percent total nitrogen and an amount of phosphorus

[1]H. D. Haskins et al., "Inspection of Commercial Fertilizer," Massachusetts Agricultural Experiment Station Bulletin 51, 1929.

depending on the kind of raw materials used. Tankage is used as a fertilizer or feed. Where inorganic sources of nitrogen produce a plant growth increase of 100 percent, process tankage has a rating of 60 percent.

Leather. Leather meal is made from scrap pieces and shavings largely from the shoe industry. Without treatment, leather meal is valueless as fertilizer: before it is used, it must be ground, steamed or roasted; treatment with 0.5 percent sulfuric acid aids its digestion. Sole leather contains 4 to 7 percent nitrogen whereas upper leather has 10 to 13 percent nitrogen. It is used as a feed or fertilizer and also in rum distilleries to give rum a dark red color. Chrome tanned leather contains about 3.5 percent Cr_2O_7, which is harmful to plants when present in solution in more than trace amounts. Where inorganic sources of nitrogen produce a plant growth increase rated 100, processed leather has a rating of about 60 percent.

Garbage-Sewage Compost. Garbage-sewage compost is garbage mixed with an equal mass of sewage sludge. Composts, to be used as fertilizer, must be processed in a way that encourages nitrification. Little nitrification occurs in garbage compost alone. Sewage sludge mixed with garbage increases nitrification and decomposes more rapidly than garbage compost alone. Where inorganic sources of nitrogen produce a plant growth increase rated 100, garbage compost has a rating of 16 percent.[2]

Fish Scrap, Acidulated. Acidulated fish scrap gave early mixed fertilizers their dark color and odor. About 120 pounds (60 kg) of 62.5 percent sulfuric acid is used per ton (0.91 m ton). Drying is then unnecessary to prevent decay, and the phosphorus is rendered more available. It contains about 7 to 10 percent nitrogen, 1 to 2 percent phosphorus, about 2 percent sulfur and traces of iodine and mercury. A small amount is now used in garden fertilizer. Early settlers in America saw Indians following the practice of fertilizing corn by placing a fish in each hill. Where inorganic sources of nitrogen produce a plant growth increase rated 100, acidulated fish scrap has a rating of 75 percent.

Fish Scrap, Dried. Dried fish scrap is prepared from nonedible fish, such as menhaden, and from edible fish, such as anchovy. Also, the offal from crab houses, fish canneries and spoiled edible fish are used. Material designated as scrap is ordinarily fit only for fertilizer whereas that designated as meal must be fit for feed. Fish meal is made by cooking raw fish with steam, then pressing to remove the oil and finally drying and milling. Pelleted fish meal, which has been treated with an antioxident, is

[2]A. L. Prince and H. W. Winsor, "The Availability of Nitrogen in Garbage Tankage and in Urea in Comparison with Standard Materials," *Soil Sci.* 21 (1926), 59.

now offered for sale to poultry producers. A composite analysis of many samples shows a total nitrogen content of about 9 percent, a total phosphorus content of about 3 percent, and a total calcium content of about 6 percent. Where inorganic sources of nitrogen produce a plant growth increase rated 100, fish meal has a rating of 75 percent.

Cottonseed Meal. High-grade meal is deep yellow; fermented meal is red brown; moldy meal is mottled; and meal ground without removing the seed coat is very dark brown in color. It is produced by pressing the seed to remove the oil and then grinding the cooked cottonseed meats. The meal is used as cattle feed because it contains gossypol, which is not suitable for feeding to poultry and swine. Some meal, particularly that which is off color or moldy, is used as fertilizer. It contains 6 to 7 percent of total nitrogen and less than 1 percent phosphorus and potassium. Although there is no scientific evidence to support its use in tobacco fertilizers, some bright leaf growers demand it, and manufacturers produce a tobacco fertilizer using a formula containing cottonseed meal as a source of part of the nitrogen in the mix. Experiment stations and agricultural workers in the tobacco belt do not recommend organics for tobacco or, for that matter, any other crop. Some pecan growers use it. Where inorganic sources of nitrogen produce a plant growth increase rated 100, cottonseed meal has a rating of 65 percent. In cotton-producing areas where the cotton bolls are stripped resulting in the harvest of hulls as well as of lint and seed, a product known as *cottonseed hull ash* is sold as a source of potassium fertilizer; it contains about 27 percent potassium.

Cocoa Shell Meal. Husks of brown cocoa seeds are ground and may be used as a conditioner for fertilizer. They are considered poisonous for animals, and the carbon-to-nitrogen ratio is too high for good nitrification. *Cocoa shell meal* contains about 2 to 3 percent nitrogen, 1 percent phosphorus and 2 percent potassium. *Cocoa tankage* is the name given to the residue resulting from chemical processing to produce chocolate. It is a defatted ground cocoa cake and contains about 4 percent nitrogen, 1 percent phosphorus, 2 percent potassium, 1 percent magnesium and 20 percent lime. Where inorganic sources of nitrogen produce a plant growth increase rated 100, cocoa meal, because of its high carbon-to-nitrogen ratio, has a rating of only about 40 percent.

Castor Pomace. Castor pomace is the ground residue of castor beans from which the oil has been extracted. Like cocoa meal, castor pomace is considered poisonous to animals and is used only for fertilizer purposes. It is a by-product of the manufacture of castor oil. It contains about 5 to 6 percent nitrogen, 1 percent phosphorus and 1 percent potassium. Castor pomace is given a rating of 75 percent relative to a 100-percent rating for inorganic sources of nitrogen.

Linseed Meal. Linseed meal is the residue obtained after extracting oil from flax seed. A light yellow, fluffy powder, it contains about 6 percent nitrogen, 1 percent phosphorus and 1 percent potassium. Because it is an excellent feed, its value prohibits its use for fertilizer. As a source of nitrogen fertilizer, it has a rating of about 65 percent relative to a 100-percent rating for inorganic nitrogen.

Other Vegetable Wastes. Besides cottonseed meal, cocoa meal, linseed meal and castor pomace, other meals and pomaces are sometimes offered to farmers and manufacturers of feed and fertilizers. Among these are apple pomace, copra cake meal, cranberry pomace, peanut meal, peach pomace, pumpkin pomace, rubber seed meal, tomato pomace, grape meal, velvet bean meal and wine lees. The water content of pomaces is high, and their composition is quite variable. The composition of seed meals is not quite so variable, but the results of research on their use should be studied carefully before using them for feeds or fertilizers.

Composts. Straw, peanut hulls, sawdust, leaves, bagasse, corn stalks and similar wastes or residues are high in cellulose and pentosan but low in nitrogen; because of this high carbon-to-nitrogen ratio, they do not decompose readily. If the nitrogen content exceeds 1.2 percent on a dry basis, they will rot rapidly under moist, slightly alkaline, aerobic conditions. The addition of 100 pounds per ton (50 kg/m ton) of $Ca(NO_3)_2$ or its equivalent in nitrogen and calcium to waste organic matter promotes rapid compost formation and organic matter decomposition. This well-decomposed, dark brown, organic matter has a high buffer capacity over a considerable range of soil pH values, and it serves to stabilize soil structure and to improve water infiltration capacity of soil. In the eastern United States, a mixture of ammonium salts and ground limestone is used to supply the calcium and nitrogen. In the West, where soils and water are alkaline, gypsum and ammonium sulfate are generally employed to supply the calcium and nitrogen. After several months in a warm moist condition, the mass will have the appearance of a humified lignoproteinate and the properties of rotted manure. The 1972 *Yearbook of Agriculture* contains a discussion on how to make a compost.

Plastics. Many of a large group of organic, synthetic, building materials contain proteinaceous or amide-type substances. They are commonly known by their trade names; an example is Formica®. Shavings of such materials are sometimes considered as sources of synthetic protein for feed or fertilizer. Those that have been tested, without further amelioration, usually have too narrow a carbon-to-nitrogen ratio to be very soluble or readily available as nutrients for plants or animals. There are laboratory methods for the study of the quality of insoluble nitrogen in plastics offered for fertilizers. Some such materials have been tested and declared crude, inert or slow acting as sources of feed or fertilizer.

Microorganisms, particularly bacteria, fungi, and actinomycetes, play a major role in the decomposition of soil organic matter, the degradation of pesticides, nitrogen transformations and the breakdown of waste amendments, such as sewage sludge, food-processing wastes, composts, crop residues and manures. Their success in decomposing these wastes and the subsequent use by higher plants of the degradation products is dependent on the proportion of nitrogen to carbon contained in the wastes.

A chemical analysis of the protein-organic waste for its carbon and nitrogen content is the first step in determining its quality as a fertilizer. The carbon-to-nitrogen ratio is very important in the decomposition process. For protein-organic wastes or crop residues with a carbon-to-nitrogen ratio greater than 20:1, nitrogen will temporarily be immobilized in microbial tissue and will create nitrogen-deficient soils for plants being grown following the addition of such highly carbonaceous residues or wastes. For wastes or residues with a carbon-to-nitrogen ratio less than 20:1, nitrogen will be mineralized, i.e., released as NH^+_4 or NO^-_3, and as such will become available for uptake by plant roots.

The following examples with clover and wheat straw residues will demonstrate this point. Wheat straw contains 45% carbon and 5% nitrogen, which is a carbon-to-nitrogen ratio of 90:1. Clover contains 45% carbon and 3% nitrogen, which is a carbon-to-nitrogen ratio of 15:1.

Assume an addition of 5,000 kg of each to a hectare of land. It is known that a carbon-to-nitrogen ratio of 20:1 is optimum. How many kg of nitrogen must be added with the wheat straw to convert the 90:1 carbon-to-nitrogen ratio to 20:1? The amount of carbon added in the straw is $5,000 \times 45\% = 2,250$ kg. Then $2,250/x = 20/1$; $x = 112$ kg nitrogen. However, the 5,000 kg of straw contained 0.5% nitrogen, and thus the straw contributed $5,000 \times 0.5\% = 25$ kg nitrogen. Therefore, the amount of nitrogen that must be added with the wheat straw in order for it to have a 20:1 carbon-to-nitrogen ratio, and therefore be decomposed, is 112 kg − 25 kg nitrogen = 87 kg.

The same calculations for clover reveal that 5,000 kg of clover with 3% nitrogen provides 150 kg of nitrogen and 2,250 kg of carbon with 45% carbon. Thus, the amount of nitrogen needed to provide a carbon-to-nitrogen ratio of 20:1 is $2,250/x = 20/1$ and $x = 112$ kg. But the clover is already providing 150 kg of nitrogen or a surplus of 150 kg − 112 kg = 38 kg after decomposition. Residues and wastes containing this carbon-to-nitrogen ratio permit microorganisms to decompose the organic matter and to mineralize the surplus NH^+_4 and NO^-_3, which is available for uptake by plant roots.

Animal manure, like clover, has a fairly low carbon-to-nitrogen ratio and decomposes rather rapidly. On the other hand, sawdust, like wheat straw, has a high and hence undesirable carbon-to-nitrogen ratio and decomposes rather slowly. However, decomposition can be speeded up by adding generous amounts of fertilizer (particularly nitrogen) to sawdust compost or to soil into which sawdust has been incorporated. The same applies, to a slightly lesser extent, to other woody, organic materials, such as ground bark, leaves and grain straw.

MANURES FOR FERTILIZER

Manures of all kinds have long been known to have substantial fertilizer value. About 800 B.C., Homer in the Greek poem the *Odyssey* writes about the "mule and cow dung that had been saved in a heap for hauling to the fields." In early agronomy and soils textbooks, the care, handling, preservation and use of animal manure was placed high on the list of good husbandry.

Human manure, sometimes referred to as "night soil" hauled and handled from "honey wagons" (names given by World War II American GI's in Japan), has long been used as fertilizer. Historically, human manure use for fertilizer in Europe and heavily populated areas of the Far East has been considered not so much as a means of disposal but as a means of supplying plant nutrients for crop production.

Bear estimated that the loss of nutrients in human sewage in 1949 in the United States was about 640,000 tons (582,400 m tons) of nitrogen, 400,000 tons (364,000 m tons) of P_2O_5 and 300,000 tons (273,000 m tons) of K_2O. In 1980, with a population of 220,000,000, this is estimated to be 960,000 tons (873,600 m tons) of nitrogen, 600,000 tons (546,000 m tons) of P_2O_5 and 450,000 tons (409,500 m tons) of K_2O. At $0.15 per pound (0.45 kg) for nitrogen, $0.10 for P_2O_5 and $0.05 for K_2O, the value is $453 million.

SEWAGE SLUDGE

One of the many successful human manure products developed as an activated sewage sludge has the trade name of Milorganite® and is manufactured in Milwaukee, Wisconsin. This material contains about 6 percent nitrogen and 1 percent phosphorus. It is popular for golf greens and other nonfarm uses. There are at least 15 cities in the United States, including the capitol, Washington, D.C., that are now producing or have produced in the past a similar product. It is estimated that less than 25 percent of the human waste in America is returned to the soil. Much of what is returned is used on turf. If toxic metal hazards were eliminated, more of this material would be used.[3]

[3]A. L. Page, *Fate and Effects of Trace Elements in Sewage Sludge When Applied to Agricultural Lands*, 670/2-74-005, United States Environmental Protection Agency, 1974.

Disposal of solid wastes is a problem for all communities. One phase of this problem is the need to dispose of large amounts of sewage sludge produced during the conventional biological treatment of sanitary and industrial wastewater. Most cities have a sewage treatment plant in which wastewater is treated to remove harmful microorganisms and suspended solids. The solids removed from the water are then subjected to a digestion process that stabilizes the wastes. This material is commonly termed "sewage sludge."

Disposal. The sewage sludge can be disposed of in several different ways, including placement in a lagoon, burning or application on agricultural land. Because of present day economics, application of sewage sludge to cropland is becoming the preferred method for many cities.

The major emphasis of research has been directed toward the benefits and potential problems that can result from application of municipal sludge on agricultural land. Considerable attention has been given to determining the chemical composition of sewage sludges produced by different cities so that application rate recommendations may be based on a valid knowledge of the components of the sludge. All sewage sludges contain appreciable concentrations of the major, essential plant nutrients nitrogen, phosphorus, potassium, sulfur and several micronutrients such as zinc, copper, manganese and iron.

A valid approach to sampling the sewage sludge from a given city is essential because the results of studies indicate that there is considerable variation in the chemical composition of sewage sludge. To insure proper use of sewage sludge in crop production, a sound sampling and analysis program is important.

Forms of Nitrogen. Both organic and inorganic forms of nitrogen are present in sewage sludges. The major inorganic form is ammonium, which is rapidly converted to nitrate in soils. Ammonium and nitrate are immediately available for plant growth. The organic forms of nitrogen in sewage sludge are not immediately available for plant uptake, but they are slowly released as ammonium through the decomposition of sludge in soils. Thus, sewage sludges contain immediately available nitrogen plus some that is slowly released to the crop throughout the entire growing period, which is different from commercially available forms of nitrogen, such as anhydrous ammonia, ammonium nitrate, ammonium sulfate or urea.

However, since the concentration of nitrogen in sewage sludges is relatively low in relation to commercial fertilizer materials, much larger amounts of sludge must be applied in order to obtain the same amount of plant-available nitrogen. This results in higher costs for application and transportation.

Rates. Annual sludge application rates are adjusted so that the nitrogen needs of the crop are balanced by the amount of available nitrogen present in the sludge. Adding more sludge nitrogen than can be removed by the crop during the growing season will promote nitrate accumulation in the soil or leaching of nitrate into ground water. Excess nitrate in soil or water represents a potential hazard to human and animal health. In addition to nitrogen, the majority of phosphorus, potassium and sulfur added to soils in sewage sludges will be available to crops during the first year after sludge application.

Sewage sludges have been demonstrated to provide the essential nutrients required for plant growth. The application of equivalent amounts of available plant nutrients in either sewage sludge or commercial fertilizer materials will result in similar yields. Furthermore, many of the physical properties of soil that enhance plant growth, such as water-holding capacity, infiltration, aggregation and nutrient-holding capacity, are improved by sludge application to soils.

Heavy Metals. One of the problems associated with land application of sewage sludge is the accumulation of metals in the soil profile because most sludges contain appreciable levels of heavy metals. Heavy metals in sludge originate largely from industrial wastewater treated in municipal sewage treatment plants.

In addition to metals that are essential for plant growth, such as zinc and copper, sewage sludges may also contain significant amounts of cadmium, lead and nickel. The latter three metals are not essential for plant growth, but they can be taken up from the soil by crops. This results in elevated concentrations of these metals in either the forage or grain of many agronomic crops.

Even though copper and zinc are essential for plant growth, they can become damaging to plants if they are applied to soils in excessive amounts. Cadmium, lead and nickel are not phytotoxic but are important principally because the introduction of these metals into the food chain has significant human and animal health implications. By far the greatest concern is centered around the contamination of food crops with cadmium. The concentrations of cadmium in soil that result in excessive cadmium levels in plant tissues are far below those that cause decreased plant yields.

One of the major thrusts in research has been to define the metal concentrations that can be tolerated by different crops and to evaluate various management practices in order to minimize the metal uptake by different types of crops. Research has been conducted at many universities to evaluate the uptake of metals from soils treated with sludges. The major agronomic crops investigated were corn, soybeans, oats, wheat, rye and Coastal Bermuda grass.

Since the metal concentration in sludges are dependent, to some extent, on the amount of industry within the area, the application of sludge from a nonindustrialized city results in little change in the metal concentration in crops. However, if significant amounts of industry are present, metal concentrations in the different plant parts may be increased.

Guidelines. The research conducted at agricultural experiment stations has resulted in the adoption of some general guidelines for the amount of metals that can be applied to different soil types. These guidelines can be utilized to prevent the excessive application of sludges on agricultural land, thus enabling the growth of any type of crop in future years. In contrast to conventional agricultural practices, it is essential that proper management techniques be utilized to insure that the availability of metals in soil is kept at a low level. This is commonly accomplished by maintaining the soil pH at 6.5 or above in order to minimize metal uptake by plants. Studies are presently being conducted to evaluate the proposed guidelines and to determine their validity for annual and long-term sludge applications.

Polychlorinated Biphenyls (PCB's). Another problem in application of sludge on agricultural lands is the presence of elevated concentrations of chlorinated hydrocarbons in a small percentage of sludges. The most prevalent group of compounds identified is PCB's. In the majority of contaminated sludges analyzed, the PCB concentration was less than 10 ppm, a level that should not alter the PCB level in food materials produced on soils treated with sludge.

In general, these types of organic compounds are not taken up by plants; the major mode of entry into animal food chains is through direct ingestion of soil or plant materials that have surface contamination of sludges containing PCB's. A management technique that can be used to prevent PCB intake by animals grazing on sludge-amended lands is to eliminate direct grazing of livestock on pastures immediately following sludge application. For the majority of sludges, the chlorinated hydrocarbon concentration is relatively low. In these cases, no overt problems should be encountered concerning the intake of PCB's or other types of organotoxins by livestock when sludges are used as a fertilizer material.

With proper management, sludge application to cropland can provide substantial benefits to farmers. However, failure to follow general guidelines developed by agronomists may lead to numerous problems in crop and animal production. Specific recommendations and guidelines have been provided by a number of universities. Examples are those published by King and Morris at the University of Georgia,[4,5] Nelson, Sommers and Botcher[6] at Purdue in Indiana and by Page at the University of California.[7]

In India and Turkey, manures are dried and used to produce methane,

[4]L. D. King, and H. D. Morris, "Land Disposal of Liquid Sewage Sludge: I. The Effect on Yield, *In Vivo* Digestability and Chemical Composition of Coastal Bermuda Grass," *J. Envir. Quality* 1 (1972), 325–329.

[5]L. D. King and H. D. Morris, "Land Disposal of Liquid Sewage Sludge: II. The Effect on Soil pH, Manganese, Zinc and Growth and Chemical Composition of Rye," *J. Envir. Quality* 1 (1972), 425–429.

[6]D. W. Nelson, L. E. Sommers, and A. D. Botcher, "Water Quality in an Agricultural Watershed," Conference Nonpoint Sources, Water Pollution Problems, Policies, Prospects, ed. H. R. Potter, Purdue Water Resources Center, 1977, pp. 91–103.

[7]Page, *Fate and Effects.*

which serves as a source of fuel for cooking and heating. Clemson University researchers produced methane from swine manure fermentation.

LIVESTOCK WASTES

In 1972, livestock in the United States produced manures and other debris equivalent to the waste of 2.0 billion people. This amounts to about 10 times more animal wastes than human wastes. Complicating the disposal problem is the fact that about half of these wastes are produced under concentrated conditions. Few treatment facilities exist to handle the livestock wastes. The nation faces the challenge of using the soil, with its remarkable abilities for self-rejuvenation, as an animal-waste-disposal medium in a way that minimizes the pollution of the soil and related resources.

When herds and flocks were smaller and scattered over wider areas, disposing of livestock wastes was relatively easy. Farmers simply spread the wastes over cropland and did little else to help nature recycle them. But old-fashioned methods of waste disposal do not work today for three reasons:

1. Environmental Protection Agency rules regulate manure use on land.
2. Economic studies indicate that the cost of handling manure makes it no longer competitive in price with chemical fertilizers.
3. Many large scale, confinement-type livestock and poultry operations do not have the cropland or pasture on which to spread manure, even if the practice were economically and environmentally feasible.

How, then, to dispose of livestock wastes? They cannot be allowed to accumulate because they give off noxious odors and are a breeding place for vermin and a source of unsavory dusts when dry and polluting runoff when wet. Before looking at some of the livestock-waste-management techniques, it is important to consider the value of manure (Table 9.2).[8]

VALUE, PRODUCTION, COMPOSITION AND MOST PROFITABLE USE

The value of an animal's manure is based on its composition, which varies greatly depending on factors such as, kind of animal; its age, condition and individuality; kind of food it consumed; the kind of litter used; and the handling and storage it receives.

[8]H. F. Perkins, M. B. Parker, and M. L. Walker, "Chicken Manure—Its Production, Composition, and Use As Fertilizer," Georgia Agricultural Experiment Station Bulletin, N.S. 123 (Based on 82 broiler and 31 hen houses), 1964.

Table 9.2
THE VALUE AND COMPOSITION OF A TON (909 KG) OF SOME ANIMAL MANURES

	Fresh Manure Horse, Steer, Cow, Swine			Air-Dried Broiler Manure			Air-Dried Hen Manure		
	Nitrogen	P_2O_5	K_2O	Nitrogen	P_2O_5	K_2O	Nitrogen	P_2O_5	K_2O
Percent	0.5%	0.25%	0.5%	1.7%	1.9%	1.5%	1.3%	2.7%	1.4%
Weight[1]	10.0 lb	5.00 lb	10.0 lb	34.0 lb	38.0 lb	30.0 lb	26.0 lb	54.0 lb	28.0 lb

[1]Lb ÷ 2.2 = kg.

Based on the composition of manures given in Table 9.2 and the values for nitrogen, P_2O_5 and K_2O of 15¢, 10¢ and 5¢ per pound (0.45 kg), fresh cattle manure is worth $2.50 per ton (0.91 m ton), air-dried broiler manure is worth $10.40/ton and air-dried hen manure is worth $10.70/ton. Based on these same nutrient prices, a ton (0.91 m ton) of 10-5-10 fertilizer is worth $50.00. It costs the same to spread a ton (0.91 m ton) of chicken manure as it does to spread a ton of 10-5-10. With the ton of hen manure, the grower gets 108 pounds (49 kg) of nutrients spread; with the fertilizer, he gets 500 pounds (228 kg) of nutrients spread for the same price.

Custom application of fertilizer by ground equipment costs $5.00 to $7.00 per ton, depending on the kind and condition of the material. Aerial application costs $1.50 to $3.00 per 100 pounds (45.5 kg) of material or up to $60.00 per ton. Because of the concentration of high-analysis fertilizer materials and spreading costs, it is more economical, and more plant food can be supplied by adding 100 pounds (45.5 kg) of 10-5-10 fertilizer by air than a ton (0.91 m ton) of cow manure by ground equipment. Usually, aerial application of fertilizer is reserved for areas inaccessible to ground equipment, such as forests, rice paddies, and crops growing on fields too wet for traction. It seems evident that the cost of handling manures in many cases is greater than the fertilizer value of the waste being disposed of.[9] It should be emphasized, however, that most poultry manure, dry-lot cattle manure and pig parlor manure must, by law, be moved from their place of origin; therefore, its value is certainly worthy of consideration. In 1978, it was calculated that poultry in South Carolina annually produce about:

$$35,890,000 \text{ lb nitrogen @ } 15¢ \text{ per lb} = \$5,383,500$$
$$24,298,000 \text{ lb } P_2O_5 \text{ @ } 10¢ \text{ per lb} = 2,429,800$$
$$10,923,000 \text{ lb } K_2O \text{ @ } 5¢ \text{ per lb} = \underline{546,150}$$
$$\$8,359,450$$

To purchase in the form of commercial fertilizers the nitrogen, phosphorus and potassium that are present in all animal manures produced annually in South Carolina would cost approximately $100 million. This is about as much as is spent annually in the same state for commercial fertilizers. Manure is a most valuable by-product of livestock farming, having in the case of cows, a fertilizer value equal to about 4 to 5 percent of the value of the milk produced.

Amount of Manure Produced. As a general average, a mature beef cow voids about 70 pounds (31.8 kg) of excrement daily; of this, about 20 pounds (9.1 kg), approximately 30 percent, is urine and about 50 pounds

[9]F. P. Miller, "Land Conservation and the Environment," in *Our Land and Its Care* (Washington, D.C.: Fertilizer Institute, 1977).

(22.7 kg), or approximately 70 percent, is feces. This is somewhat more than a ton (0.91 m ton) per month. However, due to appreciable losses under practical conditions, a ton (0.91 m ton) per cow monthly is a commonly accepted figure for beef cows. Maturing milking cows like the Holstein excrete two tons (1.8 m ton) or more per month. Water makes up about 85 percent of the feces and 93 percent of the urine.

Compared with manure production by beef cows, on an equal live-weight basis, horses produce about 25 percent less, hogs 25 percent more and sheep and chickens about 50 percent less. One of the main reasons for these differences is a marked variation in the proportion of liquid to solid.

In general, a ton (0.91 m ton) of manure can be returned to the land for each ton of dry feed harvested from the land. With average crop yields, this provides about 10 tons per acre (22.4 m tons/ha) of manure every four years.

For practical considerations, the following assumptions can be made. The excrement of farm animals contains three-fourths of the fertilizer elements present in the feed consumed. Of this three-fourths, the feces contain one-half of the nitrogen, 95 percent of the phosphorus, and 20 percent of the potassium; the urine, of course, contains the remainder of what is excreted. About one-half of the nitrogen and the phosphorus and nearly all of the potassium in the total excrement is water soluble and thus subject to loss by leaching. In addition, nitrogen is subject to loss by volatilization.

Composition. The composition given in Table 9.2 compares fresh and undried cattle (horse, steer, cow, swine) manure with a special air-dried poultry manure, which is produced in Georgia. However, a more realistic comparison of the composition of the various animal manures, all on a fresh (moist) basis, is given in Table 9.3.

The nitrogen, phosphorus and potassium contents of cow and horse manure are quite similar, especially when calculated to the same moisture content. In terms of the fertilizer formula, cow and horse manures are

Table 9.3
COMPOSITION OF FRESH ANIMAL MANURE, INCLUDING LIQUID AND SOLID

Kind of Animal	Average % of Water	Pounds in One Ton of Manure[1]		
		Nitrogen (N)	Phosphoric Acid (P_2O_5)	Potash (K_2O)
Cow	78	9 to 10	3 to 7	7 to 10
Horse	63	10 to 15	4 to 10	11 to 18
Hog	74	10 to 17	8 to 17	7 to 15
Sheep	63	13 to 34	8 to 12	11 to 28
Chicken	58	9 to 25	14 to 20	5 to 12

[1]Lb per ton × 0.5 = kg/1000 kg.

approximately 0.5-0.25-0.5 on a N-P_2O_5-K_2O basis. On this basis, a ton (0.91 m ton) contains 10 pounds (4.5 kg) of nitrogen, 5 pounds (2.2 kg) of phosphate and 10 pounds (4.5 kg) of potash, respectively, or the same amounts as in a 100-pound (45-kg) bag of 10-5-10 fertilizer, costing at present prices about $2.50. The manure, however, contains considerable calcium, magnesium and sulfur, appreciable amounts of all of the micro-nutrient elements, and also supplies organic matter. Thus, the total value of a ton (0.91 m ton) of average cow or horse manure, based on fertilizer prices, is close to $3.00. The value of the crop increases produced by the manure, or by the 10-5-10, may be several times this figure.

How to Get Full Value from Manure. Getting full value from manure requires close attention to the following:

1. In the barn: Conserve the urine by providing tight floors and gutters and adequate bedding. If manure in loose-run pens and stalls becomes dry, a large loss of ammonia nitrogen to the air will occur.

2. From the barn to the field: Under South Carolina conditions, it is usually best to haul the manure directly to the field and to spread it as it is being produced.[10] This reduces losses to a minimum, keeps the premises clean, and utilizes labor to good advantage. The soluble fertilizer constituents of manure are quickly fixed when they come into contact with soil. When land to be manured is so steep that loss by runoff is risked, it may be desirable to withhold application until just before plowing or, if on pastures, until a time of lush green growth.

3. In the field: When *fermented manure* (manure in loose-run barns, stalls, and old piles) is spread and it dries out before being rained upon or worked into the soil, loss of nitrogen, as ammonia, occurs. This may be prevented by spreading during rainy periods or by immediately discing or plowing down after spreading. When *unfermented manure* (fresh manure as produced) is applied, there is little if any danger of loss in this manner.

4. Storing manure: When storage of manure is necessary, it is best to build a compact pile with rather vertical sides in order to lessen both drying out and leaching. In many cases, it is profitable to provide a concrete manure pit or bin for the storage of manure in order to prevent loss by leaching or runoff.

Losses Are Heavy. The annual loss of fertilizer elements, due to various causes, from manure in South Carolina in terms of the market value of

[10]T. C. Peele, H. P. Lynn, C. L. Barth, and J. N. Williams, "South Carolina Guidelines for Land Application of Animal Waste," South Carolina Agricultural Experiment Station Bulletin 570, 1973.

these elements probably falls in the range of 20 to 35 million dollars. One-half of this loss could probably be prevented (Figure 9.1).

MANURES FOR ANIMAL FEEDS

The use of manures as a source of protein in feeds has been studied in some detail. In multistock, dry-lot feeding operations, beef cattle have been fed corn; hogs in the same dry lot eat the cattle droppings; and poultry in the same enclosure consume the hog and cattle droppings. Because different housing is required for different species, this is not feasible at present in livestock operations in the United States.

Poultry manure, that has been ensiled, has been used as a source of protein for other animals with some success. However, disease organisms can create problems. The Food and Drug Administration in the United States declares that feeds containing poultry litter must be labelled "adulterated."

According to some animal nutritionists, processed animal wastes may well be worth more as a feed than as a fertilizer. If this is so, the long-discarded farm practice of running hogs behind cattle may eventually be reenacted with a modern twist.

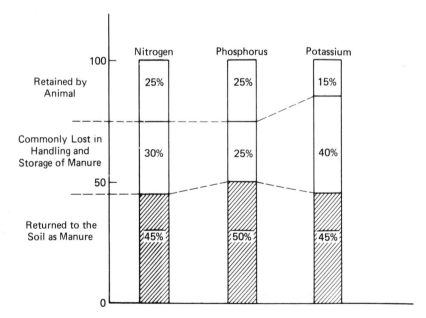

Figure 9.1 The disposition of nutrients commonly incurred in the process of converting feed consumed into body weight and excrement.

From the previous discussion, it is evident that the thriving market for manures and other by-product protein-organics that prevailed in the early part of this century no longer exists. The interest in these materials today is focused on the pollution and waste-management viewpoint rather than on the use of food-processing wastes and manures as fertilizer. The most relevant questions seem to be:

1. What are the economic and environmental costs and benefits of the waste-utilization operation, and how can processing and handling be improved to reduce the costs?

2. How much protein-organic waste can be tolerated in feeds before they become too "adulterated"?

3. How much protein-organic waste can the soil assimilate without endangering the crop yield or polluting the soil beyond its rejuvenation capacity?

Answers to questions 1 and 2 will be the object of much research by agronomists, soil scientists, engineers and animal scientists of the future. The answer to number 3 is perhaps the greatest challenge faced by soil scientists. It will be a basic problem of the remaining two decades of this century.

WASTE MANAGEMENT OF MANURES AS FERTILIZERS

Some parameters have already been established to plan the utilization of the livestock wastes in the United States that have been estimated as being equivalent to the wastes of 2.0 billion people. It was suggested that ½ acre (0.20 ha) of corn is needed to accommodate the manure from 1,000 broilers, 100 laying hens, 10 hogs (30 to 200 lb or 13.6 to 90.9 kg), 2 feeder cattle (400 to 1,100 lb or 181.8 to 500 kg) or 1 dairy cow (1,200 lb or 545.5 kg). A maximum of 300 to 400 lb/acre (336 to 448 kg/ha) of nitrogen per year from the manures was postulated.

This same basis of 300 to 400 pounds per acre (336 to 448 kg/ha) of nitrogen per year is also suggested as a maximum for the utilization of food-processing wastes having nitrogen similar to that in manures.

Here are some of the manure waste-management methods currently in use or under investigation.

Land Disposal.   When the land is available (and far enough away from nonfarm communities so that no one objects), spreading manure on cropland or pasture is still an excellent way of recycling wastes.[11] Manure contains many essential plant nutrients, especially nitrogen, phosphorus and potassium, as well as some trace elements not generally found in chemical fertilizers. Animal wastes also help to build and maintain soil fertility and tilth and to cut down on erosion.

Agricultural engineers have been working on ways of improving land-disposal techniques—with special emphasis on cutting labor requirements, improving stockpiling procedures, and controlling runoff when it rains.

A big step forward has been the adaptation of the liquid manure system, popular in Europe, to U.S. farming technology. Basically, this system calls for flushing the manure into collection pits or tanks and holding it until it is time for disposal. The manure-water mixture is then spread on the land by a tank wagon, through a sprinkler irrigation system or through irrigation ditches.

Lagoons. Another method of recycling livestock wastes involves lagoons, which are ditches in which wastes are dumped to allow aerobic decomposition (breaking down manure with oxygen-using bacteria). The decomposed material can be redistributed on cropland and pastures with irrigation equipment.

The big advantage of the lagoon system is its simplicity and low cost. There are lots of drawbacks, too. It must be fed new waste every two days; a poorly designed lagoon can, and too often does, leak fertilizer nutrients into nearby streams, ponds or rivers; overloaded lagoons produce noxious gases and vile odors as the bacteriological decomposition changes as a result of oxygen depletion.

Agricultural engineers are working on ways to improve the lagoon-disposal system. One group is testing modifications of the Pasveer Oxidation Ditch. Developed in Holland to handle human wastes, the Pasveer treatment calls for mechanical stirring and aeration to supplement natural oxidation.

Engineers are also testing ways of purifying lagoon runoff while making maximum use of fertilizer nutrients. The engineers have built a series of three basins downhill from a lagoon. The lagoon runoff flows through the basins where hydroponically grown grasses strain out the solid matter and purify the runoff. The grasses can be harvested as a forage for livestock.

Dehydration. A few very large feed lots have begun dehydrating manure, bagging it and selling it to home gardeners. However, large amounts of manure must be handled to justify the investment in machinery and relatively high operating costs.

---

[11]C. L. Barth, J. T. Ligon, and C. L. Parks, "Using Animal Manure as Fertilizer," Clemson Extension Circular 578, 1977.

Dehydration may not be too practical for poultry manure because the product reeks when wetted. Some poultry producers are nevertheless dehydrating manure with electric fans on endless belts and selling it to growers and gardeners.

Toxic Effects.   Hog manure contains copper, zinc and arsenic. Beef cattle manure contains 1 percent or more NaCl. Poultry manure sometimes contains a coccidiostat, which acts like a herbicide. Human manures often contain large quantities of lead, zinc and cadmium. Some contain PCB's. When manure containing these kinds of chemicals is applied to the land at very heavy rates, seed germination and seedling growth may be reduced.

The application of manure to the soil in very heavy amounts, unless immediately covered and completely buried, may also create conditions favorable for flies and odors. Of all pollution problems, odor is the one people complain about the most.

Weed seeds also create problems when manure is put on the land.

WASTE MANAGEMENT OF MANURES AS FEEDS

Some examples of work being done on recycling livestock and poultry wastes as feeds follow.

In Alabama, one researcher has gotten good feed efficiency with a beef ration containing 40 percent cattle manure, 48 percent whole shelled corn and 12 percent ground hay. He merely scrapes the manure off the feeding floor, blends it with the hay and corn, and ensiles it in a small bunker silo. According to him, fermentation destroys any pathogens and degrades residues of pesticides, antibacterials and growth promotants that might be present in the manure.

Similar results, using broiler litter silage in the beef ration, has been observed by Clemson University (South Carolina) scientists.

At the University of Illinois, agricultural engineers centrifuged hog wastes from an oxidation ditch. They report that the wastes contain up to 75 percent crude protein (on a dry-weight basis) and have up to 10 times as much lysine as normal corn.

Scientists at several institutions are feeding dry or ensiled poultry wastes to cattle, sheep, hogs or back to the birds themselves. One scientist doing this type of research notes that caged layer manure contains anywhere from 30 to 45 percent protein—about half of which is true protein while the remainder is uric acid. One dreaming ecologist forecasts the day when agriculture will no longer be needed because man will supply his nutritional needs with his own wastes.

A start has been made, and much research that has been done in the past is certainly relevant. However, food-processing effluents containing large amounts of soda ash (NaHCO$_3$), such as peach-cannery wastes, those containing large amounts of lye (KOH), such as turkey-canning wastes, as well as many others need careful study. Some success in waste management of poultry- and peach-canning wastes has been the distribution of the liquid alkaline waste onto pastures through sprinkler irrigation systems. These effluents should be scrutinized closely for the presence of animal and plant pathogens before being used for feed or fertilizer.

Each situation needs to be evaluated on its own merits and each situation is different. It is clear that soil science and all of its ramifications are involved in any waste-management consideration where wastes are left in or on the land. There is very little pollution, whether of air, water or soil, that does not start out or end up on the land.

UTILIZATION OF ANIMAL MANURES AND SEWAGE SLUDGES FOR FOOD AND FIBER PRODUCTION

A task force of scientists representing agricultural economics, agricultural engineering, agronomy, animal science, dairy science, poultry science, sanitary engineering, soil science, veterinary pathology, veterinary toxicology and water resources was appointed by the Council on Agricultural Science and Technology (C.A.S.T.) to prepare a report which is the subject of this section. Individual members of the task force were named in Volume 2, No. 1 of NEWS from C.A.S.T. Dr. Parker Pratt, chairman of the Department of Soil Science and Agricultural Engineering at the University of California at Riverside was chairman of the task force. The task force met in Kansas City and prepared a draft of the report. The draft was subsequently revised by the chairman, reviewed by members of the task force, edited and reproduced. A part of the summary of the report as given in Volume 3 of NEWS from C.A.S.T. follows:[12]

> The utilization of animal manures and sewage sludges as fertilizers and soil amendments and the use of properly processed animal wastes as ingredient for animal feeds offer a number of advantages to society. Proper utilization in agriculture, forestry and land reclamation will improve water quality as a result of decreased disposal of wastes into rivers, lakes and oceans and will improve air quality as a result of decreased incineration. Savings in use of natural resources, including fossil fuels, and in livestock production are potentially important advantages.
>
> Application of animal manures and sewage sludges to agricultural and forest land is influenced by physical, chemical, microbiological, sociological, economic and political or legal factors. The limitations or obstacles to the present use or increased efficiency of use of these materials include the following:

[12]P. F. Pratt et al., "Utilization of Animal Manures and Sewage Sludges in Food and Fiber Production," NEWS from C.A.S.T., Vol. 3, 1975, 23–25.

1. Animal manures and sewage sludges, as sources of plant nutrients, are bulky, low-grade fertilizers of variable composition. The total plant nutrient content is only 10 to 20 percent of that of most commercial chemical fertilizers. Concentrations of nutrients, soluble salts, trace elements and water vary tremendously and are seldom known. Optimum rates of application are thus difficult to predict.

2. As low-grade fertilizers, these materials cannot be transported more than a few tens of miles (kilometers) before the costs of transportation exceed their fertilizer value. Many animal confinement facilities and many large cities lack agricultural land within economic transportation distances of the point of production.

3. Difficult on-farm management problems are created by the physical properties of these materials. Application techniques are inefficient and time consuming. When liquid sludges are applied to the surface of the soil, a drying interval is required, and this loss of time can result in costly delays in seeding. The production of animal manures and sewage sludges is continuous, whereas the needs for fertilizers are seasonal. Thus, storage is required by either a distributor or the farmer. The products, particularly sewage sludges, are difficult to store on the farm.

4. Animal manures and sewage sludges both contain soluble salts that can create problems in their use as fertilizers, particularly in the irrigated lands of arid regions where soluble salts are already present from irrigation waters. In arid, irrigated regions, applications of manures and sludges must be limited because of the detrimental effect of the salt on crops when large quantities are applied.

5. The undesirability of leaching of soluble salts, particularly nitrate, to ground waters may limit the rates of use of animal manures and sewage sludges on lands.

6. Sewage sludges contain heavy metals that are retained in soils and can accumulate to the point where they are toxic to some plants and thus restrict the types of crops that can be grown. As a result of uptake by crops, they may also be toxic to animals and humans, so this may limit the use of the harvested crop. So little is known about this hazard at present that it cannot be properly evaluated. Of primary concern as potential toxicants are zinc, copper, nickel and especially cadmium.

7. Sewage sludges contain pathogenic bacteria, viruses and parasites that represent a public-health risk to farm workers and public via the food chain. The degree of risk depends on the processing of the sludge and on its management on the farm, but experiences with application of sludges to land in a number of locations suggest that the risk is low.

8. Odors and associated nuisances, both real and imagined, create conflicts between urban residents and the farmers of adjacent croplands who could advantageously use the wastes. These conflicts add transportation or application costs for land application of sewage sludge or of animal manures from dairy or animal-feeding operations located near cities.

9. One of the serious obstacles to the use of sewage sludges on cropland is sociological resistance that stems from fear of pathogens, odors and nuisances and possible environmental deterioration. Once resistance

and fear have been generated in a community, the people have great difficulty in accepting the fact that the risk of creating any of these problems is extremely low in a properly managed system of applying sludges or manure to land.

10. Another obstacle to the use of sewage sludges from crop production is the requirement by environmental protection agencies that both crop and water quality be monitored. Farmers who would otherwise receive and use this material cannot or will not pay the additional costs of monitoring for possible environmental side effects. Monitoring is not required if chemical fertilizers are used.

11. One obstacle to the use of processed animal manure as a feed ingredient has been the concern for possible transmission of pathogens and undesirable organic residues into food products. Because of this concern, Federal and state government agencies have been slow to approve the practice.

12. The lack of meaningful guidelines that can be interpreted on a regional basis or in terms of local situation and needs has been an obstacle to the use of sewage sludges on cropland and also to the use of animal manures as feed ingredients. Investments in systems to apply sewage sludges to land or in processing plants to produce feed ingredients from animal manures cannot be expedited until such guidelines are established.

Actions are needed that will promote the beneficial use or increased efficiency of use of sewage sludges and animal manures. As a society with limited resources, we need to stop thinking in the negative terms of waste disposal and instead should think in the positive terms of appropriate use. Research and development activities are needed that will concentrate on the beneficial-use concept to overcome some of the disadvantages and obstacles discussed. It would be helpful if government agencies at all levels would take actions to favor rather than to restrict the reasonable use of these resources.

SUMMARY

1. Frequently, the cost of handling is greater than the value of the waste material being used.
2. The greatest interest in waste protein-organics is from a pollution-control, waste-management viewpoint.
3. The protein-organic fertilizers and feedstuffs, with the exception of urea, are derived from plant and animal sources.
4. Urea is used interchangeably as a protein source in ruminant feeds and as a nitrogen source in fertilizers.
5. Many wastes can be used interchangeably as feedstuffs and fertilizers. In either use, toxic metal and PCB hazards must be considered.

6. Protein-organics are used much less extensively for fertilizer and more extensively for feeds today than formerly.

7. Fertilizers and feedstuffs are end-use products.

8. Manure and food-processing by-products are the principal protein-organic wastes.

9. Food-processing wastes are used primarily as ingredients of feeds, but progress has been made in the use of peach- and poultry-processing effluents as fertilizer for grasses.

10. Manures are used mostly for fertilizers, but research has demonstrated the use of poultry and cattle manure as feed. The Food and Drug Administration labels such feed adulterated. The Environmental Protection Agency has rules regulating manure use on land.

11. One problem in processing food wastes and manures for practical use is treatment of the material to prevent it from being a carrier of diseases.

12. The quality of the waste material and the cost of processing and handling it are the factors influencing the efficient utilization of protein-organics as fertilizer or feed.

13. The challenge to soil scientists is how to use the soil, with its capacity for self-rejuvenation, as an animal-waste-disposal medium in a way that minimizes the pollution of it and related resources.

14. The most practical way to get organic matter into the soil is by growing higher yields per acre (ha) and incorporating the crop residue.

15. The decomposition and mineralization of wastes and residues is dependent on the carbon-to-nitrogen ratio. A carbon-to-nitrogen ratio greater than 20:1 will cause nitrogen-deficient systems.

16. Conservation, in an immediately available form, of the nitrogen that volatilizes from sewage sludges and animal manures would greatly increase the fertilizer value of many of the products.

QUESTIONS

True or False:

1. ___Protein-organic sources of nitrogen are superior to inorganic sources for the production of tobacco, ornamentals and vegetables.

2. ___A principal limitation to the use of land for the disposal of large amounts of animal manure is the possibility of ground water contamination with nitrate.

3. ___Water drained from irrigated land usually has a higher salt content than the irrigation water applied to the land.

4. The disposal of animal manure on land is more a problem today than previously because:
   (a) The manure is now appreciably higher in nitrogen than it was formerly.
   (b) Land for disposal is now less accessible to centers of manure production than it was formerly.
   (c) Disposal on land has become increasingly more hazardous to human health.

5. The concentration of nitrogen in nitrate form that is considered by the United States Public Health Service to make water unsafe for human consumption is:
   (a) 10 ppm.
   (b) 100 ppm.
   (c) 1,000 ppm.

Completion:

6. The factors influencing the efficient utilization of protein-organics as feed or fertilizer are _____ and _____.

7. The most practical way to get organic matter into the soil is by growing higher yields per acre (ha) and _____.

8. The greatest interest in protein-organic wastes is from the standpoint of _____ and _____.

9. A bench mark of _____ to _____ pounds of nitrogen per acre per year is suggested as the maximum amount that soils and crops can tolerate in the form of manures.

10. Probably the greatest limitation to the use of sewage effluents for irrigation and fertilization of land is that the supply is _____.

Essay:

11. Are there any benefits to be derived from disposing of sewage effluents on land?

chapter 10

The Fertilizer Situation in the United States, 1973–1985*

Welded barge for transporting ammonia. (Courtesy of Ingalls Shipbuilding Corporation.)

*This chapter consists for the most part of a report written for presentation to the United States Senate Agriculture Committee by a task force appointed by

Agricultural production in the United States has become increasingly dependent on commercial fertilizers; an estimated 30 to 40 percent of production is attributable to fertilizer use. The demand for fertilizer materials in the United States has risen sharply during the past three and a half decades, from 8.6 million tons (7.83 million m tons) in 1940 to 48.9 million tons (44.0 million m tons) during 1975–1976.[1] The demand for fertilizers accelerated rapidly during 1973 as a result of United States Department of Agriculture action eliminating set–aside acreage (hectarage) requirements for wheat, feed grains and cotton in 1974; increased world demand for food; and an increase in fertilization rates per acre (ha) in response to expected higher prices for farm produce.

Fertilizers in late 1973 were in critically short supply, having moved from a surplus situation during the late 1960s and early 1970s that resulted from overproduction and led to depressed fertilizer prices. The industry virtually stopped all plant and equipment expansion. In addition, devaluation of the U.S. dollar in 1973 had the effect of lowering the price of U.S. products to foreign customers, creating an increase in export demand including fertilizer.

As fertilizer producers try to evaluate their future plans for expansion, several new unknowns loom on the horizons. Pollution control laws and environmental controls will limit operations in older plants and will add substantially to investment and operating costs of new plants. There is a shortage of raw materials, particularly natural gas which is the basis of most nitrogen fertilizers. The energy crisis could have a very significant impact on all aspects of agriculture: the use of farm machinery, transport of agricultural products, irrigation, fertilizer and pesticide production and other management tools.

The purpose of this brief report is to discuss fertilizer needs, problems related to supplying these needs and possible alternatives for solving these problems.

---

the Council for Agricultural Science and Technology. Task force members were C. H. Davis, J. R. Douglas, E. A. Harre, V. J. Kilmer and L. B. Nelson, all of the Tennessee Valley Authority; M. H. McVickar of Chevron Chemical Company; J. F. Reed of the Potash Institute; W. C. White of the Fertilizer Institute and U. S. Jones of Clemson University. This report was slightly revised by the author in 1978.

[1]*Fertilizer Trends*, compiled by E. A. Harre, M. N. Goodson and J. D. Bridges, Muscle Shoals, Alabama: National Fertilizer Development Center, Tennessee Valley Authority, 1976.

Nitrogen, phosphorus and potassium are termed macronutrients because relatively large amounts of these elements are required by crops. Elements needed in smaller amounts by plants (secondary nutrients) are sulfur, calcium and magnesium. Plants use very small amounts of boron, copper, iron, manganese, molybdenum, zinc and possibly chlorine and cobalt. These latter elements are designated as micronutrients.

Although all of these elements are vital to plant growth, this report will deal entirely with nitrogen, phosphorus and potassium. The production and marketing of these three major nutrients constitute a very large proportion of the fertilizer business. The demand for nitrogen, phosphorus and potassium is expected to increase by about 5% annually or 50% within the next decade, according to Tennessee Valley Authority estimates.

Nitrogen Needs.   Most crops require more nitrogen than either phosphorus or potassium. A lack of sufficient nitrogen in soils is the most frequent nutritional factor limiting plant growth.

The annual "on-farm" nitrogen requirements in the United States are estimated to be 20.0 million tons (18.2 million m tons). Of this 20.0 million tons (18.2 million m tons) of nitrogen, about 10.0 million tons (9.1 million m tons) are supplied by commercial fertilizers.[2] An additional 10.0 million tons (9.1 million m tons) must come principally from native soil supplies and legumes.

Nitrogen provides the building blocks for plant and animal protein. An estimated 18.7 million tons (17.0 m tons) of nitrogen, or about 90 percent of the total nitrogen used in agriculture in the United States, is used to produce animal–based protein. Approximately 70 percent of our protein intake, amounting to 0.22 lb (100 g) per capita per day, is in the form of meat, poultry, milk, eggs and cheese. This diet requires about 190 pounds (86 kg) of farm-site nitrogen per capita per year.

The farmer uses nitrogen fertilizers in solid form (ammonium nitrate, urea, ammonium phosphates), as liquid nitrogen solutions (urea, ammonium nitrate, aqua ammonia), and as a gas (anhydrous ammonia). About 40 percent of the nitrogen used in the United States is applied in the fall,[3] mainly to cool–season grasses and grain crops. The remainder is applied during the spring and early summer. Fall application of nitrogen for spring-seeded crops is generally not recommended in the warmer areas of the United States because of nitrogen losses that occur during the absence of a crop.

[2]"Consumption of Commercial Fertilizers," United States Department of Agriculture Statistical Reporting Service, Columbia, S.C. 1976.

[3]N. L. Hargett, *Fertilizer Summary Data* (Muscle Shoals, Alabama:National Fertilizer Development Center, Tennessee Valley Authority, 1977).

329

Crop plants take up about 50 percent of the nitrogen that is applied during the same year that the crop is grown. Biological transformations and losses by leaching and as nitrogen gas to the atmosphere are the primary causes of this low rate of utilization,

Nitrogen Production Capacity, 1973–1985.[4] The fixation of atmospheric nitrogen as synthetic ammonia is required for the production of all synthetic nitrogen fertilizers. Nitrogen constitutes about 80 percent of the atmosphere and is for all intents and purposes inexhaustible.

During the middle 1960s, ammonia production in the United States kept pace with capacity levels, and the industry was able to maintain an average operating rate of over 85 percent of capacity. By 1966, however, production from new large-scale plants began to reach the market, and operating rates declined to as low as 76 percent in 1969. Even with this relatively poor performance, which led to the closing of many small units, inventory levels continued to grow. Not until 1969, when a trade surplus of over 900,000 tons (810,000 m tons) of nitrogen was achieved, did the situation begin to improve.

Since 1969, demand has been catching up to potential supply, the trade balance has declined, and the industry in 1972 returned to an 85-percent level of operation. In 1976, producers approached a 90-percent level as they attempted to meet both a strong domestic demand and a relatively high level of export demand.

For 1975, total nitrogen-production capacity reached 14.6 million tons (13.3 million m tons) of nitrogen. Total use and net exports exceeded 13.1 million tons (11.9 m tons), an operating level of 90 percent. With this level of operation, however, net exports were minus 647,000 tons (588,770 m tons) compared with the high of plus 904,000 tons (822,640 m tons) in 1969. By 1980, a 30-percent increase in total capacity is anticipated, compared with a 17-percent gain in demand. As a result, the United States is moving rapidly toward becoming a net nitrogen exporter again with a net export requirement of 389,000 tons (354,000 m tons) of nitrogen for 1980. For the remainder of the decade, no greater capacity is needed to meet the growing domestic demand. If the gas feedstock supply situation does not deteriorate, it appears that the United States will be in a position to contribute to expanded world trade in nitrogen fertilizers. However, this comes at a time when many other nations are also expanding capacity, with the idea that the United States will be their primary market.

To indicate the supply situation up to 1985, two levels of production were selected for calculating the supply-demand position and the trade level needed to keep the market in balance. The two levels selected were 90 percent and 80 percent of total capacity. The 80 percent was included

[4]Prepared by the Tennessee Valley Authority for presentation at the CENTO symposium on the Mining and Beneficiation of Fertilizer Minerals, Istanbul, Turkey, November 19–23, 1973. Data revised in 1976.

to show the effect on supply of a 10-percent reduction in production caused by feedstock supply shortages.

These market forecasts are shown in Table 10.1. It compares the assumed total supply level under the two levels of operation with the total demand. From the table, it can be seen that a 90-percent operating rate did not meet domestic demand after 1973 and that increased imports were needed. However, with new plant construction after 1975 and the 90-percent rate, by 1985 a trade surplus of 385,000 tons (364,000 m tons) of nitrogen is indicated. If the United States had not increased ammonia capacity, the trade deficit could have exceeded 30 percent of total production, well above any previous trade deficits that occurred in the nitrogen industry in the United States.

Table 10.1
THE NITROGEN SITUATION IN THE UNITED STATES AND FUTURE SUPPLIES[1]

Fiscal year	Total ammonia capacity	Total ammonia production	Average operating rate	Other demand[3]	Fertilizer demand	Net trade or calculated trade balance
	(1,000 tons nitrogen[2])		%	(1,000 tons nitrogen[2])		
1962	5,086	4,450	87	1,224	3,370	−103
1963	6,500	5,126	79	1,328	3,929	−148
1964	6,701	5,895	88	1,654	4,353	−149
1965	7,468	6,655	89	2,063	4,639	−78
1966	9,945	8,079	81	2,558	5,326	17
1967	11,460	9,435	82	2,851	6,027	80
1968	12,061	10,115	84	2,934	6,694	370
1969	13,249	9,986	75	2,483	6,958	904
1970	13,538	11,014	81	3,048	7,459	473
1971	13,759	11,299	82	2,989	8,134	148
1972	13,783	11,753	85	3,200	8,016	189
1973	13,848	12,024	87	3,360	8,339	288
1974	14,266	12,839	90	3,600	9,471	−232
1975	14,614	13,153	90	3,820	9,980	−647
			Forecast supply level 1			
1980	19,680	17,712	90	4,800	12,523	389
1985	19,826	17,843	90	4,837	12,621	385
			Forecast supply level 2			
1980	19,680	15,744	80	4,800	12,523	−1579
1985	19,826	15,860	80	4,837	12,621	−1598

[1]From the records of the Fertilizer Distribution Economics Section, Test and Demonstration Branch, Division of Agricultural Development, Tennessee Valley Authority, Muscle Shoals, Alabama.

[2]Ton × 0.91 = m ton.

[3]Adjusted for net change in producers' stock; includes industrial uses, processing losses, and the net change in unreported field inventory.

Any further increase in production, even at the 90-percent level, would necessitate increases in exports by 1985. The capacity expansion is well above that needed to meet the expected demand for nitrogen for fertilizer and nonfertilizer purposes.

There is a strong argument for not allowing this country to be dependent on another country for a product vital to our nation's agriculture. However, the potential trade deficits and surpluses at the 80- and 90-percent operating rate are shown to indicate the need for market research before additional ammonia capacity is planned beyond that which has already been announced.

Phosphate Needs. A little over 5.2 million tons (4.7 million m tons) of phosphorus (P_2O_5) was used in fertilizers by farmers in the United States during 1976. Phosphorus needs of plants are much smaller than those of either nitrogen or potassium, but phosphorus is vital to all life and growth.

Phosphorus is highly immobile in soils because it readily forms slowly soluble compounds with a number of elements commonly found in soils. As a result, crop plants utilize about 20 percent of the phosphorus that is applied during the same year that the crop is grown. The residual phosphorus remains in the soil, and small portions of it become available to subsequent crops. However, the slow rate at which this residual phosphorus becomes available to crops makes repeated additions of this element necessary for sustained high yields.

Phosphate rock is the starting material for all but a small fraction of the phosphate fertilizers used in commercial agriculture. The rock is processed into soluble phosphate fertilizers consisting mainly of ammonium and superphosphates. The practice of applying phosphates as fertilizer solutions is also growing rapidly.

Phosphate Production Capacity, 1973–1985.[5] The United States is the major world supplier of phosphate fertilizers, and recent capacity announcements for the next few years indicate that it will retain this position. Total supply capability for the United States by 1985 could exceed 10 million tons (9.1 million m tons) of P_2O_5 while demand is expected to be about 5.7 million tons (5.2 million m tons). This allows this nation to meet domestic demand increases and to double its 1973 level of export of 1.1 million tons (1.0 million m ton) of P_2O_5 for the next few years even with the operating levels at only 80 percent of total phosphoric acid capacity.

From Table 10.2, either of two calculations becomes apparent if the large-scale increase in capacity comes into production as scheduled. If plants continue to operate at 90 percent of capacity or better, the United States will need to triple its phosphate trade balance between now and

[5]Prepared by the Tennessee Valley Authority for presentation at the CENTO symposium, on the Mining and Beneficiation of Fertilizer Minerals, Istanbul, Turkey, November 19–23, 1973. Data revised in 1976.

Table 10.2
THE PHOSPHATE SITUATION IN THE UNITED STATES AND FUTURE SUPPLIES[1]

Year	Phosphoric acid capacity	Phosphoric acid production	Average operating rate	Nonacid contribution to supply	Total phosphate supply[3]	U.S. demand	Net trade or calculated trade balance
	(1,000 tons P_2O_5)[2]		(%)		(1,000 tons P_2O_5)[2]		
1962		1,469		1,668	3,046	2,807	196
1963		1,852		1,674	3,329	3,073	158
1964	2,725	2,126	78	1,764	3,846	3,378	211
1965	3,097	2,522	81	1,687	4,002	3,512	334
1966	4,830	3,315	69	1,742	4,696	3,897	316
1967	5,375	3,810	71	1,800	5,172	4,305	622
1968	5,575	4,187	75	1,654	5,371	4,453	976
1969	5,501	4,122	75	1,432	5,096	4,666	812
1970	5,602	4,458	80	1,285	5,520	4,574	572
1971	5,387	5,013	93	1,165	5,736	4,803	615
1972	5,641	5,316	94	1,278	6,174	4,873	776
1973	6,370	5,750	90	1,260	6,429	5,100	1,126
1974	7,182	6,464	90	965	6,783	5,098	1,685
1975	8,692	7,823	90	865	7,906	4,511	3,395
			Forecast supply level 1				
1980	9,324	8,390	90	865	8,477	5,313	3,164
1985	10,131	9,118	90	1,028	9,234	5,722	3,512
			Forecast supply level 2				
1980	9,324	7,459	80	865	7,578	5,313	2,265
1985	10,131	8,105	80	1,028	8,322	5,722	2,600

[1]From the records of the Fertilizer Distribution Economics Section, Test and Demonstration Branch, Division of Agricultural Development, Tennessee Valley Authority, Muscle Shoals, Alabama.
[2]Ton × 0.91 = m ton.
[3]Nonacid contribution to supply plus 90% of total phosphoric acid production.

1985. Without the increased trade levels, reductions in operating rates or plant closures will occur. Any decline in our phosphate trade balance or domestic demand growth will bring rates of operation down just that much faster.

Forces are at work that could make expansion of the phosphate industry an orderly process. This is especially true if more attention is paid to construction scheduling and if the forces of supply, demand and price are allowed to function properly. Among the factors reducing production capacity are stricter pollution controls that may force closure of older plants. Phosphate rock production is now adequate to supply all of the new phosphate plants and to expand exports. Rock prices are up, and the price at which phosphate rock is available has a direct bearing on the number of P_2O_5 producers and their level of production. About 10 million tons (9.1 million m tons) of new mining capacity was needed to supply these new units and rock producers were faced with meeting this growing market under most stringent circumstances. Environmental restrictions, competition for land for other uses and lower P_2O_5 content of the rock reserve all lead to increased rock prices and some difficulty in obtaining needed supplies.

Still another factor that could, in the long term, help to keep the phosphate expansion within bounds is on the demand side. Economic pressures continue for the substitution of wet-process phosphoric acid for thermal acid in the industrial–use market. With some improved technology, there could by 1980 be a significant tonnage of wet-process acid moving into industrial markets.

One result of the tight supply of phosphoric acid in the United States was the curtailment in the growth of fluid mixed fertilizers. After increasing from less than 500,000 tons (455,000 m tons) in 1960 to 3,456,000 tons (3,144,900 m tons) in 1974, a growth rate of almost 20 percent per year, in 1975 tonnage of fluid mixtures decreased to 3,170,000 tons (2,884,700 m ton).[6] It appears that the short supply of superphosphoric acid will limit further increases for several years. The effect of this reduced supply was felt as early as 1972 when only 17,000 tons (15,470 m tons) of additional P_2O_5 was distributed as liquids, compared with 55,000 tons (50,050 m tons) the previous year.

Two factors have contributed to the shortage of supplies for liquid mixtures. High-quality phosphoric acid, normally supplied from thermal phosphoric acid producers, has been going to the industrial market, and no excess has been available for fertilizers. Secondly, the new superphosphoric acid capacity that was scheduled to begin operation in 1973 experienced start-up difficulty and poor levels of operation. It will take several years to correct those problems, making it 1980 before supplies will again be adequate and liquids can continue their rapid increase in the share of the mixed fertilizer market that they hold.

It is possible that the rush of new capacity scheduled in the United States may lead to an oversupply position once again, especially if trade levels are not maintained because of stockpiling by importing countries or construction of new plants. This could delay additional capacity, result in some plants closing down and bring about an orderly expansion program through the decade.

Potash Needs. Crop requirements for potassium are quite high, often approaching or even exceeding nitrogen requirements in some crops. Unlike nitrogen and phosphorus, potassium does not form an integral part of plant components, such as protoplasm, fats and cellulose. The function of potassium in plant growth appears to be catalytic in nature.

[6]*Fertilizer Trends*, compiled by E. A. Harre, M. N. Goodson and J. D. Bridges.

Application of potash to U.S. and Canadian crops amounted to more than 5.2 million tons (4.7 million m tons) of K_2O in 1976.[7] Potassium chloride is the least expensive form and is by far the most widely used material to supply potassium in fertilizers. In 1976, 55 percent of the potassium in the United States was used in mixed fertilizers that contain nitrogen and/or phosphorus. The remaining 45 percent was applied directly to soils, chiefly in the form of potassium chloride. Corn receives the most potassium fertilizers among field crops, followed by cotton, wheat and soybeans.[8]

Under most conditions, very little potassium leaches from the soil. Nevertheless, potassium recovery by crops is on the order of 45 percent of that applied during the same year that the crop is grown.

Potash Production Capacity, 1973–1982. Underground potassium salt deposits are the major raw material source for potassium fertilizers. World potash reserves are large enough to take care of needs for many centuries. Total North American potash capacity is presently more than twice domestic demand. Nearly half of the estimated 11 billion tons (12.1 billion m tons) of K_2O reserves are in North America. World potash demand has increased appreciably in recent months. This increase in demand has come both from outside North America and from within the United States. Producers in the United States cannot meet domestic needs, but Canadian capacity is high, and imports from Canada make up more than 60 percent of total potash usage in the United States.

Canadian producers in 1973 announced a 5–year deal for the sale of up to 3.57 million tons (3.25 million m tons) to Japan and a 494,505–ton (450,000–m ton) sale to India. Moreover, Israeli shipments of potash have been held up by a blockade of the Straits of Bab el Mandeb, causing other world exporters to experience a temporary increase in demand.

Even so, United States and Canadian production capacity is more than adequate to supply needs, if it were not for a tremendous shortage of rail cars. This has led to an inability to meet both overseas and domestic demand. Grain, especially, competes with potash for shipment space. So, although there are inventories of potash at the production plants in the United States and Canada, potash shortages could be felt at the fertilizer manufacturers' plants, at the dealers' warehouse and on the farm. This transportation problem could continue for some time and hamper shipment to major markets in the United States.

Projected world potash consumption to 1982, as well as the North American share of about 23 percent, is shown in Table 10.3. Figure 10.1

---

[7]*Fertilizer Trends*, compiled by E. A. Harre, M. N. Goodson and J. D. Bridges.

[8]S. A. Barber, R. D. Munson, and W. B. Daney, "Production, Marketing, and Use of Potassium Fertilizers," in *Fertilizer Technology and Use,* 2nd ed. (Madison, WI: Soil Science Society of America, 1971).

Table 10.3
PROJECTED WORLD POTASH CONSUMPTION

	1976–77	1977–78	1978–79	1979–80	1980–81	1981–82
	(1,000 tons K_2O_5)[1]					
North America	6,060	6,270	6,480	6,790	7,000	7,410
Latin America	1,130	1,180	1,275	1,375	1,500	1,620
Asia	2,350	2,600	2,800	3,000	3,200	3,350
Africa	410	435	460	500	510	520
Oceania	290	300	330	360	400	420
W. Europe	5,400	5,600	5,700	5,900	6,100	6,200
E. Europe	10,100	10,600	11,000	11,500	11,800	12,200
World Total	25,740	26,985	28,045	29,425	30,510	31,720
Percent Change		4.8	3.9	4.9	3.7	3.9

[1]Tons × 0.91 = m tons.

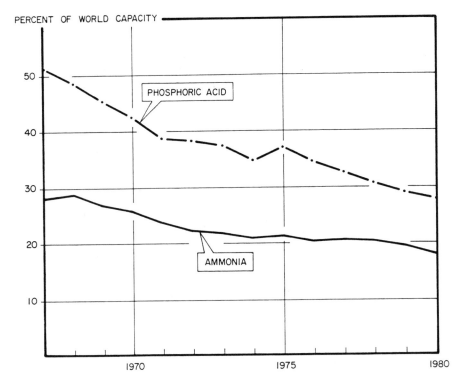

Figure 10.1   North American share of world production capacity.

shows the North American share of phosphoric acid production capacity
at about 28 percent and of ammonia at about 18 percent in 1980.

Nitrogen, phosphate and potash have one thing in common—natural gas or fuel oil is required for their manufacture. About 60 percent of total gas consumption is for feedstock; process heat is the principal non-feedstock use. Natural gas and fuel oil requirements per ton of product are shown in Table 10.4. Natural gas accounts for an estimated 70 to 80 percent of the total energy consumed in fertilizer manufacture. Overall, agriculture's fossil fuel energy expenditures now exceed the captured solar energy represented in the foods consumed by the human population of the United States.[9] Agriculture's total energy expenditure is about 3 percent of the U.S. supply.[10]

Total energy requirements for fertilizer production are distributed among the three major nutrients as follows:

Nitrogen—88 percent
Phosphates—5 percent
Potash—7 percent

Table 10.4
NATURAL GAS AND FUEL OIL REQUIREMENTS FOR PRODUCTION OF SELECTED FERTILIZERS[1]

		Natural Gas		Fuel Oil gal/ton[3]
	Feedstock	Nonfeedstock ft³/ton[2]	Total	
Anhydrous ammonia (centrifugal compressors)	22,219	15,908	38,127	9
Concentrated superphosphate		565	565	3.7
Diammonium phosphate		1,000	1,000	2.5
Wet-process phosphoric acid		344	344	
Muriate of potash		1,890	1,890	129
Sulfate of potash		4,173	4,173	96
Sulfur (Frasch process)		6,000	6,000	33

[1]Adapted from William C. White, 1973. "Fertilizer-Food-Energy Relationships," presented at meeting of the American Chemical Society, Chicago, Illinois, August 28, 1973.
[2]Ft³/ton × 0.031 = m³/m ton.
[3]Gal/ton × 4.16 = l/m ton.

[9]P. R. Stout, "Energy Requirements in Food Production," Proceedings, 1973 Plant and Soil Conference, Sacramento, California.

[10]David Pimentel et al., "Food Production and the Energy Crisis," *Science* (1973), 182.

Fixation of atmospheric nitrogen as ammonia is the vital initial step for the production of virtually all nitrogen fertilizers. Ammonia is synthesized from a 3:1 (by volume) mixture of hydrogen and nitrogen at an elevated temperature in the presence of an iron catalyst. The hydrogen may be obtained from natural gas, water, oil or coal. All of the nitrogen used is obtained from the air.

There is presently no satisfactory substitute for natural gas (essentially methane) as a source of hydrogen for ammonia production. Other hydrogen sources are either too expensive, too dirty or of limited availability.

From 35,000 to 40,000 cubic feet (980 to 1,120 m³) of natural gas is required to produce a ton (0.91 m ton) of anhydrous ammonia. During 1972 in the United States, 11.4 million tons (10.4 million m tons) of anhydrous ammonia was produced, requiring 456 billion cubic feet (12.8 billion m³) of natural gas. This usage accounted for 2 percent of the 22 trillion cubic feet (0.6 trillion m³) of gas consumed that year. In 1976, 13.2 million tons (12.01 million m tons) of anhydrous ammonia was produced, requiring 527 billion cubic feet (14.8 billion m³) of gas.

Practically all of the phosphatic fertilizer used in the United States is derived from phosphate rock, and all except single superphosphate require phosphoric acid for their production. Single superphosphate production is decreasing rapidly; ammonium phosphates dominate the current fertilizer market. Phosphoric acid, made by treating phosphate rock with sulfuric acid (wet–process), is the starting step for phosphate fertilizer production. This initial step accounts for over 80 percent of the energy consumed in the production of phosphate fertilizers. This is an "indirect charge" to the energy account, the chief expenditure being the natural gas used to mine the sulfur (Frasch process) to make the sulfuric acid used in making wet–process phosphoric acid.[11]

The emphasis on natural gas should not obscure the serious electrical shortages phosphate producers have experienced in Florida. Without electricity, draglines and pumps remain still. Power interruptions in Florida during 1973 reduced phosphate production as much as 5 percent.[12]

Future Fossil Fuel Requirements. Two major factors are seen in the steadily increasing demand for agricultural commodities, our rising population numbers and external demands. Our exports of agricultural commodities amounted to $5.7 billion in 1968, $8.1 billion in 1972 and $11.1 billion in 1973. The United States Department of Agriculture has projected $15.1 billion in exports of farm commodities by 1980 or even sooner.

[11]C. H. Davis, and P. A. Corrigan, "Energy Requirements for Alternative Methods for Processing Phosphate Fertilizers." Presented at annual meeting, American Society of Agronomy, Las Vegas, Nevada, 1973.

[12]*Farm Chem.*, November 1973, p. 14.

The accelerated production of agricultural commodities requires increased fertilizer production and usage. Fertilizer production in turn is wholly dependent on fossil fuels, as are many other agricultural inputs and operations. Farming uses more petroleum than any other single industry.[13]

By 1980, total nitrogen demand is estimated to increase to 17.32 million tons (15.76 million m tons) annually, an increase of 5.46 million tons (4.97 million m tons) or 35 percent over 1973 demand (Table 10.1). This assumes that the United States will depend solely on domestic supplies. This will require a 35-percent increase in natural gas consumption over a 7-year period, amounting to an annual increase of 5 percent. Thus, the estimated 475 billion cubic feet (13.3 billion m³) of natural gas consumed in anhydrous ammonia production in 1973 is expected to increase to an annual consumption of 641 billion cubic feet (17.94 billion m³) by 1980.

While energy requirements for phosphate and potash production are much less than those required for nitrogen, increased requirements for fossil fuel and electricity will result as demand for these fertilizers increase. Annual phosphate requirements are expected to increase by 12 percent by 1985 (Table 10.1). Phosphoric acid production in 1972 required 4.2 million tons (3.82 million m tons) of sulfur, which in turn required 25.2 billion cubic feet (0.71 billion m³) of natural gas to mine. Other steps in phosphorus pentoxide production required an additional 20 percent, making the total consumption of natural gas 30.3 billion cubic feet (0.85 billion m³) in 1972. A 12-percent increase by 1985 would increase the annual gas consumption to 34 billion cubic feet (0.95 billion m³).

Natural gas requirements for potash production are also expected to increase at the rate of about 2 percent per year. The 4.4 million tons (4.0 million m tons) of potash produced in 1973 required an estimated 8.4 billion cubic feet (0.24 billion m³) of gas, mainly for process heat. Annual gas requirements for potash production are expected to be 9.5 billion cubic feet (0.27 billion m³) by 1980.

Energy/Food Relationships. The energy required to produce fertilizers translates directly into energy required to produce food. Many people have begun to calculate the extent of energy use in our food-production system since it is clearly evident that our present system of agriculture depends entirely on fossil fuel and electricity. For example, the single largest energy input for corn production is the energy required to produce fertilizer nitrogen. This energy amounts to over 30 percent of the estimated total 2.9 million kilocalories required to produce 81 bu/acre (5,443 kg/ha) of corn, using 112 lb/acre (125 kg/ha) of nitrogen fertilizer.[14] Since 25 cubic feet (0.7 m³) of natural gas is required for each pound (0.45 kg) of nitrogen in fertilizer, 34 cubic feet (0.95 m³) of gas equals 1 bushel (27.3 kg) of

[13]Committee on Agriculture, House of Representatives, p. 20 (92nd Congress), 1971.

corn. A relationship of 39 cubic feet (1.1 m^3) of natural gas per bushel (27.3 kg) of wheat has been calculated by other workers (Table 10.4).

Similar energy-fertilizer-food relationships hold for all agricultural production.

TRANSPORTATION SITUATION

During 1977, more than 51 million tons (46.4 million m tons) of fertilizer materials were distributed to farmers in the United States and Puerto Rico, the primary market being the midwestern states. The locations of production plants for the three major plant nutrients, however, are the Gulf States, Florida, and New Mexico; therefore, the industry must rely heavily on the nation's transportation system to deliver its products to the farmer. Fertilizers move by rail, barge and truck, with approximately 60 percent of all movement being by rail and 15 to 20 percent being by barge. The remainder is relatively short-haul by truck.

Producers attempt to ship materials to local and regional storage during all months of the year; however, some years they are hampered by a lack of rail cars, and deliveries get behind schedule. The large increase in grain exports during 1973 led to congestion and delays at port facilities, and rail cars badly needed for fertilizer were tied up for extended periods of time by grain shipments. The 1973 mass movement of coal in open cars, due to the energy crisis, added to the severe shortage of rail cars.

It is estimated that about 50,000 to 60,000 new rail cars will be built soon, but this will still mean tight supplies to meet the needs of the fertilizer industry during the peak spring season. No relief was in sight as projected levels of agricultural exports for 1974 showed an increase from $12 million to $19 million.

Rail cars may not be the only limiting factor. While the number of barges for bulk handling of fertilizers should be adequate for the spring season, a lack of fuel for the towboats could slow delivery. This is also true of fuel for the diesel engines used by the railroads. Trucks for local delivery to the farm should be adequate, especially when the number of trucks owned by farmers that can be used for fertilizer is considered. Gasoline shortages and the lower speed limits have had some effects on delivery times.

ALTERNATIVES

1. Viable alternatives to our present dependence on fertilizers for food are by no means plentiful. Significant changes in fertilizer demands can be brought about only by drastically changing our dietary habits.

[14]Pimentel, *Science*, 1973, 182.

For example, a larger part of the daily per capita protein requirement could be met with plant protein and a concurrent reduction in the consumption of animal-based protein. This could result in a corresponding decrease in the demand for fertilizer nitrogen and the fossil fuel required for its preparation.

2. At present, 1979, the only feasible substitute for natural gas as a hydrogen source for fertilizer production is coal gasification. However, coal gasification technology is not well developed, and a major research and development effort will be required.

3. Better use can be made of organic wastes, including animal and poultry manures, sewage sludges and municipal composts. However, even the fullest utilization of these materials will not have a very great impact on commercial fertilizer demands. Organic wastes are relatively low in plant nutrients, and hence tons per acre (m tons per ha) rather than pounds/acre (kg/ha) are required. The low economic value of these materials precludes transport over long distances. Organic wastes may also contain contaminants that are toxic to crops and humans.

4. An increase in the practice of crop rotation with legumes versus monoculture could be encouraged. Although fixation of atmospheric nitrogen by legumes is very expensive relative to synthetic fixation as anhydrous ammonia, it is possible to obtain part of our nitrogen requirement in this way.

5. Agricultural commodity exports in 1980 are forecast at $15.1 billion. Production of these commodities will require over 2.7 million tons (2.47 million m tons) of nitrogen, phosphorus and potash combined. Fertilizer demands in the United States could be reduced by curtailing exports of agricultural commodities, as well as fertilizers.

SUMMARY

1. Fertilizer use accounts for at least one-third of the agricultural production in the United States.

2. During 1977, more than 51 million tons (46.4 million m tons) of fertilizer materials were used by farmers. This demand is expected to increase by a compounded 5 percent annually.

3. This steady increase is brought about by two major factors, rising population and an increase in export demands.

4. Fertilizer production is entirely dependent on fossil fuels, particularly fuel oil and natural gas. The latter is used both as feedstock and process heat.

5. During 1978, the amount of natural gas consumed in fertilizer production was less than 2.5 percent of the total consumption in the United States.

6. Coal gasification appears to be the only promising alternative to natural gas.

7. The energy requirements for fertilizer production must be considered as a part of the energy requirements to produce food.

8. The high quality of the food diet in the United States is the major factor in the high demand for fertilizers.

QUESTIONS

Completion:

1. Significant changes in fertilizer demands can be brought about only by drastically changing our _____ habits.

2. One change that would reduce fertilizer demand would be the substitution of our daily per capita animal protein requirement with _____ protein.

3. The only feasible substitute for natural gas as a hydrogen source for fertilizer production is _____ gasification.

4. Fertilizer demands in the United States could be reduced by curtailing exports of agricultural _____, as well as _____.

True or False:

5. ___ Viable alternatives to our present dependence on fertilizers for food are plentiful.

6. ___ The full utilization of organic wastes, including animal and poultry manures, sewage sludges and municipal composts, will have a very great impact on commercial fertilizer demands.

7. ___ The fixation of atmospheric nitrogen by legumes is very expensive relative to synthetic fixation as anhydrous ammonia.

Essay:

8. Discuss the transportation situation as related to fertilizer and agricultural commodity distribution in the United States.

9. Should the United States permit a level of nitrogen production that would result in a dependence on imports to supply our nation's agricultural needs?

chapter 11

Fertilizer Basics—
An Autotutorial
Laboratory Exercise

The primary function of fertilizers is to supply available plant nutrients to the soil. The nutrient elements are supplied by various carriers, and the value of the fertilizer depends on the composition and solubility of these plant nutrient carriers.

The fertilizer analysis gives the percent nitrogen as nitrogen, the percent phosphorus as P_2O_5 and the percent potassium as K_2O. A 4-12-12 contains 4 percent nitrogen, 12 percent P_2O_5 and 12 percent K_2O. In a ton, each 4-12-12 has 4 units of nitrogen, 12 units of P_2O_5 and 12 units of K_2O, or a total of 28 units of available plant nutrients. The numbers 4-12-12 are referred to as the grade of fertilizer. A unit is 1 percent of a ton or 20 pounds.

The ratio of the available nutrients can be figured from the fertilizer analysis. For instance, a 3-9-18 and a 6-18-36 both have their available nutrients in a simple ratio of 1-3-6. A 5-10-10 and a 10-20-20 have a 1-2-2 ratio.

If a ton of fertilizer of a certain analysis is to be mixed and the chosen carriers do not weigh a ton, some inert material, called a filler, such as gypsum, ground limestone or organic trash, may be added to make up the difference. Water is used as a filler for liquid fertilizers. Fillers were formerly chosen to give solid fertilizer mixtures a good physical condition and to neutralize any residual acidity that may be produced when the fertilizer reacts with the soil. Limestone, tankage, gypsum and rice hulls are excellent conditioners for powdered fertilizers. Ground limestone is an excellent filler, since it will reduce soil acidity, but granular limestone, which is commonly used in granular fertilizers, is too coarse to reduce very much soil acidity. Powdered fertilizers have been largely replaced by granular ones in the United States.

Equivalent acidity is defined as the acidity developed in the soil by the fertilizer, measured in terms of the calcium carbonate required for its neutralization. Equivalent basicity refers to the basic residue left in the soil by the fertilizer, expressed as equivalent calcium carbonate.

In most cases, fertilizer mixtures on the market today are acid forming, i.e., they are compounded in such a way that the resultant equivalent acidity is near the theoretical figure of about 1.8 pounds of lime equivalent per pound of nitrogen. For example, 500 pounds per acre of a 10-10-10 fertilizer mixture will develop an acid residue in the soil requiring about 90 (1.8×50) pounds of lime to neutralize.

344

The laboratory work consists of an examination of the common carriers and also the mixtures of these carriers to form various analyses or grades. The solution of problems that involve fertilizer formulas will be an important part of this practice.

For your convenience, a fertilizer-mixing sheet form is on page 346.

This fertilizer package contains fertilizer samples, a visual presentation (35 mm slides), narration on a cassette tape with a sound tone to indicate when to change slides and a problem exercise to be completed after you have seen the slides and listened to the cassette tape.[1]

The primary reason for the addition of fertilizer to soil is to supply plant nutrients. Because of the nature of fertilizer nutrients, they cannot be supplied in a pure state but must come from carriers that contain a certain known amount of the nutrient. This laboratory is designed to familiarize you with fertilizers and fertilizer blending.

After completing this lab, you should be able to:

1. Recognize what is in a fertilizer container and know how to calculate plant nutrient content of fertilizer materials on an elemental and oxide basis.

2. Recognize the different kinds of fertilizer materials or "carriers" available.

3. Understand the chemical constituents of different fertilizer materials and their reaction in aqueous solutions and in soils.

4. Calculate the amount of fertilizer required when given a recommendation for a certain amount of actual nutrients to be added to the soil.

5. Calculate the cost per pound of a nutrient when given the cost per ton of fertilizer.

[1]Cassette tape and slides are available at cost from the Communications Center, Clemson University, Clemson, S.C. 29631.

MIXING SHEET FORM

Analysis of Fertilizer

	To supply (lb)			Lb material needed/ton of fertilizer	Equivalent acidity or basicity for the amount of each carrier used, lb CaCO$_3$
	N	P$_2$O$_5$	K$_2$O		
1. NaNO$_3$ (16 percent nitrogen)					
2. (NH$_4$)$_2$SO$_4$ (21 percent nitrogen)					
3. NH$_4$NO$_3$ (33.5 percent nitrogen)					
4. Urea (45 percent nitrogen)					
5. Cottonseed meal (7.0 percent nitrogen)					
6. (NH$_4$)$_2$HPO$_4$ (18 percent nitrogen - 46 percent P$_2$O$_5$)					
7. Superphosphate (20 percent P$_2$O$_5$)					
8. Concentrated superphosphate (46 percent P$_2$O$_5$)					
9. Muriate of potash (60 percent K$_2$O)					
10. Sulfate of potash (52 percent K$_2$O)					
11. Limestone (100 percent CaCO$_3$)					
12. Total materials					
13. Filler, sand					
14. Grand total					

Cassette tape
Slides
Box of fertilizer samples
Fertilizer exercise sheets

Table 11.1
THE NITROGEN CONTENT AND EQUIVALENT ACIDITY OR BASICITY OF COMMON FERTILIZER
MATERIALS

		Equivalent acidity or basicity		
	Percent nitrogen	Per unit of nitrogen	Per 100 lb of carrier	Per pound of nitrogen
NITROGEN CARRIERS				
Ammonium sulfate	21.0	107	112	5.35
Ammonium nitrate	33.5	36	60	1.80
Ammo-pho®	11.0	107	59	5.35
Anhydrous ammonia	82.2	36	148	1.80
Nitrate of soda	16.0	36B	29B	1.80B
Nitrate of potash	13.0	36B	23B	1.80B
Uramon®	42.0	36	76	1.80
Urea	45.0	36	81	1.80
Sewage sludge (dried)	6.0	36	11	1.80
Cocoa shell meal [1]	2.5	20B	2B	1.00B
Cottonseed meal	7.0	29	10	1.45
Dried blood	13.3	36	24	1.80
Tankage	9.0	36	16	1.80
Tobacco stems	1.8	86B	8B	4.30B
PHOSPHATE CARRIERS				
Ammonium phosphate (18 percent nitrogen, 46 percent P_2O_5)	18.0	107	119	5.35
Superphosphate (20 percent P_2O_5)			0	
Concentrated superphosphate (46 percent P_2O_5)			0	
Rock phosphate			10B	
Steamed bone	2.0		25B	
Raw bone	3.8		20B	
POTASH SALTS	All neutral, except some of the nitrogen-potash salts and by-product potash from cement and other industries, which commonly are basic.			
MISCELLANEOUS MATERIALS				
Dolomitic limestone			90–100B	
Marl			50– 90B	

B indicates equivalent basicity instead of equivalent acidity.

[1]The values for most vegetable organics like cottonseed meal and cocoa shell meat are fairly constant. The values for animal tankages, fish products and certain other materials are variable, but the values given are representative.

347

PROBLEMS

1. Rank the nitrogen fertilizers in Table 11.2 in increasing order on the basis of cost of nitrogen per unit (1 percent of a ton).(Consider the cost of application of solutions, and bag and bagging costs of solids as being equal.)
    1.
    2.
    3.
    4.
    5.
    6.

2. Compute the amounts of materials necessary for a ton of 4-12-12 using the following carriers:

    Ammonium sulfate   21 percent nitrogen
    Superphosphate   20 percent P_2O_5
    Muriate of potash   60 percent K_2O
    Filler   Ground limestone

    a. Is the ground limestone filler enough to make this a nonacid-forming fertilizer?

3. a. Compute the amounts of materials necessary for a ton of 4-12-12 using the carriers listed below.

    b. Compute the equivalent acidity and supply enough limestone to make this a nonacid-forming fertilizer.

    Ammonium sulfate   21 percent nitrogen
    Superphosphate   46 percent P_2O_5
    Muriate of potash   60 percent K_2O
    Limestone   100 percent $CaCO_3$ equivalent

    c. If the limestone filler was not used, what higher analysis 1-3-3 ratio could be produced from the same ingredients?

    d. How much sulfur would a ton of this higher analysis material contain if the ammonium sulfate contains 24 percent sulfur?

4. Compute the cost of the ton of 4-12-12 fertilizer in Problem No. 2, using the price list in Table 11.2. Make no charge for the filler.

5. Compare the cost of a ton of factory-mixed 4-12-12 fertilizer ($108.00) with a ton of 4-12-12 computed in Problem No. 4.

6. Compare the cost of a ton of factory-mixed 5-10-10 ($82.00) with a ton of factory-mixed 6-12-12 ($91.00). Which is the cheaper per ton? Which is the cheaper when considering the analysis? Which one should a farmer buy as a rule, based on the cost of plant nutrients?

7. a. You have a recommendation to apply 800 pounds of 3-9-18 per acre. How many pounds of 6-18-36 per acre would be required to be equivalent in nutrients to the original recommendation?

Table 11.2
APPROXIMATE COST TO CONSUMERS

Material	Nutrient Percent	Cost per Ton in 50-lb bags,[1] $	Cost of Nutrient per Unit[2] of Nutrient $
Cottonseed meal	6 nitrogen	125.00	20.80
Sodium nitrate	16 nitrogen	135.00	8.44
Nitrogen solution (applied)	19 nitrogen	100.00	
Nitrogen solution (applied)	30 nitrogen	130.00	
Calcium nitrate	14 nitrogen	132.00	
Ammonium sulphate	21 nitrogen	110.00	
Ammonium nitrate	33.5 nitrogen	150.00	
Urea	45 nitrogen	200.00	
Anhydrous ammonia	82 nitrogen	250.00	
Superphosphate	20 P_2O_5	100.00	
Superphosphate	46 P_2O_5	160.00	
Muriate of potash	60 K_2O	100.00	
Sulfate of potash	52 K_2O	90.00	
Ground limestone		35.00	
Ammonium phosphate	18 nitrogen–46 P_2O_5	270.00	
Ammonium polyphosphate (applied)	10 nitrogen–34 P_2O_5	135.00	

[1]Except for nitrogen solutions, anhydrous ammonia and ammonium polyphosphate.

[2]1 percent of a ton.

349

    b. A ton of 3-9-18 costs $82.00, and a ton of 6-18-36 costs $137.00. How much money per acre would the farmer save by using 6-18-36 instead of 3-9-18 in Problem 7(a)?

8. A soil must be fertilized with an amount per acre of 100 pounds of nitrogen, 40 pounds of P_2O_5 and 80 pounds of K_2O. Available fertilizers are:

    1. 5-10-20 grade (1-2-4 ratio)

    2. Ammonium nitrate (33 percent nitrogen)

How many pounds per acre of each fertilizer will be required to supply the nitrogen, P_2O_5 and K_2O? Use the 5-10-20 grade to supply the P_2O_5 and K_2O and the ammonium nitrate to supply the part of the nitrogen not supplied by the 5-10-20.

VISUALS	NARRATION
1. A fertilizer bag showing the grade numbers. The first number is nitrogen (N). The second number is phosphorus P_2O_5. The third number is potassium K_2O.	If you look at a fertilizer bag, you will find numbers on it. The first three numbers are always, in order, the percent by weight of nitrogen, phosphorus and potassium. All three of these components of fertilizer *must* be designated; other fertilizer components *may* be designated. Fertilizer is labelled according to state standards.

The first number represents the percent by weight of available nitrogen—the percent of nitrogen the plant may use. Phosphorus, however, is expressed in terms of P_2O_5, and potassium is expressed in terms of K_2O. Neither nitrogen nor the oxides of phosphorus and potassium are found in fertilizer. The use of the oxide expression originated in a custom first used by chemists in the nineteenth century. They found that, after analyzing rocks, minerals or soils for chemical elements and expressing them as the oxides, the sum of all the oxides usually was about 100 percent. This is not surprising in view of the fact that about 20 percent of the atmosphere is oxygen and that the environment will oxidize most elements on the earth's surface. Nevertheless, scientists in scholarly journals now report fertilizer analyses on the elemental basis.

Since 1950, efforts have been made by some members of the industry to make the conversion from oxide to elemental basis, but the public has not accepted it yet; customs and habits are hard to change. Therefore, care should be taken in establishing new customs or habits to make sure they are good ones and will be long lasting.*

2. N = N
    P_2O_5 x 0.44 = actual phosphorus
    K_2O x 0.83 = actual potassium

The numbers on the bag, therefore, may need to be converted to actual phosphorus and actual potassium. This is accomplished easily if you understand that P_2O_5 is 44 percent phosphorus and K_2O is 83 percent potassium. The numbers on the bag may be multiplied by 0.44 and 0.83 to calculate actual phosphorus and potassium, respectively.*

3. $(NH_4)_2SO_4$ = molecular weight 132
    nitrogen molecular weight 14
    percent nitrogen $= \left(\dfrac{2(14)}{132}\right) \times 100 = 21\%$

After reading the bag, one should understand what is inside. You have before you several samples of fertilizer carriers. Ammonium sulfate, for instance, contains about 21 percent nitrogen; this is calculated as follows. The formula for ammonium sulfate is $(NH_4)_2SO_4$. If the atomic weight of nitrogen is 14, sulfur is 32, oxygen is 16 and hydrogen is 1, the molecular weight of ammonium sulfate is 132. The weight of all of the nitrogen in a molecule of ammonium sulfate is 2 × 14 or 28. The percentage of nitrogen is calculated as shown.*

4. Sources of N $\equiv$ carriers of nitrogen
    $2NH_3 + H_2SO_4 \rightarrow (NH_4)_2SO_4$
    $NH_3 + H_3PO_4 \rightarrow NH_4H_2PO_4$
    $2NH_3 + CO_2 \rightarrow CO(NH_2)_2 + H_2O$
    $NH_3 + HNO_3 \rightarrow NH_4NO_3$

Nitrogen, however, does not come only from ammonium sulfate. As can be seen, several sources or carriers of nitrogen are available.

The symbol for ammonia is NH_3, and it is the basic nitrogen raw material and is also the least expensive per unit of nitrogen. More nitrogen is added to the soil as ammonia than any other source. Under normal temperature and pressure, it is a colorless, pungent, basic gas; when in contact with nitric acid, it forms solid ammonium nitrate containing 33 percent nitrogen. When dissolved in water, ammonium nitrate is only 19 percent nitrogen. The discussion of nitrogen carriers continues with the second reaction from the top of the chart.

Ammonium sulfate is formed by reacting ammonia with sulfuric acid. We have already seen that it contains about 21 percent nitrogen.

Ammonia plus phosphoric acid, H_3PO_4, forms mono- or diammonium phosphate, depending on

*Indicates sound tone.

351

the conditions of the reaction. Ammonium phosphates contain 11 to 21 percent nitrogen and in addition contain 46 to 53 percent P_2O_5, a second essential element for plant growth.

Ammonia plus carbon dioxide produces urea, a 45 percent nitrogen material. Because carbon dioxide production is inherent in the manufacture of ammonia, no additional raw material is required to make urea. It is considered a nonprotein, organic nitrogen material and is the highest analysis *solid* nitrogen material on the market. Because it is widely used as a protein supplement for ruminants, as a fertilizer for forests such as Douglas fir, and as a favorite nitrogen material for rice, urea usage is destined to surpass ammonium nitrate, which in 1975 was the second most widely used source in nitrogen.

A widely used liquid nitrogen fertilizer is a mixture of about 42 percent ammonium nitrate, 33 percent urea and 25 percent water. This fluid fertilizer contains 30 percent nitrogen.

Finally, a mixture of ammonia and water, the same as household ammonia, is sometimes used as a carrier of nitrogen for fertilizer. Like its anhydrous counterpart, it exerts an appreciable vapor pressure and, unless put in the ground and covered, will be lost to the atmosphere.*

5. Chemical reactions illustrating sources of phosphorus
   1a. $Ca_3(PO_4)_2 + 2H_2SO_4 \rightarrow Ca(H_2PO_4)_2 + 2CaSO_4$
   1b. $Ca_3(PO_4)_2 + 3H_2SO_4 \rightarrow 2H_3PO_4 + 3CaSO_4$
   1c. $Ca_3(PO_4)_2 + 4H_3PO_4 \rightarrow 3Ca(H_2PO_4)_2$
   2. $2NH_3 + H_3PO_4 \rightarrow (NH_4)_2HPO_4$

Phosphorus, also, has many sources or carriers. It is found as an ingredient of rock phosphate, the chemical formula which is expressed here as tricalcium phosphate, $Ca_3(PO_4)_2$. The rock is finely ground and treated with sulfuric acid, H_2SO_4, as seen in 1(a) on the left, to produce superphosphate containing 20 percent P_2O_5. An excess of sulfuric acid, as noted in 1(b), may be added to rock phosphate, resulting in the formation of the products phosphoric acid, H_3PO_4, and calcium sulfate, $CaSO_4$. Calcium sulfate, also called gypsum, is centrifuged, and liquid phosphoric acid is made for fertilizer and other products, including soap powder.

Item 1(c) demonstrates the reaction of liquid phosphoric acid with rock phosphate to form a concentrated superphosphate, largely monocalcium phosphate, which contains 46 percent P_2O_5.

VISUALS	NARRATION

In item 2, the basic raw material for nitrogen, anhydrous ammonia, is bubbled through phosphoric acid to produce ammonium phosphate, a high-analysis multinutrient material. This is the main ingredient of most blended solid and liquid fertilizers used on farms today.*

6. Sources of potassium
   KCl
   K_2SO_4
   KNO_3
   $K_2SO_4 \cdot 2MgSO_4$

Muriate of potassium or KCl is mixed or blended with liquid or solid ammonium phosphate. In this way, multinutrient fertilizers, such as 5-10-10 liquid and 13-13-13 solid fertilizers, containing nitrogen, phosphorus and potassium can be made.

There are several potassium carriers; however, most potassium comes from muriate of potash, KCl. Others are sulfate of potash, K_2SO_4, nitrate of potash, KNO_3, and sulfate of potash-magnesia. Look at the samples of fertilizer carriers on the table. Do not attempt to memorize the samples by sight; you cannot identify the analysis of a fertilizer merely by looking at it.*

7. Multiple grade fertilizers
   100 lb 6-18-36 = 200 lb 3-9-18

You also have before you several samples of multinutrient fertilizers. You will notice that the percentages of nutrient are very low—seldom totalling more than 50 percent. Also note near the bottom of the slide that 100 pounds of 6-18-36 = 200 pounds of 3-9-18. This phenomenon is caused in part by the fact that the carriers themselves are not pure nutrient.*

8. Various kinds of fillers

To improve physical condition, to reduce soil acidity or from habit the fertilizer buyer may choose a fertilizer diluted with filler. For example, he may require the manufacturer to mix 1,000 pounds of 6-24-24 with 1,000 pounds of filler and to make 2,000 pounds of 3-12-12 because he thinks 3-12-12 is a good fertilizer. Filler may include anything that may be added to a fertilizer to make it less concentrated. Fillers include rice hulls, sand, water and limestone. Water is used as filler in all fluid fertilizers. Limestone is the most widely used filler for solid fertilizers, but this practice is fast fading out of the picture for reasons that we will now discuss.*

9. Eventually all sources of nitrogen convert to HNO_3 in soil.
   $2HNO_3 + CaCO_3 \rightarrow Ca(NO_3)_2 + H_2CO_3$

Nitrogen fertilzer carriers do leave an acid residue in the soil because, in time, bacteria convert all nitrogen carriers to nitric acid, HNO_3.

VISUALS	NARRATION

NARRATION

Plants take up nitrogen for the most part as nitrate, NO_3^-, which is the form found in nitric acid. Limestone in the soil neutralizes nitric acid and forms calcium nitrate, $Ca(NO_3)_2$. This is taken up by the plant roots or, if not, is readily leached through the soil profile to the water table. In either event, the end result of adding nitrogen fertilizer to the soil is loss of lime, $CaCO_3$, by leaching or crop removal, and the soil thus becomes more acid. This is a real problem in modern soil fertility. Almost all fertilizers containing nitrogen are acid-forming; they cause the soil pH to fall. To avoid the problem of too-low soil pH, limestone, which is largely $CaCO_3$, may be added with the fertilizer. You have in your lab packet a table showing equivalent acidity or basicity of common fertilizer materials. Take this out and turn to Table 11.1.*

10. Equivalent acidity
(100 lb) (0.21) = 21 lb of nitrogen
(21 lb N) (5.35 lb $CaCO_3$/lb N) = 112.3 lb $CaCO_3$

Using, for example, a strongly acid-forming fertilizer, such as ammonium sulfate, one may calculate the amount of limestone needed to counteract the fertilizer. According to Table 11.1 in the packet, ammonium sulfate requires 5.35 pounds of calcium carbonate to neutralize every pound of actual nitrogen used. This means that if 100 pounds of ammonium sulfate is added to a field, 112 pounds of calcium carbonate must be added to counteract the acidity produced. The lime may be used as a filler to make the fertilizer non-acid forming, or a more economical way is to add it later with a lime spreader after testing the soil.*

11. Cost per unit
Unit = 1 percent of a ton = 20 lb.
$NaNO_3$ = 16 percent nitrogen.
1 ton $NaNO_3$ = \$135.
cost/unit = (cost of fertilizer per ton/nutrient percent) =
$\dfrac{\$135}{16}$ = \$8.44/unit of nutrient

After deciding what he wants to buy, the grower must also be able to calculate the cost of the fertilizer. You have in the lab packet a table entitled "Approximate Cost to Consumers." Sodium nitrate, for example, is 16 percent nitrogen and costs \$135 per ton. A unit of fertilizer is 1 percent of a ton or 20 pounds. Fertilizer is priced on the basis of so many dollars per unit. Thus, every ton of sodium nitrate is 16 percent nitrogen, and it contains 16 units of nitrogen. The cost per unit of sodium nitrate is therefore the cost of the fertilizer per ton divided by the percent of nutrient. In this case, the nutrient costs \$8.44 per unit. Complete the column marked "Cost of Nutrient per Unit of Nutrient" in Table 11.2. In this manner, you can compare the cost per unit of the various sources of nitrogen.*

VISUALS	NARRATION

12. Need 150 lb nitrogen, 54 lb P_2O_5 and 108 lb K_2O. Have 6-18-36 and anhydrous ammonia = 82 percent nitrogen

If a grower has a soil-test recommendation to supply a certain number of pounds of individual nutrient per acre, he must know how to calculate how much of what analysis to buy. Assume an acre field must be supplied with 150 pounds of nitrogen, 54 pounds of P_2O_5 and 108 pounds of K_2O. Assume also that the fertilizer dealer has only 6-18-36, which is a 1-3-6 ratio, and anhydrous ammonia which is 82 percent nitrogen. How much of each should he buy?*

13. $\dfrac{108 \text{ lb}}{36} K_2O = 3$ $\dfrac{54 \text{ lb}}{18} P_2O_5 = 3$

100 lb × 18 percent = 18 lb
300 lb fertilizer = 54 lb P_2O_5
100 lb 6-18-36 = 36 lb K_2O
300 lb 6-18-36 = 108 lb K_2O

Notice that the percentage of both P_2O_5 and K_2O divide evenly into the amounts needed. Thus, if 100 pounds of 6-18-36 contains 18 pounds of P_2O_5, then 300 pounds will contain 54 pounds of P_2O_5. Similarly, if the fertilizer is 36 percent K_2O, then the same 300 pounds of it will contain the necessary 108 pounds of K_2O.*

14. Nitrogen
Anhydrous ammonia = 82 percent nitrogen
$0.82x = 132$ lb
$x = \dfrac{132}{0.82} = 161$ lb anhydrous ammonia

Nitrogen, however, does not come out quite so even. The purchase of 300 pounds of 6-18-36 for one acre of ground will supply only 18 pounds of nitrogen. The other 132 pounds of nitrogen must be added with anhydrous ammonia. Because anhydrous ammonia is only 82 percent nitrogen, some amount greater than 132 pounds must be added. If 82 percent of that amount is 132, then the formula may be arranged as shown. Eighty-two percent of some weight of anhydrous ammonia is equal to 132 pounds. Solving for x reveals that the producer must use 161 pounds of anhydrous ammonia.*

15. Sodium nitrate — 16 percent nitrogen
Superphosphate — 20 percent P_2O_5
Muriate of potash — 60 percent K_2O
Filler — Limestone

Suppose a fertilizer plant manager wanted to formulate a 5-10-5 fertilizer. He will do this with the help of carriers. If the fertilizer is to contain sodium nitrate, 20 percent superphosphate and muriate of potash, along with ground limestone as filler, compute the amount of each carrier needed to produce a ton of 5-10-5.*

16. 1 ton 5-10-5 = 100 lb nitrogen, 200 lb P_2O_5 and 100 lb K_2O
$0.16x = 100$ lb (nitrogen)
$0.2x = 200$ lb (P_2O_5)
$0.6x = 100$ lb (K_2O)
$x = 635$ lb $NaNO_3$
$x = 1,000$ lb Superphosphate
$x = 166$ lb KCl

Calculation is similar to the procedure used for anhydrous ammonia earlier. Sodium nitrate is 16 percent nitrogen so that the equation is set up as shown. Similarly, superphosphate is 20 percent P_2O_5; the equation is similar, $0.20x = 200$. A similar procedure is followed for K_2O. One ton of the fertilizer must contain 100 pounds of nitrogen, 200 pounds of P_2O_5 and 100 pounds of K_2O. The amounts of carrier used will be 625 pounds of sodium nitrate, 1,000 pounds of 20 percent superphosphate and 166 pounds of muriate of potash.*

VISUALS	NARRATION

17. 625 lb NaNO₃

1,000 lb superphosphate

<u>166 lb KCl</u>

1,791 lb total

2,000 lb − 1,791 lb = 209 lb lime filler

The carriers do not total 2,000 pounds, so the balance must be filled with some material. In this case, we will add ground limestone as filler. To achieve a ton or 2,000 pounds of material, 209 pounds of filler must be added.*

18. NaNO₃ = 1.80 lb B; not acid-forming; lime not needed to neutralize it

As you see from Table 11.1, sodium nitrate is not acid-forming; however, other nitrogen fertilizers are, and the amount of neutralizing value supplied by the filler may or may not be sufficient.*

19. 6-12-6 = 24 percent nutrient

5-10-5 = 20 percent nutrient

1,791 lb carriers

$\dfrac{20}{1,791} = \dfrac{24}{x} = 2{,}150$ lb

x is lb of carrier for a 6-12-6;

2,150 lb is more than a ton;

no higher analysis available for 1-2-1 than 5-10-5

If the limestone filler had not been used, a higher analysis fertilizer could have been formulated. The ratio of 5-10-5 fertilizer is 1-2-1. Other analyses of this ratio are possible, such as 8-16-8. With this in mind, what higher analysis 1-2-1 ratio fertilizer is possible with the same carriers? As can be seen, the next higher analysis is 6-12-6. If 5-10-5 is 20 percent nutrient and 6-12-6 is 24 percent nutrient, the analysis may be compared by simple algebra. If 20 percent nutrient required 1,791 pounds of fertilizer carriers, then 24 percent nutrient would require a proportionally larger amount of fertilizer, in this case 2,150 pounds. Because 2,150 pounds is more than a ton, this analysis cannot be formulated with these carriers, and so the highest analysis possible with these carriers is 5-10-5.*

20. NaNO₃ = 27 percent sodium

625 lb NaNO₃/ton of fertilizer

(0.27)(625 lb) = 169 lb sodium

Suppose a producer is growing a crop, such as sugar beets, that is helped by sodium, and he wants to know how much sodium a ton of this 5-10-5 contains. Sodium nitrate is 27 percent sodium. This fertilizer contains 625 pounds of sodium nitrate per ton of fertilizer. Thus every ton contains 27 percent of 625 pounds or about 169 pounds of sodium.*

21. (5 units nitrogen) ($8.44/unit) = $42.20

(10 units P₂O₅) ($5.00/unit) =   50.00

(5 units K₂O) ($1.67/unit) =   8.35

Total $100.55

After the fertilizer is formulated, a price may be placed on the product. Using the "Approximate Cost to Consumer" list enclosed in your packet, calculate a price for the 5-10-5 fertilizer just formulated. As shown, a ton of 5-10-5 has 5 units of nitrogen; at $8.44 per unit, this is $42.20. If the costs of the other nutrients are calculated similarly, the total price is the sum of $42.20 for nitrogen, $50 for phosphorus and $8.35 for potassium, or a total of $100.55*

356

VISUALS	NARRATION
22. 6-12-6 = $110/ton. ($110/ton)(1 ton/24 units of nutrient) = $4.58/unit. 5-10-5 = $101/ton (110/ton)(1 ton/20 units of nutrient) = $5.05/ unit	If factory-mixed 6-12-6 cost $110 per ton and the 5-10-5 costs $110 per ton, which fertilizer would be the better buy? If the 6-12-6 contains 24 units of nutrient, then, as shown at $110 per ton, this is $4.58 per unit of nutrient. Similarly, the 5-10-5 is 20 percent nutrient and cost $5.05 per unit. The 6-12-6 is a better buy.*
23. 3-12-12 (1-4-4 ratio) 6-24-24 (1-4-4 ratio) 600 lb 3-12-12 = 300 lb 6-24-24	If you decide that you need 600 pounds of 3-12-12 per acre and the salesman has only 6-24-24, how much of the higher analysis material can be used to get an equal response? Both fertilizers are in a ratio of 1-4-4. Because 6-24-24 is twice as concentrated as 3-12-12, you need only ½ as much, not to mention the amount of time and labor saved by using half as much material to achieve the same result.*
24. 3-12-12 = $85/ton 6-24-24 = $140/ton 2 × $85 = $170; $170 − $140 = $30 3-12-12 (600 lb/acre)($85/2,000lb)= $25.50/acre 6-24-24 (300 lb/acre)($140/2,000 lb) = $21.00/acre $25.50/acre 21.00/acre savings $ 4.50/acre	Not only labor is saved; usually, the higher the analysis of fertilizer, the more economical it is. Two tons of 3-12-12 = 1 ton of 6-24-24. If the 3-12-12 costs $85 per ton and the 6-24-24 costs $140 per ton, the saving in favor of the higher analysis fertilizer is thirty dollars. If the 3-12-12 is applied at the rate of 600 lb/acre, it will cost $25.50 per acre, as shown. If the 6-24-24 is applied at 300 lb/acre, it will cost only $21.00 per acre, a savings of $4.50.*
25. I am nitrogen	As you have seen, fertilizer production, marketing and use is a complicated but very important industry. Without it the war against hunger cannot be won. Several kinds of fertilizer are available. The fertilizer you buy may be made up of several kinds of carriers—each of which has its own characteristics. Complete the questions in the packet and turn them in as directed by your instructor. Rewind this tape and place it and all other A—T lab materials neatly where you found them.

Index

221E0012